Water Pollution
Microbiology

Water Pollution Microbiology

Edited by

RALPH MITCHELL

**Gordan McKay Professor of Applied Biology
Division of Engineering and Applied Physics
Harvard University, Cambridge, Massachusetts**

WILEY-INTERSCIENCE

a Division of John Wiley & Sons, Inc.
New York · London · Sydney · Toronto

*We seek for slumbering trout
And whispering in their ears
Give them unquiet dreams.*

W.B. Yeats

Preface

An increasing number of courses are taught both in biology departments and in engineering schools which attempt to assess the role of microorganisms in water pollution and their control. Too often these courses use out-of-date information and concepts. However, it is quite difficult to provide current information and ideas about two such apparently divergent fields as microbiology and water pollution. The objective of this book is to provide a text for advanced courses in which modern microbiological concepts are applied to the solution of problems in water pollution control. For my own teaching I have relied on original publications to bridge the gap between the microbiologist and the engineer. I attempted in my selection of contributors to this volume to provide the reader with both the information available in those original articles and a critical assessment of that information by the person most competent in that field.

Obviously it was not possible to be comprehensive. I would have liked to include chapters on general ecological principles, energy transfer, and many other topics. However, the economic limitations dictated that only the most pertinent subjects be included.

I hope that this volume will stimulate the student to initiate research into the microbiology of both polluted and unpolluted waters. Only by obtaining a clear understanding of the fundamental nature of aquatic ecosystems and the effect of perturbations on them can we hope to develop new concepts which will ultimately yield effective and economic pollution control.

I express thanks to the Environmental Protection Agency for their support in the preparation of this book.

The editor wishes to thank the Macmillan Company for permission to quote from *The Collected Poems of W. B. Yeats.*

RALPH MITCHELL

Cambridge, Massachusetts
June 1971

Contents

x   Contents

Water Pollution
Microbiology

1 Sources of Water Pollution

Ralph Mitchell, Laboratory of Applied Microbiology, Division of Engineering and Applied Physics, Harvard University, Cambridge, Mass.

1-1. Introduction

Pollution is defined by Cairns and Lanza in Chapter 10 of this volume as the appearance of some environmental quality for which the exposed community has inadequate information and is thus incapable of an appropriate response. In healthy waters, the level of pollution is sufficiently low that the native microbial community contains the information necessary to neutralize the negative effects. The ability to reverse the effects of pollution or to maintain natural waters in a healthy state is dependent on the capacity to understand the ecological processes responsible for the transfer of this information. Our knowledge of these processes is meager indeed.

The relationship between microbiology and pollution control of many of the subjects discussed in this book is self-evident. However, some sources of pollution whose origin has been recognized more recently, including eutrophication, thermal pollution, and oil pollution, require some introduction and will be discussed below.

1-2. Pollutants

The term eutrophication is derived from the Greek *eutrophos*, which means "corpulant." Eutrophic waters have been enriched with nutrients which support excessive algal photosynthesis. The microbial degradation

1

of these algae results in oxygen depletion of natural waters. Secondary effects include extensive fish kills, decline in recreational value, and foul taste in drinking water.

Lake Erie provides a classic example of accelerated eutrophication (1). The total concentration of phosphorus in the lake doubled between 1942 and 1958. The dissolved oxygen concentration in 1959 was less than 3 ppm in 75% of the lake bottom.

At that time the fish population had completely changed. Blue pike, which were among the dominant fish prior to 1950, began to decline in the late 1950s and had completely disappeared by the mid-1960s. The white fish and walleye population collapsed at the same time. Commercial production of these fish was approximately 10 million pounds per year for each fish prior to the mid-1950s. By 1965 less desirable fish began to predominate and commercial production had declined dramatically. The data shown in Figure 1.1 are typical of highly eutrophic situations. Similar conditions can be found in impounded waters in most countries of the developed world (2).

Municipal and agricultural wastes account for most of the nutrients entering natural waters. Less than 30% of the population of the United States is served by sewage treatment plants. Most plants in use depend on secondary treatment which removes much of the available organic matter, but only 30% of the phosporus and 20% of the nitrogen. Thus the conventional sewage treatment plant releases high concentrations of nutrients into surface waters. The remaining phosphorus and nitrogen must be removed by tertiary treatment if eutrophication is to be reversed.

Animal wastes and fertilizers are an important source of eutrophication in agricultural areas. The animal population in the United States in 1970 was estimated at 564 million head with a waste level equivalent to 2 billion people. In areas where animals are concentrated, the potential eutrophication is enormous. Similarly, in areas of intense crop production the sharp rise in the use of fertilizers in the past decade has dramatically increased the rate of eutrophication of local waters.

The role of phosporus in eutrophication is discussed by Stumm and Stumm-Zollinger in Chapter 2, and the role of nitrogen is discussed by Goering in Chapter 3.

More than 10,000 miles of streams in the United States are contaminated by acid drainage from abandoned coal mines. The passage of water through the mines is accompanied by microbiological action in which pyrites are oxidized to yield sulfuric acid. The pH usually declines to 4 or lower, and there are no fish living in these waters. However, dense mats of algae develop. Acid mine water cannot be used for drinking by humans or livestock without treatment.

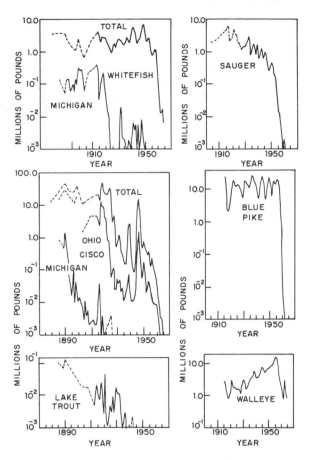

Figure 1.1 Commercial production of blue pike, cisco (lake herring), lake trout, sauger, walleye, and whitefish in Lake Erie. Broken lines represent production during periods when annual data were not available (1).

It would be desirable to treat either the source waters or the mine itself to prevent the development of acidity. However, our knowledge of the processes occurring in the mine is insufficient for the development of adequate control measures. Neutralization by lime is widely practiced. Prevention of oxidation by sealing mines has been attempted without significant success. More information about the biochemistry and ecology of the microorganisms responsible for acid production in mines undoubtedly would lead to the development of adequate control measures.

The microbial processes controlling the production of acid mine wastes are discussed by Lundgren, Vestal, and Tabita in Chapter 4.

We have only recently begun to realize that vast quantities of oil find their way into the sea, accumulate there, and cause extensive damage to the biota. The most obvious result of oil pollution is the death of large numbers of birds and fish. However, the most destructive effects occur in the intertidal zones and are not immediately obvious to the untrained observer. These are the zones of intense biological activity where invertebrates breed and where there are high concentrations of algae. When hydrocarbons settle on these sediments, they kill both invertebrate larvae and algae. Processes controlling degradation of hydrocarbons in the sea are discussed by Floodgate in Chapter 7.

The disturbance of biological communities is a potential danger of oil pollution. Many aquatic organisms use chemical signals to communicate with each other. These signals are essential for reproduction, aggregation, and detection of prey by predators. The male lobster, for example, is attracted to the female by a chemical exuded by the female. The gravid female exudes another chemical which repels the male. The ability of starfish to find their prey, the oyster, is facilitated by exudation of a chemical by the oyster which the starfish can detect. The chemical detection of prey by predators is not confined to animals in the aquatic environment. Microbial predators can also chemically detect their prey. These microbial predator–prey detection systems are totally inhibited by crude oil and by hydrocarbons. Blumer (3) demonstrated inhibition of the starfish–oyster predation by crude oil. The implications of this form of inhibition for marine ecology are obvious.

The hazards to man from hydrocarbons in oil lie in their concentration in the food chain. One would expect to find a situation analogous to DDT, that is, apparently innocuous concentrations of hydrocarbons in the water concentrating to hundreds or thousands of ppm in fish. Many of the aromatic hydrocarbons which are extremely slow to degrade probably find their way into the food chain in this way. The fact that many of these compounds are carcinogenic should make determination of accumulations in marine fish a priority for research. Kaufman and Plimmer describe the hazards to man of pesticide hydrocarbons in Chapter 8.

The use of electricity in the United States is increasing at a rate of approximately 7% per annum. Power plants usually depend on surface water as coolants. The development of nuclear sources of power, which require approximately 50% more water for each temperature rise than fossil fuel plants, will markedly increase the quantity of surface water required for cooling purposes in the next few years. A projection of that development is shown in Figure 1.2. It has been estimated that if the trend in power use continues, the power industry will be using more than 20% of the fresh runoff water by 1980 (4).

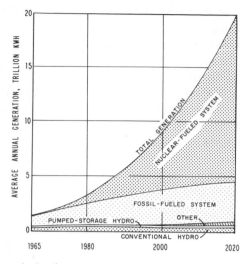

Figure 1.2 Projected development of sources of electrical power in the United States (4).

The effects of increased temperature on the aquatic environment are striking. Higher temperatures serve to de-aerate the water. Thus the river might have a strong capacity to assimilate organic matter at the ambiant temperature of 23°C. If waste heat increased the temperature to 30°C, the dissolved oxygen concentration would fall below the minimum required to satisfy existing stream standards (4). In impounded waters an input of heat would cause destratification with the consequent upwelling of nutrients, and a further decline in water quality. In Chapter 10 Cairns and Lanza describe the detrimental effects of thermal pollution on algal and protozoan communities.

For almost a century water pollution control has been based on two principles: (*a*) infinite dilution of the wastes and (*b*) secondary treatment. In areas of low population density it has also been assumed that both organic matter and pathogens would be diluted in the infinite sink of the biosphere. In recent years urbanization has forced the development of facilities for domestic wastes where local outfalls were acceptable in earlier times. The quantities of domestic waste flowing into rivers, lakes, and the sea were simply too great for the natural waters to be considered an infinite sink. The negative effects of eutrophication are not sufficiently realized in many areas. Secondary treatment facilities are being built to overcome the problem of the infinite sink. However, this form of treatment, while disposing of the organic matter and pathogens in the sewage, allows high concentrations of nutrients to be returned to the natural water.

The fallacy inherent in the consideration of our natural waters as an infinite sink is most apparent when we deal with industrial wastes, pesticides, oil, and waste heat. We are only beginning to understand that chemicals may accumulate in the biosphere and cause widespread, deleterious ecological changes. Our rapidly increasing use of power could result in a change in the biological equilibrium in many of our rivers and lakes.

H. T. Odum, in his monograph on the relationship between the environment, power, and society (5), points out that industrial man has upset the equilibrium of the biosphere by utilization of fossil fuels. This puts a great stress on the biosphere as the absorbent of wastes with low levels of energy. The resultant perturbations adversely affect both energy and material flow through the system. Table 1.1 shows Odum's view of the effect of stress

Table 1.1   Species Diversity in Some Systems (5)

System	No. of Species Found When 1000 Individuals Are Counted
Systems with few species in stressed environments	
Mississippi Delta, low salinity zone	1
Brines	6
Hot springs	1
Polluted harbor, Corpus Christi, Texas	2
Complex systems in stable environments	
Stable stream, Silver Springs, Florida	35
Rain forest, El Verde, Puerto Rico	75
Ocean bottom	75
Tropical sea	90

on the diversity of species in natural populations. It is apparent from this table that in the absence of perturbations homeostasis is maintained in biological systems, and there is a maximal level of diversity. Under conditions of stress, diversity declines and the whole ecosystem becomes more unstable. A more detailed discussion of H. T. Odum's approach to disturbances of energy flow by auxiliary energy sources is discussed by Stumm and Stumm-Zollinger in Chapter 2.

The one statement that can be made with certainty about these ecological disturbances is that we are only beginning to diagnose pollutants which either accumulate in the food chain or have a deleterious effect on the biota in extremely low concentrations. We are virtually ignorant of the ecological processes and interrelationships occurring in either healthy or polluted natural waters.

Our capacity to restore the quality of our natural waters is not simply dependent on legislation or on current technology on waste treatment.

Water pollution is analogous to a newly discovered group of diseases. Before these diseases can be adequately controlled, we must study extensively the nutrition and physiology of the patient.

Information gained from these studies will enable us to determine which materials disturb the ecology of natural waters or are potentially hazardous to man. Hopefully we can develop new concepts which we can use to prevent these materials from entering our natural waters. An understanding of the process of self-purification of natural waters may facilitate the development of new standards for materials which are already accepted as pollutants, but whose effect either on the environment or on man is as yet largely unknown. This information may also lead to the development of processes by which pollutants already in a natural water, for example, nutrients or oil, can be neutralized.

REFERENCES

1. A. M. Beeton, in *Eutrophication, Causes, Consequences, Correctives,* U.S. National Academy of Sciences, Washington, D.C., 1969.
2. W. Rodhe, in *Eutrophication, Causes, Consequences, Correctives,* U.S. National Academy of Sciences, Washington, D.C., 1969.
3. M. Blumer, Woods Hole Technical Report No. 70-20 (1970).
4. P. A. Krenkel and F. L. Parker, in *Thermal Pollution* (P. A. Krenkel and F. L. Parker, Eds.), Vanderbilt Univ. Press, Nashville, 1967.
5. H. T. Odum, *Environment, Power, and Society*, Wiley, New York, 1971.

Part I
Microbial Changes Induced by Inorganic Pollutants

2 The Role of Phosphorus in Eutrophication

Werner Stumm, Federal Institute for Water Resources and Water Pollution Control, Swiss Federal Institute of Technology, Zurich, Switzerland.

Elisabeth Stumm-Zollinger, CH 8700 Küsnacht-Zurich, Switzerland.

One of the most important problems in the pollution of inland waters is the progressive enrichment of waters with nutrients concomitant with mass production of algae, increased water productivity, and other undesirable biotic changes. Such a progressive deterioration of water is referred to as eutrophication.

2-1. The Problem: Conflict between Resource Exploitation and Water Pollution Control

There is a continuous flux of energy to and from the earth's surface. Living matter extracts radiation energy from the sun and uses this energy to organize the living system; that is, the input of solar energy is necessary to maintain life. The flux of energy through the system is accompanied by cycles; cycles of water, nutrients, and other elements (hydrogeochemical cycles) and by cycles of life through different trophic levels. An *ecological system* may be considered as a unit of the environment that contains a biological community (primary producers, various trophic levels of consumers and decomposers) in which the flow of energy is reflected in the trophic structure and in natural cycles (1).

The energy fixed by all the plants in the ecosphere is between 10^{17} and 10^{18} kcal/year. Man as a human animal plays a relatively minor role in the physiology of the ecosphere; the energy involved in his metabolism is about 10^{15} kcal/year. On the other hand, man as an inventive intellectual being, with his capacity to dominate and to manipulate nature, dissipates from 10 to 20 times in the United States, and in Europe from 50 to 100 times, as much energy as he requires for his metabolism (2). The energy utilized by our society for its own advantage imposes a stress upon the environment, because most of it ultimately causes a simplification of the ecosystem, specifically a reduction of the food web and a shortening of the food chain. The less complex a natural ecosystem, the less stable it is and the more liable to perturbations and to catastrophe.

Figure 2.1*a* shows a diagram of the exponential growth of the human population, of energy consumption, and of fertilizer production (3). Basically, it is this growth which causes progressively rising degrees of ecological disorder. Pollution is inextricably interrelated with the use that is made of energy resources. Figure 2.1*b* illustrates that fertilizer consumption is related to population density. In order to feed the increasing human population, we are dependent on a *highly productive agriculture*. This creates a conflict between resource exploitation and pollution control because the activities necessary to maintain a productive soil monoculture (fertilizing, plowing, seeding, weeding, controlling pests, etc.) are incompatible with and counteract measures designed to keep surface waters in a nonproductive, nonpolluted state.

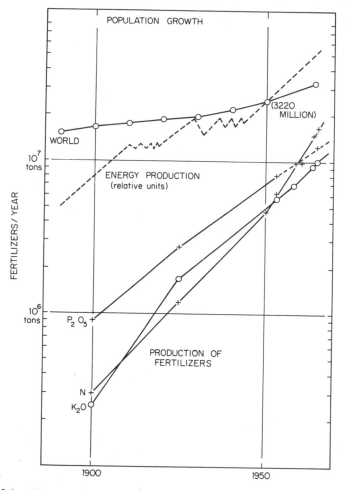

Figure 2.1a Exponential growth of population, energy utilization, and fertilizer production (3).

2-2. Nutrients and the Balance between Photosynthesis and Respiration

In a balanced ecological system a mutual interaction between the activities of producer and consumer organisms, a steady state of photosynthesis and respiration, that is, between production and destruction of organic material as well as between production and consumption of oxygen, seems to be maintained. This balance provides a constant surplus of oxygen. The

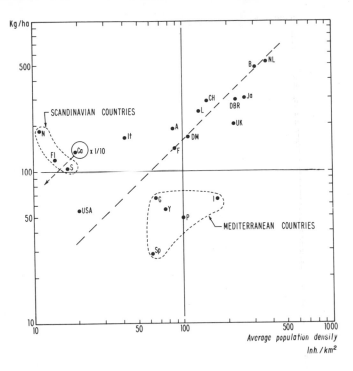

Figure 2.1b Fertilizers used $(N + P_2O_5 + K_2O)$ per unit of arable land surface in relation to population density (3). Letters refer to countries; DBR = German Federal Republic, A = Austria, CH = Switzerland, N = Norway, Fi = Finland, S = Sweden, Ca = Canada, USA = United States of America, F = France, DM = Denmark, L = Luxemburg, UK = England, Ja = Japan, B = Belgium, NL = Netherlands, Sp = Spain, G = Greece, Y = Yugoslavia, I = Italy, P = Portugal.

stationary state between photosynthetic production, $P = dp/dt$ (rate of production of organic material, p = algal biomass) and heterotrophic respiration, R (rate of destruction of organic material) (Figure 2.2) can be characterized in terms of a simplified stoichiometric chemical equation (4, 5).

$$106CO_2 + 16NO_3^- + HPO_4^{2-} + 122H_2O + 18H^+$$
$$(+ \text{ trace elements; energy})$$
$$P \downarrow\uparrow R. \hspace{3em} (1)$$
$$\{C_{106}H_{263}O_{110}N_{16}P_1\} + 138O_2$$
$$\text{algae protoplasm}$$

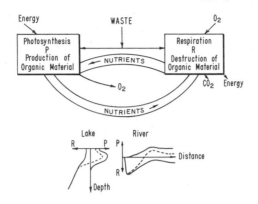

Figure 2.2   Balance between photosynthesis and respiration. A disturbance of the *P–R* balance results from vertical (lakes) or longitudinal (rivers) separation of *P* and *R* organisms. An unbalance between *P* and *R* functions leads to pollutional effects of one kind or another: depletion of O_2 if $P < R$, or mass development of algae if production rates become larger than the rates of algal destruction by consumer and decomposer organisms $(R < P)$.

A.   The Regulation of the Nutrient Composition of Natural Waters. Figure 2.3 shows the correlation between dissolved nitrate, phosphate, and carbonate carbon for waters of the Western Atlantic (4). For other oceans approximately the same constant correlations are found. The mole ratio is $\Delta N \div \Delta P \div \Delta C \approx 16 \div 1 \div 106$ and reflects the atomic compositions of algal protoplasm. In the photosynthetic zone, phosphate, nitrate, and carbonate carbon are eliminated from the water and used for algal growth in a ratio of $1 \div 16 \div 106$. In the deeper water layers, settled algae are mineralized whereby phosphate nitrate and carbonate carbon are released into the water in a similar ratio. Because oxygen participates in the photosynthesis and respiration reactions, corresponding correlations between phosphate and dissolved $O_2(\Delta O_2/\Delta P = 138 \div 1)$ and between nitrate and $O_2(\Delta O_2/\Delta N \approx 9)$ are observed. Table 2.1 shows that the observed correlations are indeed in accord with the stoichiometric composition of plankton.

It is remarkable that the summation of the complicated processes of the *P–R* dynamics, brought about by so many different organisms, results in such simple stoichiometry. The stoichiometric formulation of Reaction 1 reflects in a simple way *Liebig's law of the minimum*. Figure 2.3 shows that, as a result of photosynthetic assimilation, seawater becomes exhausted in dissolved phosphorus and in nitrogen simultaneously. Not considering temporary and local deviations, one infers that both nitrogen and phos-

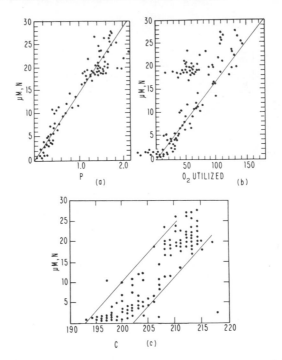

Figure 2.3 Stoichiometric correlations between soluble nitrate, phosphate oxygen and carbonate carbon in the western Atlantic. The correlations can be explained by the stoichiometry of reaction 1. (*a*) Correlation between nitrate nitrogen and phosphate phosphorus in waters of western Atlantic. (*b*) Correlation between nitrate nitrogen and apparent oxygen utilization in the same samples. The concentrations are μM. Phosphorus corrected for salt error (4). (*c*) Correlation between nitrate nitrogen and carbonate carbon in waters of western Atlantic. Concentrations in μM (4).

Table 2.1   Ratios of the Elements Involved in the Oxidation of Organic Matter in Seawater at Depth and Those Present in Plankton of Average Composition, by Atoms (after Richards and Vaccaro, 1956)[a]

Seawater Analyses	ΔO	ΔC	ΔN	ΔP	Ref.
Northwest Atlantic	−180	105	15	1	Redfield (4)
Cariaco Trench,					Richards and Vaccaro
upper layers	−235	—	15	1	(1956)
Plankton analyses	−276[c]	106	16	1	Fleming (1940)[b]

[a] Richards, F. A. and R. H. Vaccaro: *Deep Sea Research* **3**, 214 (1956).
[b] Fleming, R. H., *Proc. 6th Pacific Sci. Cmgr. Calif.* **3**, 535 (1939).
[c] Estimated assuming 2 atoms $O \leftrightharpoons$ 1 atom C and 4 atoms $O \leftrightharpoons$ 1 atom N; as quoted in [32].

phorus determine the extent of organic production in the western Atlantic. One might speculate that originally phosphorus (*e.g.*, from apatite) was the minimum nutrient. In the course of evolution the concentration of nitrogen was adjusted by nitrogen fixation and by denitrification to the ratio presently found. Alternatively, one could also argue that the stoichiometry of living matter is an evolutionary reflection of the composition of seawater.

Similarly as in the oceans, simple stoichiometric relations can be observed in some lakes [Figure 2.4 (5)]. The molar ratios may differ somewhat from the ratios postulated in Reaction 1, but systematic analytical errors, difficult to avoid in the analysis of soluble phosphorus, may influence the slope of the curves.

The stoichiometric relations of Reaction 1 are not readily observable, of course, in systems where the mixing is fast in comparison to the transport of plankton. In rivers and estuaries, "pockets" frequently contain progres-

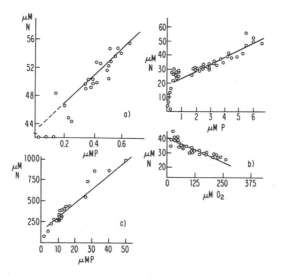

Figure 2.4 Correlation between concentrations of soluble nitrate, phosphate, and oxygen in eutrophic lakes. Data of Lake Constance (*a*) [from results in (24)], Lake Zurich (*b*) [*cf.* (5)] during summer stagnation, and of Lake Norrviken (*c*) [*cf.* (5)] (winter stagnation). Data given are for samples taken at various depths and at various times; for the correlation with dissolved O_2 only data from the hypolimnion were considered. P and N are eliminated (photosynthesis in epilimnion) from and released (respiration in hypolimnion) into the water in constant proportion. Hence the difference in [P] and [N] results from disturbance between photosynthesis and respiration. Nearly all [P] is eliminated when the water still contains substantial [N]; hence in these lakes, P is a major productivity controlling factor.

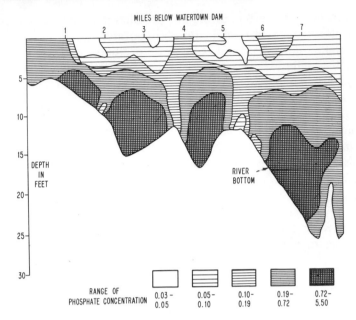

MILES BELOW WATERTOWN DAM

DEPTH IN FEET

RIVER BOTTOM

RANGE OF PHOSPHATE CONCENTRATION

| 0.03 - 0.05 | 0.05 - 0.10 | 0.10 - 0.19 | 0.19 - 0.72 | 0.72 - 5.50 |

Figure 2.5 Phosphate study profile in the Charles River basin. Data by Process Research Inc., Cambridge, Massachusetts.

sively accumulating biodegradation products of algal protoplasm (Figure 2.5).

B.   Pollution as a Departure from the Balance between Photosynthesis and Respiration.   A steady state between photosynthesis and respiration is a prerequisite for the maintenance of a constant chemical composition in the water. Photosynthesis and respiration play an important role in the self-purification of natural waters. A temporally or spatially localized disturbance of the stationary state between photosynthesis and respiration leads to chemical and biological changes, that is, to pollution (5, 6). The condition $P > R$ is characterized by a progressive accumulation of algae which ultimately leads to an organic overloading of the receiving waters. When $R > P$ the dissolved oxygen may be exhausted (biochemical oxygen demand) and ultimately NO_3^- SO_4^{2-} and CO_2 may become reduced to $N_2(g)$, NH_4^+, HS^-, and $CH_4(g)$. The balance between P and R is necessary to maintain a water in nonpolluted and aesthetically pleasing condition. When $P \approx R$ the organic material is decomposed by respiratory (heterotrophic) activity as fast as it is produced photosynthetically; O_2 produced by photosynthesis can be used for respiration (Figure 2.2). A departure from $P–R$ balance results when a natural water receives an excess of organic heterotrophic nutrients or inorganic algal nutrients. The $P–R$ balance may also be disturbed when P and R organisms become physically separated.

C. Lakes. In a stratified lake a vertical separation of P and R results from the fact that algae remain photosynthetically active only in the euphotic upper layers. After having settled under the influence of gravitation, the organic algal material that was synthesized with excessive CO_2 or HCO_3^- in the upper layers of the lake becomes biochemically oxidized in the deeper layers. Most of the photosynthetic oxygen escapes to the atmosphere and does not become available to the R organisms in the deeper water layers (Figure 2.2). An excessive production at the surface of the lake ($P >> R$) is paralleled by anaerobic conditions at the bottom of the lake ($R >> P$). This progressive enrichment in nutrients and the enhancement of productivity may ultimately lead to the filling up of a lake with sediments. Since the last ice age many lakes have so disappeared.

D. Rivers, Estuaries, and Fjords. The characteristics of hydraulic circulation influence the distribution of the biochemically important elements markedly. In rivers, a longitudinal separation of the P and R functions may frequently be observed (Figures 2.2 and 2.5). Countercurrent systems are particularly effective in producing changes in the distribution of nutrients along the direction of flow; they show nutrient accumulation in the direction from which the surface current is flowing (7).

In estuaries and in most fjords the circulation creates a trap in which nutrients tend to accumulate. Algae grown from nutrients which were carried seaward in the surface outflow eventually settle and become mineralized. The mineralization products are then carried landward by the countercurrent of more dense seawater that moves in to replace the water entrained in the surface outflow.

A reversal of the currents characteristic of the esturaine circulation, *an antiestuarine circulation*, leads to an impoverishment in nutrients. The *Mediterranean*, which is the most impoverished large body of water known, may serve as an example. The antiestuarine circulation results from the fact that in the Mediterranean evaporation exceeds the accession of fresh water by about 4% (7). Hence a much larger volume of water flows in through the Strait of Gibraltar than is required to replace the loss by evaporation. Nutrients that tend to accumulate in the deeper layers are continuously being pumped out toward the Atlantic and toward the Black Sea.

E. Phosphorus as a Limiting Nutrient. A complex set of conditions must be fulfilled to allow optimum algal growth. In addition to the prin-

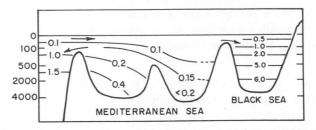

Figure 2.6 Influence of circulation upon distribution of phosphorus (4). Distribution of phosphorus in Black Sea, Mediterranean, and off-lying Atlantic Ocean (diagrammatic). Arrows indicate direction of currents in the Bosphorus and Strait of Gibraltar. Contours, phosphate phosphorus in mg atoms/m³. Depths in meters.

cipal elements such as C, N, P, K, and S that are needed for the synthesis of organisms, the nutritional requirements embrace a long list of quantitatively minor substances including Fe, Mn, Cu, Co, Zn, B, and Mo. Furthermore, organism growth may be stimulated by small quantities of organic growth factors, for example, thiamine, niacin, biotin, and vitamin B_{12}.

An approximate estimate of the nutritional requirements of phytoplankton can be derived from its elemental composition (Equation 1 and Table 2.1). In most natural waters carbon (in the form of CO_2 of HCO_3^-) is abundantly available.* The total reserve of trace metals (Mn, Mo, etc.) available in a natural body of water usually exceeds by far the requirements of the aquatic plants. In most cases nitrogen or phosphorus are growth limiting. These two elements are thus potential growth promoting factors for photosynthetic organisms.

It may be pointed out for the lakes considered (Figure 2.4), that when the concentration of phosphorus approaches zero (intersection of correlation line with ordinate), nitrate is still present in substantial concentrations. In these particular lakes, therefore, the assumption is well justified that, besides localized and time-transient deviations, phosphorus, rather than nitrogen, limits biomass synthesis (5). Using the entire lake as a "batch system" for determining the algal growth potential offers some advantage over the nutrient-enrichment experiments carried out in enclosed bottles. In the lake there is a continuous nutrient replenishment by a variety of routes, including transport from deeper waters and sediments. Thus, for example, certain trace elements (Mn, Mo, etc.) often may limit the productivity in bottles, but rarely become limiting in the lake with its sediments. Deficiency

---

* Recent claims (8) that carbon dioxide limits productivity are based on experiments in essentially noncarbonate bearing waters; because such waters are very rare in nature judgments drawn from these experiments cannot be applied to most natural waters.

of trace elements in a lake usually occurs only as a temporal or spatial transient. Inorganic or organic "growth factors," especially hormones and chelators, may affect the composition of the algal community rather than the biomass. A distinction must be made between the qualitative and quantitative roles of such growth factors. As pointed out by Vollenweider (3), there are no experimental data to prove conclusively that these substances may be determinant in the process of eutrophication.

For most inland waters phosphorus appears to play a major role in influencing productivity. In some estuaries and in many coastal waters, nitrogen appears to be more limiting. Under almost all circumstances phosphorus is a key element in the fertilization of natural bodies of water. One need also consider that any incipient deficiency in nitrogen can be obliterated by nitrogen fixation. The ability to assimilate or fix nitrogen, for example, by certain blue-green algae, has been shown to be of importance in natural waters.

2-3. Limnological Transformations and Phosphorus Loading

Phosphorus occurs in various forms of water. The chemistry of aquac phosphorus has been reviewed recently (9). Table 2.2 gives data on phosphorus concentrations typically encountered in natural and wste water. It is evident that in many waters only a fraction of the total phosphorus present is available as soluble orthophosphate. There is still uncertainty in the analysis of phosphate, especially in the subdivision of total phosphorus into individual fractions (dissolved inorganic P, disolved organic P, particulate P). Rigler (10), for example, has presented interesting evidence that most standard chemical determinations may yield data that are significantly in error.

Table 2.2   Representative Concentration Range of Phosphorus (13)

	Total Phosphorus mg/l as P	Percent Present as Soluble Orthophosphate
Domestic wastewater	5–20	15–35
Effluents from secondary treatment plants	3–10	50–90
Agricultural drainage	0.05–1.0	15–50
Lakes, unpolluted	0.01–0.04	10–30
Lakes, eutrophic	0.03–1.5	5–20
United States rivers	0.01–1.0	
Oceans mean value	0.07	
Rainwater	0.004–0.03	

A. Phosphorus Sources and Loading of Surface Waters. Vollen-weider (3) has reviewed the various nutrient sources extensively and has attempted to analyze them. His report should be consulted for his detailed accounting of the various factors involved. Obviously, the quantity of phosphorus introduced into a surface water from a given drainage area is dependent on density of population and livestock, on the methods and intensity of fertilization, on the type of cultivation (e.g., forests, grassland, cropland), on the pedological characteristics of the soil, and on the type of sewage and waste treatment system involved. The daily per capita excretion of P is 1.5 g. Other loadings in municipal waste can be attributed primarily to phosphates in detergents. A reasonable figure for the total municipal discharge in the United States is *ca.* 3 g P/capita/day; or related to the population density on an annual basis, 1 mg P/(year)m²/ population density (capita/km²). Table 2.3 (Part I), gives representative estimates on the phosphorus runoff from a representative area with a population density of 150 inhabitants/km². The phosphorus loading of lakes can then be estimated by relating the drainage area to the lake surface (Table 2.3, Part II). Ohle (11) has shown that there is indeed a correlation between the "surrounding factor" (drainage factor = drainage area/lake surface) and the primary production of such lakes.

B. Phosphorus Removal in Waste Treatment. Present day aerobic biological waste treatment mineralizes substantial fractions of bacterially oxidizable organic substances but is usually not capable of eliminating more than 50% of phosphate components. This inefficiency of biological treatment with respect to removal of algal nutrients can be understood by comparing the elemental composition of domestic sewage with the stoichiometric relation between C, N, and P in bacterial sludges (12). Most municipal wastes are nutritionally unbalanced for a heterotrophic enrichment process in the sense that they are deficient in organic carbon (Figure 2.7).

An estimate of the extent of possible effects of algal nutrients (potential fertility) is obtained from the schematic stoichiometry equation of production and respiration (Equation 1). If phosphorus is the limiting factor, 1 mg P allows the synthesis of approximately 0.1 g algae biomass (dry weight) in one single cycle of the limnological transformation. This biomass, after settling to the deeper layers, exerts a biochemical oxygen demand of approximately 140 mg for its mineralization. This simple calculation demonstrates that the organic material that is introduced into the lake with domestic wastes (20–100 mg organic matter/liter) may be small in comparison to the organic material that is biosynthesized from fertilizing constituents (3–8 mg P/liter which can yield 300–800 mg organic matter/ liter).

Table 2.3   Phosphorus Runoff and Areal Loadings of Lakes from a Representative Area of Population Density 150 in./km² [a]

	Phosphorus, g/year m² (drainage area)	
	Lower Estimate	Higher Estimate
I. Runoff (drainage area)		
Sewage		
Human wastes	0.08	0.08
Detergents	0.04	0.04
Runoff from highways	0.01	0.01
Industrial wastes	0.01	0.01
	0.14	0.14
Agricultural wastes and drainage from forests		
Arable	0.01	0.05
Meadows and grassland	0.01	0.05
Forests	0.01	0.01
	0.03	0.11
Total	0.17	0.25
	Phosphorus, g/year m² (lake area)	
II. Loading (lake surface)		
For surrounding factor (= drainage area/lake surface)		
5	0.85	1.25
10	1.70	2.50
20	3.4	5.0
Assuming complete removal of P in sewage		
5	0.17	0.57
10	0.32	1.12
20	0.62	2.22

[a] From Vollenweider (3).

More complete phosphate removal can be accomplished readily by treating wastes with lime, iron, or aluminum salts, leading to chemical precipitation (Figure 2.8) of phosphate. A high degree of control over the P elimination can usually be achieved by pH adjustment. For a recent review on chemical phosphate elimination, see Leckie and Stumm (13). However, even if it were possible to eliminate all phosphorus discharged from the sewage system, substantial nutrient loading from less controllable sources would always remain.

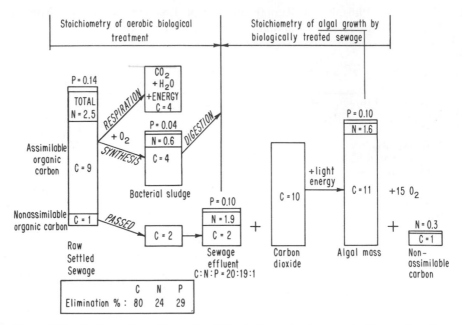

Figure 2.7 Stochiometry of aerobic biological waste treatment. Comparing the relative composition of "average" domestic sewage with the mean stoichiometric relations between C, N, and P in bacterial cells shows in a schematic way that only a fraction of the phosphate and nitrogenous material can become incorporated into the sludge. The inorganic nutrients as released by biologically-treated sewage effluents can be converted into algal cell material. CO_2 (or HCO_3^-) and other essential elements are usually available in sufficient quantities relative to nitrogen and phosphorus.

C. Limnological Transformations.   The concentration of phosphorus in a surface water cannot be predicted merely on the basis of supply and of hydrographic conditions. Figure 2.9a, adapted from Phillips (14), schematically depicts some of the phosphorus locations in a body of water and the exchange processes between the various locations.

Table 2.4 compares the nutrient budget for the ocean (15) with that of a hypothetical lake. According to the estimates given, phosphorus is present in the ocean at concentrations 100 times larger than those eliminated by phytoplankton; the annual use, however, far exceeds the annual contribution to the ocean. Accordingly, the major proportion of the nutrients must be regenerated from organic debris. In the hypothetical lake the annual contribution to the lake may amount to a significant fraction of the annual use by the photoplankton. But even here the cycle of phosphorus is determined largely by regeneration of phosphorus from the biota. The residence

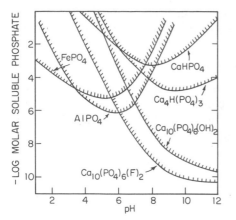

Figure 2.8 Solubility diagrams for solid phosphate phases. The diagrams have been calculated by considering the solubility equilibria together with the appropriate acidity and complex formation constants. The calcium phosphate phases were calculated assuming that 1×10^{-3} M free Ca^{+2} is maintained and that F^- is controlled by $CaF_2(s)$. The aluminum and iron phosphate phases were computed assuming that Al^{+3} and Fe^{+3} are controlled by the respective hydoxide solids.

time in fresh water of *dissolved* phosphorus, that is, the average time phosphorus atoms remain in solution, varies from 0.05 to 200 hours (16). A system having a short residence time may be low in dissolved phosphate, as in the sea, or it may be very active biologically, as during algal blooms. When both conditions occur simultaneously as in a small lake, the residence time becomes very short (16). Accordingly, the concentration of dissolved phosphate in natural waters gives little indication of phosphate availability; quite the contrary, in a highly productive system most of the soluble phosphorus has been taken up into the biomass.

The total reserve of phosphorus in a body of water (*i.e.*, the quantity of soluble, particulate, sestonic, and accessible sedimented phosphorus), however, is a pertinent gross parameter because it gives the ultimate *capacity for biomass synthesis*. The latter can be estimated from the stoichiometry of the photosynthesis reaction.

No single relationship exists between biomass and *productivity*. Produc-

Table 2.4   Phosphorus Budget in the Ocean and a Hypothetical Lake

	Ocean,[a] μg P/liter	Lake,[b] μg P/liter
Total phosphorus	70	1000
Annual use by phytoplankton	0.77	1300
Annual contribution by river, drainage, and waste	0.008	730
Annual loss to sediments	0.008	180
Annual loss through effluent	—	550

[a] From Emery, Orr and Rittenberg (15).
[b] Lake represented in model of Figure 2.9b.

tivity measures the *rate* of biomass production. Riley (17) has established the following empirical relationship between standing crops of algae [A] and primary productivity, $d[A]/dt$:

$$d[A]/dt = (p - r - h)[A] \tag{2}$$

where p, r, and h are empirically determined coefficients for photosynthesis, phytoplankton respiration, and consumption by herbivores (respiration mediated by other organisms), respectively. Because the proportionality coefficient is time and space dependent, the relationship between biomass and productivity varies more or less individually with time and from lake to lake.

A simplified steady-state model (18) depicting some significant limnological transformations of phosphorus (Figure 2.9*b*) attempts to emphasize that the dynamics of the transformations, especially the rate of regeneration of nutrients from phytoplankton, detritus, and sediments and the rate of supply of soluble phosphorus to the algae, are often more important in determining productivity than is the concentration of soluble phosphorus or the phosphorus reserve. Values given in Figure 2.9*b* for abundance of

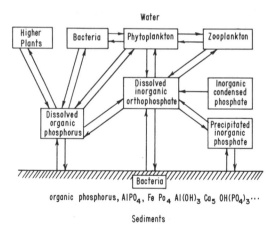

organic phosphorus, $AlPO_4$, $Fe\ PO_4$ $Al(OH)_3$ $Ca_5\ OH(PO_4)_3 \cdots$

Sediments

Figure 2.9a Phosphorus transformations (14). Schematic representation of some of the phosphorus locations in a body of water and of exchange processes between the various locations. Hayes and Phillips (22) have suggested that the turnover times (the time required for loss from a phase of as much phosphorus as is present in that phase) range from *ca.* 5 min for the exchange between dissolved inorganic phosphate and phytoplankton to many days for some of the other exchanges. The slowest processes appear to involve the water–sediment exchanges ($\sim$15 days for abiotic processes and $\sim$3 days for bacterially mediated processes). Exchange between inorganic phosphate and dissolved organic phosphorus appears to be fairly rapid ($\sim$8 hours). These figures were proposed for lake waters. They should be viewed as merely suggestive for a general orientation.

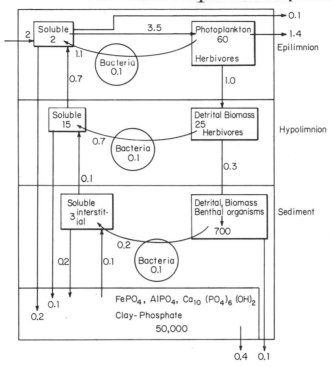

Figure 2.9b A simplified steady-state model describing important steps in the limnological transformation of P in a lake (18). The model simulates a real system by giving a hypothetical balance of the abundance of P in various forms (the numbers in the boxes are μg P/liter lake volume) and of the exchange rates (the numbers on the arrows are μg P/liter lake volume/day). The cycle of phosphorus is determined largely by regeneration of P from biota. Primary production depends to a large extent on the supply of P to the trophogenic layer. For deeper lakes the rate of supply from sediments is small in comparison to the supply by the hypolimnion and by the introduction of P from waste and drainage. A significant fraction of P introduced into the lake is irretrievably lost to the sediments.

phosphorus [P] in various forms (microgram phosphorus/lake volume, V, in liters*) and for exchange rates (microgram phosphorus/lake volume in liter/day†) have been carefully chosen so as to be compatible with real systems as reported by various investigators; residence times, $[P]/[dP]/dt$,

---

* Actual concentrations can be obtained by dividing [P] by the appropriate volume fraction, for example, the actual concentration of $P_{soluble}$ in epilimnion is given by $[P(aq)_{epil}]/(V_{epil}/V)$.

† For a lake of depth Z (cm) this can be converted into an areal exchange rate $[Z(d[P]/dt)]$, for example, for a lake of 10 m depth, 1 μg/liter day corresponds to 10 mg P/m² day.

for soluble P, for P in organisms, and for P in sediments are within the range of observed values. At best this model simulates a real system to an approximate degree; its main purpose is to offer insight into some attributes and dominant forces operating in the real system. Manipulating the figures within the proper constraints may illustrate how real systems behave.

The simplest hypothesis, of course, is to assume that the photosynthetic activity of the individual algal cell increases more or less proportionally with the concentration of limiting nutrient (assumed here to be soluble phosphorus) and that the phosphorus utilization rate depends on the concentration of algae. But the many complicating factors must not be overlooked. To mention a few examples: the growth rate is affected by temperature and light intensity; the size of the plant population is partially determined by the metabolic activity of herbivores and the flux of soluble phosphorus to the epilimnion by waste discharge and by eddy-diffusion of P, regenerated from biota and sediments in the hypolimnetic waters, and by upwelling; water replenishment is significant at steady state, the growth rate increases with increasing dilution rate, but so does the washout of algae from the system; and the vertical upward flux is inversely related to the depth of the lake and the extent of thermocline (density gradient).

Eutrophication. In stratified waters the continuous sequence of nutrient assimilation and mineralization of organic matter accompanied by the physical cycle of circulation and stagnation leads to a progressive retention of fertilizing constituents and in turn to a progressive increase in productivity.

No single collective parameter can circumscribe the relationship between nutrient loading of a surface water and the resulting enrichment process; this relationship is influenced by many factors. Vollenweider (3) has shown convincingly on the basis of data on 20 lakes that a valid correlation can be established between areal limiting nutrient loading and mean lake depth on one hand, and degree of enrichment on the other. In Figure 2.10 this relationship is plotted with phosphorus as a limiting nutrient. The demarcation line indicated in this figure gives perhaps the most relevant collective reference values on permissive phosphorus loadings.

Sawyer's experimental work (19) on the Wisconsin lakes led to the working hypothesis that aquatic blooms are likely to develop in lakes during the summer when the concentration of inorganic phosphorus at the start of the vegetation period is in excess of 10 μg P/liter. After the spring overturn the lakes have a rather uniform distribution of phosphorus and, for lakes of similar depth, the concentrations are related to the annual areal loading. Vollenweider's demarcation line in Figure 2.10 indicates that significant reductions of the input of phosphorus are necessary for the

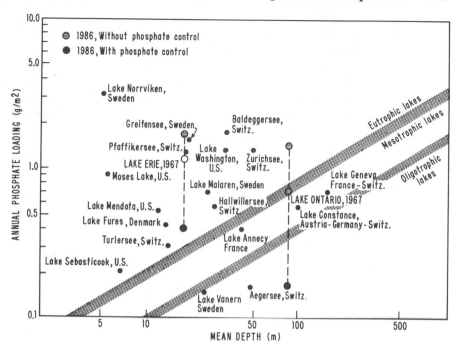

Figure 2.10 Critical phosphorus loading as a function of depth (3).

lakes considered. In Lake Erie, for example, the P input needs to be reduced at least by a factor of four.

D. Some Physical and Chemical Factors. The interaction between organisms and their abiotic environment is influenced by many physical and chemical factors. According to Mortimer (20) some of the physical factors have to do with (a) radiant energy input, (b) nutrient input and loss, (c) oxygen supply, and (d) the interaction of morphometry and motion. In view of the recent lucid discussion (20) these factors will not be reviewed in any detail here. It has already been emphasized that motion and exchange of water masses are of great relevance to the nutrient enrichment process. It has, perhaps, not always been sufficiently realized that bay and inshore water masses are frequently remarkably isolated from offshore waters. Various possible mechanisms may account for coastal entrapment. The Coriolis effect (a deflecting force resulting from the earth's rotation on the water motion) and the "thermal bar" can impede exchange between inshore and outshore water masses. As explained by

Mortimer (20) and Rodgers (21), the "thermal bar" may result during spring or fall because seasonal warming or cooling with consequent stratification takes place more rapidly in the shallow inshore water masses than in deeper isothermal waters offshore; at times when the surface temperatures of the two water masses lie on either side of the temperature of maximum density, mixing of the two masses at their boundary produces a mixture more dense than either; the sinking of this mixture develops a region of convergence that impedes, for a time, the onward movement of the inshore water in which, for a time, the inflow and pollutants from rivers and streams are trapped.

The transport of phosphorus and its release or entrapment is linked to the cycle of iron and is influenced by the redox intensity of the water and its variation with depth. The redox intensity in the deeper water layers in turn depends on the transfer of oxygen and the amount of decomposable material discharged into the lower water. At high $p\epsilon$ values, Fe(III) interacts chemically with phosphate to form insoluble compounds. Fe(III) may act as a converger for P, enhancing its sedimentation into the deeper water layers. At low $p\epsilon$ (2) and subsequent to the reduction of Fe(III) to Fe(II), these elements become soluble.

E. Phosphate Exchange with Sediments. The distribution of phosphate between the sediments and the overlying water is of considerable importance for productivity. According to Hayes and Phillips (22), the "dynamic equilibrium" of phosphorus in a surface water may be represented by

$$\text{Phosphorus in aqueous phase} \rightleftharpoons \text{Phosphorus in solid phase} \qquad (3)$$
$$\text{(small fraction of total)} \qquad \text{(large fraction of total)}$$

Significant quantities of phosphate are usually found in the sediments of surface waters, especially in the sediments of eutrophic lakes and estuaries. Inorganic solid P phases may be formed either by direct precipitation of phosphate with Ca, Al, and Fe compounds in the water column or by chemical reactions in the sediments (13). Clays are important scavengers for phosphate. The P content of a sediment increases with its clay content.

Redox potentials at the sediment water interface and in the top layers of sediments influence the affinity of the sediments for phosphorus. Heterogeneous equilibria, primarily characterized by the solubility of $AlPO_4$ (variscite), $FePO_4$ (strengite), and $Ca_{10}(PO_4)_6(OH)_2$ (hydroxyapatite) and the adsorption of phosphate on clays determine the distribution of phosphate between the aqueous phase and the solid phases (minerals, clays, etc.) in the sediments; the calculated solubility relations (Figure 2.8) have been confirmed experimentally and substantiated semiquantitatively by measurements at the sediment water interface (9, 13).

The solubility of phosphate. Heterogeneous solubility equilibria represent a buffer system for orthophosphate. As Figure 2.8 predicts, a solution in the pH range of natural waters and in equilibrium with the solid phases typically encountered in sediments maintains soluble *o*-phosphate $(H_2PO_4^-, HPO_4^{-2})$ at a molar concentration of -log P = 5.5 ± 1.0. The concentration of total soluble phosphorus, however, can be higher because of the presence of *soluble organic* phosphorus. This is illustrated by the transfer cycle:

$$
\begin{array}{c}
\text{bact.} \\
\text{particulate organic P} \rightleftharpoons \text{soluble organic P} \\
b \qquad a \qquad\qquad c \\
\text{equil.} \\
\text{solid inorganic P phases} \rightleftharpoons \text{soluble } ortho\text{-}P
\end{array}
\qquad (4)
$$

While the concentration of soluble inorganic phosphate remains defined by the solubility equilibrium, the turnover rates of reactions *a* and *c* determine the concentration of soluble organic P; that is, by incorporating phosphorus into organic compounds that do not participate directly in the solubility equilibria with solid phases, bacteria tend to increase the concentration of soluble P. The buffering action of the sediments prevents the accumulation of excess quantities of phosphorus readily available for assimilation in the overlying waters, and supplies phosphorus to the water when it becomes depleted.

Information on the chemistry of phosphate interactions at the sediment water interface is at best qualitative. The rates of interchange cannot be readily determined in the field, and the results of laboratory experiments must be used with caution to derive conclusions with regard to natural systems.

Interstitial water. In order to be able to predict the dynamics of nutrient availability from sediments, information on the nutrient concentration in the interstitial water is more important than data on the total nutrient content of sediments. Sediment samples from various eutrophic lakes in the western part of the United States contain 0.08 to 10.5 mg soluble P/ liter interstitial water (23). These concentrations may be compared with those found in the upper portion of the sediments of the less eutrophic waters of Lake Constance [0.06 to 0.15 mg/liter (24)], and those of the marine Catalina basins in the area off Southern California [0.24 to 1.2 mg/liter (25)]. Only in sediments of extremely oligotrophic lakes is the interstitial water not in saturation equilibrium with the inorganic P phases. The results of preliminary experiments carried out in this laboratory on the rate of transport through sediments indicate that at an undisturbed sediment–water interface, the rate determining step for the transfer of phosphorus from the solid phases to the overlying water is generally the diffusional transport through the interstitial water (26). Experimentally determined diffusion coefficients are $D \leq 2 \times 10^{-6}$ cm^2/sec. For a large concentration gradient of 10^{-5} M/cm a maximum diffusional transfer rate of ca. 9×10^{-6} mole/m^2 day [0.27 mg/m^2 day] may be estimated. Pomeroy's estimate (16) for Dobay Sound of 10^{-6} mole/m^2 day is reasonable and in accord with this estimate. Bacterial activity may accelerate the transfer of P from sediment to water by affecting the concentration gradient.

Sediments do not always remain undisturbed in natural waters. Ground water flow, turbulence, occurring especially in shallow waters, and the activity of burrowing organisms (tubifex or larvae of chironomides) as well as gases (CH$_4$ and H$_2$S) that may erupt in deposits undergoing rapid anaerobic decomposition and water displacement by consolidation of deposits may promote the upward transport of P.

In a productive lake the hypolimnetic phosphate concentration is subject to marked seasonal variation; one might expect that the condition $[P(aq)_{hypol}] < [P(aq)_{interstitial}]$ typically exists during and immediately following the spring and fall overturns, whereas the condition $[P(aq)_{hypol}] \gg [P(aq)_{interstitial}]$ prevails during the latter parts of the stagnation period. Hence, during the period preceding massive algal growth, that is, when nutrients are needed there would be a net addition of P to the overlying waters, while a net loss of P to the sediments would occur during the summer and the latter part of winter. Although on balance a substantial fraction of P becomes irretrievably lost in the sediments, the sequence of P release and P binding tends, especially in shallow waters, to facilitate algal blooms.

In deeper lakes, however, the quantity of P released into the water from sediments and transported into the epilimnion may be small relative to other fluxes (Figure 2.9b). Vollenweider (3) would classify, for example, a lake of 50 m mean depth as euthrophic if its areal loading exceeds 500 mg P/m^2 year. This figure [1.4 mg P/m^2 day on a daily basis] may be compared with (perhaps only valid for not too shallow lakes) the 0.27 mg P/m^2 day for the rate of transport of P from the sediments to the overlying waters, which we estimate to be an upper limit value. Hence, with increasing depth, the relative contribution of P from the sediments becomes less significant, and the beneficial effects of the sediments—the net removal of P from the water by the sediments—becomes preponderant (compare Figure 2.9b).

The removal of sediments by dredging, while perhaps occasionally of value in bays and shallow waters, is not a suitable means for lake restoration. On the contrary, most probably, such a measure would be most detrimental to the lake (18).

F. The Sedimentary Record. It is frequently inferred that the P content of sediments decreases with depth; that is, it is smaller for older deposits. This is not necessarily true, not even for lakes where eutrophication has been accelerated in recent decades by progressively increasing loading with sewage and farmland drainage. The phosphorus profile of sediments cannot serve as a simple direct memory record of the enrichment process (20). Livingston and Boykin (27) studied the phosphorus in the sediment of Linsley Pond and found that high phosphorus-binding capacity is correlated with high mineral content of the sediment. Thus, the phosphorus content of sediments decreases as the rate of mineral sedimentation declines, presumably because phosphate is being released from the sediment and recycled through the ecosystem.

Similarly, the quantity of P deposited in the sediments may depend more on the iron cycle than on the rate of supply of P from the drainage basin. As pointed out before, because of the chemical interaction of $Fe(III)$ and phosphate, P is precipitated much more efficiently under aerobic than under anaerobic conditions. At times prior to cultural enrichment, the sediment–water interface remains aerobic, and iron and manganese mobilized in the soils of the drainage area are trapped almost quantitatively (28).

2-4. Eutrophication and Ecological Succession

The introduction of organic and inorganic nutrients into surface waters affects photosynthesis and respiration (P and R), but need not necessarily disturb the P–R balance. Sewage and algal ponds are examples of systems with a very high rate of assimilation and respiration; when $P = R$, such ponds do not produce undesirable conditions. A balance between P and R is a necessary but not sufficient attribute of a nonpolluted and aesthetically pleasing condition. We may compare the algal pond with similarly productive coral reefs (29). While in algal ponds we find a simple and short producer–consumer food chain of great instability, coral reefs are stable systems made up of a complex community with a large number of energy pathways.

A. Succession of Ecosystems. All ecological systems tend in the long run to approach a steady state. The succession of an ecosystem is directed toward an increased ecosystem organization and toward a state of improved metabolic efficiency where energy is increasingly being used for mainte-

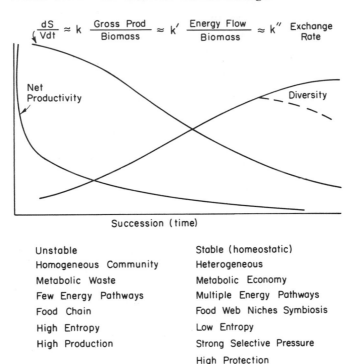

Figure 2.11 Trends in ecological succession. Ecological systems tend to proceed in the direction of a stationary state and become progressively more self-regenerating with regard to energy and materials. Most attributes listed are those of E. P. Odum (1).

nance (1, 30, 31). It culminates in a stabilized ecosystem in which maximum biomass and symbiotic function between organisms are maintained per unit of available energy flow. Common trends abstracted from observations are summarized in Figure 2.11. The figure displays qualitatively the development of ecosystems; the attributes used to describe the developmental stages are those of E. P. Odum (1). The time scale is extremely variable and depends on the type of ecosystem under consideration. Figure 2.11 also suggests a few relevant effects of man's action on ecological succession.

Energy transfer, reciprocally interrelated with *biological structure*, is perhaps the most important attribute of ecological successional stages. The input of energy causes a reduction in maturity and will "drive" an eco-system into a younger developmental stage of higher net productivity and of decreased structure. As pointed out by H. T. Odum (32), man's success in adapting some natural systems to his use has essentially resulted from applying auxiliary work circuits using fossil and atomic energy in plant and animal systems. Obviously we need to exploit ecosystems for food production. But progress in agricultural food production, essentially achieved by pumping more auxiliary energy (e.g., mechanical energy, heat and chemical energy in the form of organic and inorganic nutrients) through a system, must be paid for by destruction of homeostatic mechanisms and loss of structure. By clearing land, by planting crops, and by controlling weeds, pests, and other competitors, a monoculture of high crop productivity and of high instability is established. It appears well documented that deforestation causes erosion of primary minerals; nutrients originally locked in the biota can no longer be retained by the soil system. Not only does the soil system become less stable, but also the downhill flow of nutrients from soil to water is enhanced. As pointed out by Leopold (33), soil and water are not two organic systems, but one; a derangement in either affects the health of both. If energy input remains small, complex nutrient circuits caused by longer food chains retard the drainage and erosion of nutrients into the water and enhance their storage in soils. Dredging of sediments, cutting of reeds, or fluctuations in water level, temperature, or velocity gradients decrease the number of ecological niches and reduce the maturity of the system. Excess autotrophic or heterotrophic activity, and hence water pollution are caused by acceleration of energy flow and of exchange rates resulting from direct input of organic and inorganic waste constituents into receiving waters.

The introduction into receiving waters of materials nonindigenous to ecosystems (e.g., toxic substances, ionizing radiation, and organic substances which are not of recent biological origin such as petrochemicals) leads to a decrease in the number of pathways for energy flow and thus upsets the community structure. The ecological effects caused by such materials or agents follow the same general pattern known from other types of disturbances. Woodwell (34) in generalizing on pollution has concentrated on some of the most gross changes in the plant communities of terrestrial ecosystems and has illustrated lucidly that changes in ecosystems caused by many different types of disturbances are similar and predictable. A similarity of response has been documented for chronic irradiation, fire, sulfur dioxide, pesticides, and wastes. In the cases reported (34) the loss of structure typically involved a shift away from complex arrangements of specialized species toward the generalists; away from forests toward shrubs; away from diversity in birds, plants, and fish toward monotypes; away from tight nutrient cycles toward loose ones with terrestrial systems becoming depleted, and with aquatic systems becoming overloaded; away from stability toward instability.

B.   Aging of Lakes.   Lakes that continue to receive nutrients progressively increase their productivity and deterioration in water quality. As the process of eutrophication advances because of increased sedimentation, the lakes become shallower and eventually disappear. This culturally accelerated eutrophication has been referred to as an aging process.

How does this developmental process enhanced by human influence fit into the pattern of ecological succession? The biotic changes accompanying *cultural* eutrophication of a lake do not parallel those accompanying *natural* successions. Evolutionary succession in a hypothetical lake to which input of nutrients was curtailed to zero would be in the direction of oligotrophy. Accordingly, as pointed out by Margalef (30), oligotrophy as a more mature state should succeed eutrophy, not precede it. The development from an oligotrophic to a eutrophic state, that is, to a state of lower species diversity and lower maturity is caused by nutrients that are continuously discharged into these waters by runoff from land and by pollution. This inflow of nutrients into a eutrophic lake keeps it in a state of low maturity. As pointed out by Hutchinson (35), we should not think of oligotrophic or eutrophic water types, but of lakes and their drainage basins and sediments as forming oligotrophic or eutrophic systems. The coexistence, side by side, of a highly productive agriculture and of a nonproductive surface water is incompatible.

There is documentation that in a few instances lakes that apparently have become sufficiently insulated from nutrient supply have spontaneously developed in the direction of oligotrophy. For example, it is evidenced by the "memory" stored in sediments that Lake Zürich has gone through eutrophic episodes (36), a thousand or more years ago, as a result of events not yet known (erosion, deforestation, etc.) from which the lake could recover. This lake has, since the turn of the century, undergone extensive cultural eutrophication [Figure 2.12 (36)]. Hutchinson (35) describes the development of a small crater lake between Rome and Siena (Lago di Monterosi). In this case cultural eutrophication was initiated 2000 years ago not by artificial liberation of nutrients into the water, but by a rather subtle change in the hydrographic regime. Later the lake became rather less eutrophic. Such case histories, together with the changes observed in Lake Washington after sewage had been diverted completely from the lake (37), provide important lessons.

2-5.   Preservation and Restoration of Lakes

Can eutrophied lakes be restored? Ecological theory and the case histories referred to above indicate that prospects for preservation and restoration of lakes are quite good provided an array of remedial measures and technology is applied. What has been called the strategy of "ecosystem development" (1) provides a most important basis to evaluate measures of water quality control and lake preservation.

A.   Phosphorus Elimination.   The first and usually most important measure is to reduce the nutrient loading of the inland waters. Phosphorus elimination from point source discharges is more relevant for inland waters and more practicable than nitrogen elimination. As Part II of Table 2.3 indicates, complete removal of phosphorus from sewage (90% elimination is technologically feasible) reduces the loading to a "typical" lake by 40 to 80%. There have been arguments that because of such a low overall efficiency, phosphorus elimination from wastes does not represent an expedient remedial measure. Although such a relatively small reduction in the loading might frequently not be able to reverse eutrophication, it certainly has a beneficial influence on the ecological balance. Specifically, any reduction in the loading reduces the total reserve of P in the lake and thus reduces the ultimate capacity for biomass synthesis. Furthermore, because potential productivity depends on the rate of supply of soluble and assimilable phosphorus to the algae, algal blooms may often be prevented by reducing the P supply to the euphotic zone. As an intermediary measure (*e.g.*, prior to the installation of facilities for complete removal) phosphorus elimination during the summer months might substantially reduce the occurrence of algal blooms even if the total reservoir of phosphorus is reduced only slightly by such a measure.

B.   Measures Other than Waste Treatment.   The "strategy of ecosystem development" (1) provides a most important basis for evaluating measures of water quality control and lake preservation. Table 2.5 lists some examples of measures and how they grossly affect water quality. This brief list illustrates that water pollution control consists not only of waste treatment; many other physical and biological means of stream management need consideration. Especially important is the realization that the activities necessary to maintain a soil monoculture are counteracting measures designed to keep lakes and bays in a nonpolluted and oligotrophic state. Zoning of land in the watershed, specifically the restriction of use of land in the corridor adjoining the surface waters, with preference to

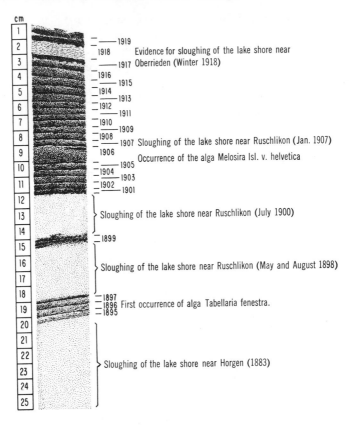

Figure 2.12 Sediment layers from Lake Zürich (1895–1919) [after F. Nipkow, Z. Hydrol. **1**, 101 (1920)]. The alternative sequence of layers result from deposition of $CaCO_3$ during the summer (oversaturation with respect to $CaCO_3$ because of pH increase due to photosynthesis) and accumulation of black Fe (II) sulfide containing sludge during winter (anaerobic conditions).

diversified ecosystems with long food chains are among the most powerful measures for water pollution control.

Some of the measures listed in Table 2.5 have dual or multiple effects and have to be assessed depending on the circumstances. For example, *mixing a body of water*, that is, destratifying a lake, has a variety of effects, and thus may either decrease or increase the lake's ecological stability. Mixing destroys ecological niches, reduces chemical activity gradients, and shortens the food chain; hence the lake becomes more dynamic with an increased energy flow per biomass, reflected in higher gross productivity and respiration. On the other hand, mixing brings P and R activities into

Table 2.5   Some Measures of Water Quality Control

Reducing Ecological Stability	Promoting Ecological Stability
Increase of energy flow By disposing nutrients for hetero-trophs and autotrophs By mixing (destratifying, sediment dredging, etc.) By heat disposal By imposing turbulence	*P–R balance restoration* By reducing waste input By harvesting or washing out of bio-mass By reducing relative residence time or by trapping of nutrients By mixing (bringing P and R to-gether) By fish management By aeration
Exploitation of adjacent soil By crop growing, seeding, weeding, and grazing By fertilizing and irrigating By deforestation By converting grassland into cropland By applying herbicides and pesticides	*Conservative land management* By reforestation By restricting monoculture produc-tivity By zoning (maintaining zones adja-cent to open waters which are kept free of fertilizers and of low net productivity) By controlling erosion By using detritus agriculture
Reduction of structure By using algicides By destruction of niches (removal of reeds) By episodic physical perturbations (flushouts, temperature discharges, shock loadings) By excessive harvesting By disposal of chemicals By interfering with chemostasis	*Enhancement of biological complexity* By establishment of ecological niches (zones, waterfront development) By seeding diverse populations and recirculating certain organisms By maintaining relatively high bio-mass compatible with energy flow By maintaining stratification By selective harvesting By maintaining high chemical buffer intensity (weathering of rocks)

closer contact and better balance, thus reducing net productivity (Figure 2.13). Furthermore, the effective relative residence time of the nutrients in the lake is reduced, thus decreasing the total nutrient reserve of the lake. Reducing the residence time of a nutrient to one-half reduces the phosphorus reserve by one-half and is thus equivalent to a 50% reduction in the nutrient input into the lake.

C. Algicides and Herbicides.   Algicides and herbicides may provide occasionally temporary relief from excessive growths of algae and macrophytes. Usually such chemicals, although they alleviate some symptoms,

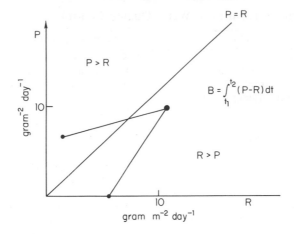

Figure 2.13 Mixing of lake may decrease net productivity. Mixing the top layers (photosynthetic functions; $P > R$) with the bottom layers (heterotrophic functions; $R > P$) increases P and R but may decrease the standing biomass, B.

may be quite harmful in the long range because they reduce the diversity and stability of the aquatic ecosystem. Methods based on control by fungi or viruses will only be successful in the long range if such a measure is accompanied by the maintenance of a diverse species composition. In this context it should be pointed out that foreign materials, for example, petro-chemicals and other refractory organic materials and toxic inorganic sub-stances, such as As and Hg, might actually enhance eutrophication because they adversely affect ecological interactions between organisms.

D.   Aquaculture.   There is no reason why in many cases we should not take advantage of high fertility (32). In some areas the ocean can be used as a sink for biological wastes. There is still room to improve the fertility of the ocean and exploit its production of harvestable food for the increasing population. We should learn to manipulate some of our aquatic environments, for example, some of the smaller lakes or some bays of larger lakes, as aquaculture enterprises. Biological wastes, if properly applied in land and aquatic ecosystems, have to be considered as potential resources; they can be recycled to useful products.

2-6.   Final Remarks

Despite technological advances we lack scientific knowledge and under-standing of how to decide which ecological balance is most desirable for man and of how to resolve the conflict between the need to feed the human population and the desire to maintain a culturally advanced civilization and a high quality of life on one hand, and the necessity to control pollution and to keep the ecosystems of our environment stabilized on the other hand. As pointed out by Dubos (38), "technological fixes are of course needed to alleviate critical situations, but generally they have only temporary usefulness. More lasting solutions must be based on ecological knowledge of the physiochemical and biological factors that maintain the human organization in a viable relationship with the environment."

REFERENCES

1. E. P. Odum, *Science* **164**, 262 (1969).
2. W. Stumm and E. Stumm-Zollinger, in *Advances in Chemistry Series* (In Press), American Chemical Society, Washington, D.C., 1971.
3. R. Vollenweider, Technical Report DAS/CSI/68.27, Organization for Economic Cooperation and Development, Paris, 1968.
4. A. C. Redfield, *James Johnson Memorial Volume*, Liverpool, 1934, p. 177, quoted from in A. C. Redfield, B. H. Ketchum, and F. A. Richards, in *The Sea*, Vol. 2 (M. N. Hill, ed.), Wiley-Interscience, New York, 1963.
5. W. Stumm and E. Stumm-Zollinger, *Chimia* **22**, 325 (1968).
6. E. P. Odum, *Fundamentals of Ecology*, Saunders, Philadelphia, 1961.
7. B. H. Ketchum, in *Eutrophication, Causes, Consequences, Correctives*, National Academy of Sciences, Washington, D.C., 1969.
8. L. E. Kuentzel, *J. Water Pollution Control Federation*, **41**, 1735 (1969).
9. W. Stumm and J. J. Morgan, *Aquatic Chemistry*, Wiley-Interscience, New York, 1970.
10. F. H. Rigler, *Limnol. Oceanog.* **9**, 511 (1964).
11. W. Ohle, *Münchner Beiträge* **12**, 54 (1965).
12. W. Stumm, *Proceedings of the International Conference on Water Pollution Research, 1963*, Vol. 2, Pergamon, New York.
13. J. O. Leckie and W. Stumm, in *Chemical Treatment* (E. Gloyna and W. W. Eckenfelder, Eds.), Univ. Texas Press, Austin, Texas, 1970.
14. J. E. Phillips, in *Principles and Applications in Aquatic Microbiology* (H. Heukelekian and N. C. Dondero, Eds.), Wiley, New York, 1964.
15. K. O. Emery, W. L. Orr, and S. C. Rittenberg, *Essays in Natural Science in Honor of Captain Allan Hancock*, Univ. South Carolina Press, 1960.
16. R. L. Pomeroy, *Science* **131**, 1173 (1960).
17. G. A. Riley, in *The Sea,* Vol. 2 (M. N. Hill, Ed.), Wiley-Interscience, New York, 1963.
18. W. Stumm and J. O. Leckie, *5th International Water Pollution Research Conference, 1970*, Pergamon, London, 1971.
19. C. N. Sawyer, *J. Boston Soc. Civil Engrs.* **53**, 49 (1966).
20. C. H. Mortimer, in *Eutrophication, Causes, Consequences, Correctives*, National Academy of Sciences, Washington, D.C., 1969.

21. G. K. Rodgers, *Pub. 15 Great Lakes Division*, Univ. Michigan, Ann Arbor, Mich. 1966, p. 369.
22. R. Hayes and J. E. Phillips, *Limnol. Oceanog.* **3**, 495 (1958).
23. A. R. Gahler, *Proceedings of the Eutrophication Biostimulation Workshop*, U.S. Dept. of Interior, Berkeley, Calif., 1968, p. 243.
24. G. Wagner, *Schweiz. Z Hydrol.* **30**, 75 (1968).
25. S. C. Rittenberg, K. O. Emery, and W. L. Orr, *Deep-Sea Res.* **3**, 23 (1955).
26. Y. S. Chen, and W. Stumm (Unpublished Data), Harvard University, 1970.
27. D. A. Livingston and J. C. Boykin, *Limnol. Oceanog.* **7**, 57 (1962).
28. F. J. H. Mackereth, *Phil. Trans. Roy. Soc. London* Series A **230**, 156 (1966).
29. H. T. Odum, *Limnol. Oceanog.* **1**, 102 (1956).
30. R. Margalef, *Perspectives in Ecological Theory*, Univ. Chicago, Chicago, 1968.
31. H. T. Odum and R. C. Pinkerton, *Amer. Sci.* **43**, 331 (1955).
32. H. T. Odum, *Pollution and Marine Ecology* (T. A. Olson and F. J. Burgess, Eds.), Wiley-Interscience, New York, 1967.
33. A. Leopold, *Symposium on Hydrobiology*, Univ. Wisconsin Press, Madison, Wis., 1941, p. 17.
34. G. M. Woodwell, *Science* **168**, 429 (1970).
35. G. E. Hutchinson, in *Eutrophication, Causes, Consequences, Correctives,* National Academy of Sciences, Washington, D.C., 1969.
36. H. Züllig, *Schweiz. Z. Hydrol.* **18**, 6 (1956).
37. W. T. Edmondson, in *Eutrophication, Causes, Consequences, Correctives,* National Academy of Sciences, Washington, D.C., 1969.
38. R. Dubos, *Reason Awake*, Columbia Univ. Press, New York, 1970.

3 The Role of Nitrogen in Eutrophic Processes

John J. Goering, Institute of
Marine Science, University of
Alaska, College, Alaska.

3-1. Introduction

The universal presence of nitrogen in all living matter explains the intimate association of the environmental chemistry of this element with biological systems. The biological transformations of nitrogen in aquatic ecosystems appear to be similar qualitatively in most respects to those occurring in the better known soil ecosystems. The various kinds of transformations are thus well understood, but a thorough understanding of the rates and mechanisms controlling these reactions in the diverse global aquatic environments (i.e., streams, lakes, and oceans) are lacking. Research is urgently needed to accurately assess the influence of man's activities which add nitrogen to these systems.

Nitrogen, with five valence electrons, forms bonds that are almost exclusively covalent in character. The N—N triple bond is one of the strongest bonds known. This triple bond has a length of 1.098 Å and a heat of dissociation of 224.5 kcal/mole [$cf.$, $(HC)_2$ 199.6; O_2 117.8; CO 256.2;

NO 150.0 (1); and H_2 102.5 (2)]. The character of this bond is largely responsible for the inert nature of nitrogen gas (1). The expenditure of large amounts of energy is required to convert molecular nitrogen into compounds which can be utilized for nutritional purposes. These compounds, referred to as fixed nitrogen compounds, contain nitrogen much less strongly bound than nitrogen bonded in nitrogen gas. Therefore, the process of nitrogen fixation in actuality does not fix nitrogen, but decreases its strong fixation in molecular form.

The fixed nitrogen in aquatic environments is readily oxidized and reduced by biologically catalyzed reactions, and it can exist in nine different oxidation states (-3 to $+5$). The most abundant form of nitrogen in unpolluted aquatic systems is molecular nitrogen with an oxidation state of zero and it generally exceeds the other nitrogen compounds by about twenty times. The most plentiful forms of reduced nitrogen are generally ammonia (-3) or ammonium ion and organic nitrogen [i.e., NH_2 (-2), NH (-1)] in dissolved and particulate matter. Nitrate ($+5$) is normally the most abundant form of oxidized nitrogen, and nitrite ($+3$) at times is also present.

Many excellent reviews of the cycling of nitrogen through the marine and lacustrine environments are available (3–5), and such a review will not be repeated here. Rather, some pathways of the nitrogen cycle which appear to be more important when considering so-called nuisance, cultural, or artificial eutrophication (e.g., N uptake by phytoplankton and by macrophytes, N regeneration) will be emphasized. The author will attempt to summarize the information now becoming available which seems especially appropriate to the eutrophication problem and also to delineate deficiencies in our knowledge of nitrogen transformations in aquatic ecosystems.

3-2. Nitrogen Compounds in Waste Water

Discharges of domestic and industrial waste water are generally regulated only from a health point of view. The abundant plant nutrients contained in these wastes are normally not removed by the conventional treatment processes and their elimination is generally not considered when designing treatment systems. This neglect often leads to over-fertilization of the receiving waters and can accelerate their rate of eutrophication.

Municipal sewage, the primary source of nutrients in wastes, is usually turbid water containing particulate, finely dispersed, colloidal, and dissolved material. Its total chemical composition will vary somewhat from municipality to municipality, primarily the result of differing industrial activity.

However, the primary source of nutrients in waste water generally originates from homes, and types and amounts of these compounds are therefore more uniform in municipal wastes.

The main sources of nitrogen compounds in municipal sewage are the end products of nitrogen metabolism in humans. The major nitrogen compound in human waste is urea, but urea is readily hydrolyzed to ammonia and CO_2 by the enzyme urease. Bacteria which possess this enzyme are abundant in sewage (*e.g.*, *Proteus vulgari, P. mirabilis*). Ammonia is, therefore, generally the most abundant form of nitrogen in municipal wastes. Eighty to ninety percent of the total nitrogen present is in the form of ammonia and urea (6). Nitrate and nitrite can also result from nitrifying activity if oxidizing conditions prevail.

Amino acids and proteins, along with protein degradation products such as proteoses, peptones, and peptides, are also present in municipal sewage. A list of the types and amounts of amino acids present in municipal sewage is given by Knott and Ingerman (7). Other organic nitrogen compounds which have been identified in sewage are uric acid, creatinine, free amines, amides, and nucleic acids.

Industrial wastes are a minor source of nitrogen in municipal sewage. Casein, a conjugated protein, occurs in dairy wastes. Meat packing plant wastes can also be a possible source of nitrogen. However, in the average municipality, industrial wastes do not contribute substantial amounts of nitrogen.

The two major forms of nitrogen, ammonia and urea, in municipal sewage are nitrogen sources available to aquatic plants and bacteria for growth. Ammonia, in particular, is a good nitrogen source for almost all plants, and urea is adequate for some. The various organic nitrogen compounds, such as amino acids and uric acid, which are present can also serve as sources of nitrogen for growth of some species.

3-3. Nitrogen Transformations Induced by Aquatic Organisms

The paths of nitrogen transformations in aquatic environments are shown diagrammatically in Figures 3.1 and 3.2, where nitrogen is represented in different compartments, for example, phytoplankton, bacteria, zooplankton, and dissolved inorganic and organic fractions. For simplicity in discussion, the transformations associated with phytoplankton and bacteria have been depicted separately. The size of each compartment at any given time is identical to the amount of nitrogen contained, usually expressed on a unit volume or unit surface area basis. The arrows represent the transport of nitrogen from one compartment to another. The rates

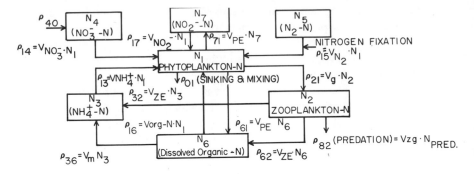

Figure 3.1 Circulation of nitrogen associated with phytoplankton.

are expressed by ρ with two subscripts, the first representing the compartment receiving, the second the compartment contributing the nitrogen. $V_{NO_3^-}$, $V_{NH_4^+}$, etc., represent fractional uptake rates which are essentially growth rates expressed in terms of nitrogen and have units of grams of nitrogen taken up/[(grams N) (time)], or simply unit/time.

Nitrogen has no radioactive isotope with a long enough half-life to be of practicable use in isotope tracer experiments. Therefore, techniques using heavy nitrogen, ^{15}N (abundance generally measured by mass spectrometry), have been developed and used successfully in the study of nitrogen metabolism in organisms (8) and the cycling of the element through lacustrine and marine environments (9–11). An example of how V and ρ can be obtained when isotope tracer techniques are used is given

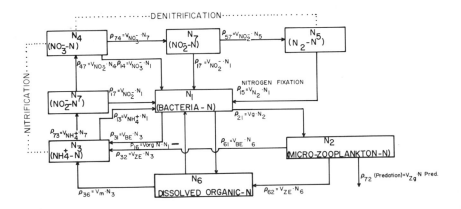

Figure 3.2 Circulation of nitrogen associated with bacteria.

by Sheppard (12). When using ^{15}N, the variable $V_{NO_3^-}$ and $V_{NH_4^+}$ are obtained after mass spectrometry from the following computation:

$$V_{NO_3^-} = \frac{\rho_{14}}{N_1} = \frac{da_1/dt}{a_4 - a_1} \tag{1}$$

where ρ_{14} is the rate of transport of nitrate from the labeled compartment 4, into the initially unlabeled compartment 1, the phytoplankton nitrogen in this case. N_1 is the concentration of nitrogen in the phytoplankton, a_4 is the atom percent ^{15}N in compartment 4. a_1 is the atom percent ^{15}N in compartment 1. Rearrangement of Equation 1, gives the rate expression used in Figures 3.1 and 3.2:

$$\rho_{14} = V_{NO_3}N_1 \tag{2}$$

A. Uptake of Nitrogen Compounds by Phytoplankton

NITROGEN LIMITED GROWTH. Although the role of nutrients (*e.g.*, N, P, Si) in limiting the production of organic matter by phytoplankton is widely acknowledged, few attempts have been made to construct mathematical models describing nutrient-growth relationships in natural populations of algae. Dugdale (13) has proposed such a model based upon the flow of a limiting nutrient through a compartmental system. This nutrient-oriented compartmental model may prove useful in predicting nitrogen-limited phytoplankton growth in lakes and in the open sea and is thus presented here. Dugdale and Goering (14) have discussed the applicability of this model to predict the growth of algae resulting from nutrients added to the marine environment by man's activities; and Dugdale and Whitledge (15) have incorporated nutrient limited growth in a computer simulated model of phytoplankton growth near a marine sewage outfall. If indeed growth and species composition of the phytoplankton population resulting from nutrient additions can be controlled, then systems could be designed to provide commercially valuable species of plants and animals. When designing such systems it must be remembered that other variables, such as light and vitamins, as well as nutrients can limit the growth of photosynthetic organisms.

The Dugdale model is intended for application where conditions of steady state or transient nutrient limitations apply. Although some interaction may be expected, regulation at steady state is essentially dependent upon one primary nutrient. For example, if the rate-limiting step is the uptake of nitrogen, the uptake of other required nutrients (P, Si, Fe, etc.) will be controlled to some degree by the uptake of nitrogen.

For simplicity it can be assumed that phytoplankton growth is regulated by one type of nitrogen compound as diagrammed in Figure 3.3. ρ_{40} repre-

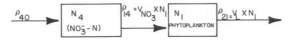

Figure 3.3 Simplified flow diagram of a major nutrient in the euphotic zone with regeneration pathways omitted (13).

sents the rate of supply of limiting nutrient (*i.e.*, nitrate); N_1 is the phytoplankton-N concentration; $V_{NO_3^-}$ is the growth rate of the phytoplankton in terms of nitrate-N; and V_L is the total loss rate of the phytoplankton, mainly from grazing, sinking, and mixing.

At steady state, all ρ's must be equal (*i.e.*, $\rho_{40} = \rho_{14} = \rho_{21}$), and thus $V_{NO_3} = V_L$. The total loss rate of the phytoplankton will therefore control the uptake of nitrate (*i.e.*, the growth rate).

To obtain real solutions for this model, valid mathematical expressions describing each of the important pathways are needed. The pathways are: (*a*) supply of nutrients (in this case, nitrate) to the nutrient pool [N_4]; the processes important here are *in situ* mineralization of organic matter containing nitrogen, addition of nutrients originating outside of the euphotic zone (*e.g.*, influents, sewage, rain, influx from deep water by mixing); (*b*) uptake of nutrients, and (*c*) losses of phytoplankton from the euphotic zone by grazing, sinking, and mixing.

SUPPLY OF NUTRIENTS ORIGINATING OUTSIDE OF THE EUPHOTIC ZONE. Theoretically, nitrogen compounds can enter the euphotic zone in four main ways: (*a*) In surface influents, (*b*) by precipitation on the water surface, (*c*) from deep water by mixing, and (d) nitrogen fixation.

The relative importance of each of these processes will vary in different aquatic systems. In unpolluted lakes most of the nitrogen compounds are derived from the soil and enter in the influents, and the major loss is usually by the effluent (5). In the open ocean the influx of dissolved nutrients from deep water is generally of prime importance.

In a two layered system, the rate of nutrient supply by mixing is (16)

$$\rho_{40} = m(N_0 - N_4) \tag{3}$$

where m is the mixing cofficient between the nutrient-limited region and that occurring just below, N_0 is the concentration of limiting nutrient just below the nutrient-limited region, and N_4 is its concentration within the nutrient-limited region.

The significance of nitrogen fixation by some species of blue-green algae in supplying combined nitrogen to lakes is well documented (17–19). The ability to fix nitrogen is believed to account for the observation that although blue-green algae have a high nitrogen content on a unit weight basis, they are often found in waters containing little combined nitrogen. Bacteria are also apparently at times important nitrogen-fixers in lakes (20).

Ammonia and nitrate are generally preferred over molecular nitrogen as sources of nitrogen for growth by phytoplankton and bacteria. Therefore, in water abundant with these nitrogen compounds, such as in polluted streams, lakes, and estuaries, nitrogen fixation would be expected to play a minor role in the overall nitrogen economy of these systems. However, there is indirect evidence to the contrary. Wilkinson (21) discovered that the amount of nitrogen converted into organic forms within a polluted estuary greatly exceeded the amount of inorganic nitrogen in the entering water. The source of this extra nitrogen was thought to have been supplied by nitrogen-fixing bacteria. There is also evidence of nitrogen fixation occurring in sewage lagoons (22, 23). Nitrogen fixation may thus play a role in providing combined nitrogen to further supplement the fertility of aquatic systems to which municipal wastes or other wastes abundant in plant nutrients have been added.

LOSSES OF PHYTOPLANKTON FROM THE EUPHOTIC ZONE.

The loss of phytoplankton from the euphotic zone is controlled by zooplankton grazing, sinking rates of phytoplankton, and vertical mixing. As previously stated, Dugdale's model of nutrient limitation is intended for application under conditions of steady state or transient nutrient limitation. At steady state, the growth rate of the phytoplankton is determined by the sum of the loss rates due to sinking, grazing, and mixing. The sinking rate of phytoplankton appears to be a function of species, age, and physiological state. Different phytoplankton-zooplankton communities will probably stabilize in different relative proportions to give varying values of V_{zg} (Figure 3.1).

B. Uptake of Nitrate and Ammonia by Phytoplankton. The uptake of nitrate and ammonium by pure cultures of phytoplankton is known to obey the Michaelis-Menten kinetics equation (24, 25). The expression has also been shown to hold for heterogenous populations of marine phytoplankton (26), and thus presumably will also be shown to describe the uptake of nitrate and ammonium by heterogenous populations of lake phytoplankton. The equation of the Michaelis-Menten expression for enzyme catalyzed nitrate uptake is

$$V_{NO_3^-} = V_{(NO_3^-)max} \frac{N}{N + K_{T(NO_3^-)}} \qquad (4)$$

where $V_{NO_3^-}$ is the fractional uptake rate of nitrate by algae, $V_{NO_3^-max}$ is the maximum fractional uptake rate, N is the ambient concentration of

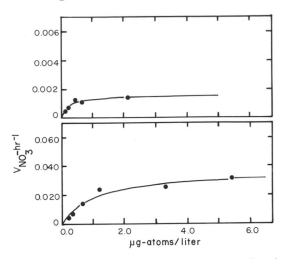

Figure 3.4   Representative response curves for $V_{NO_3^-}$ as a function of nitrate concentration, showing the characteristices of Michaelis-Menton kinetics (26).

nitrate, and $K_{T(NO_3^-)}$, is the half-saturation transport constant (*i.e.,* the concentration of nitrate at which $V_{NO_3^-} = \frac{1}{2}V_{NO_3^--max}$). The Michaelis-Menten expression for ammonium uptake would be similar. When $V_{NO_3^-}$ for natural populations of marine algae is plotted against ambient nitrate concentrations, a rectangular hyperbola is obtained (Figure 3.4), showing that Michaelis-Menten kinetics are obeyed.

It would be extremely useful in eutrophication abatement projects to be able to predict which species will result from various ambient nitrate and ammonium concentrations. One important aspect of nutrient limitation theory is its potential use in predicting phytoplankton succession resulting from competition for limiting nutrients. An example of variations in the specific growth rate of four species of marine phytoplankton with different nitrate and ammonium concentrations and two irradiance levels are given by Eppley *et al.* (27) and are reproduced in Figure 3.5. At low light levels (∿5% of surface sunlight) with either nitrate or ammonium as the substrate, the calculated specific growth rates suggest that *Coccolithus huxleyi* can grow faster at all nitrate and ammonium concentrations than can the other species. At the higher light levels (∿20% of full sunlight) the calculated specific growth rates suggest that *Coccolithus huxleyi* grows the fastest when nitrate remains about 0.8 µM or less and that at higher substrate concentrations *Skeletonema costatum* would grow the fastest.

Very few values of $K_{T(NO_3^-)}$, $K_{T(NH_4^+)}$, $V_{(NO_3^-)max}$, or $V_{(NH_4^+)max}$ for diffcrent species of algae have been determined. Eppley *et al.* (27) have

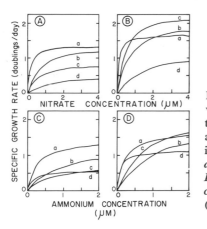

Figure 3.5 Calculated specific growth rates versus nitrate and ammonium concentration at two irradiance levels (approximately 1/20, *A* and *C*, and 1/5, *B* and *D*, of surface sunlight irradiance for wave lengths of 400–700 nm). *a—Coccolithus huxleyi* (coccolithophore), *b— Ditylum brightwellii* (diatom), *c—Skeletonema costatum* (diatom), *d—Dunaliella tertiolecta* (green flagellate) (27).

determined $K_{T(NO_3^-)}$ and $K_{T(NH_4^+)}$ for 16 species of cultured marine phytoplankton at 18°C. The $K_{T(NO_3^-)}$ ranges from about 0.1–10 μM and the $K_{T(NH_4^+)}$ from about 0.1–5 μM with more of the values falling at the lower end of the range and in agreement with the K_T values for both compounds measured by MacIsaac and Dugdale (26) for natural oceanic phytoplankton populations. Evidence also exists which shows that K_T's of marine phytoplankton are species specific and are temperature dependent, but are not influenced by irradiance or the other external factors which influence growth rate (28). The half-saturation constant appears, therefore, to be an extremely useful parameter in predicting the relative ability of different phytoplankton to use low levels of nutrients.

Nutrient interactions must be considered when applying kinetic constants to natural growth situations. For example, the uptake of nitrate by some marine phytoplankton is known to be reduced in the presence of ammonium ion (25). Inhibition of nitrate uptake begins when ammonium ion concentrations exceed 1 μM and approaches zero at concentrations of 2 μM. Experiments using ^{15}N as a tracer indicate that the $V_{(NO_3^-)max}$ is reduced in a linear fashion by increasing concentrations of ammonium, and that the $K_{T(NO_3^-)}$ is unaffected (14). These results suggest a repression of the enzyme nitrate reductase which must be present for growth on nitrate.

Values of the half-saturation constants for nitrate and ammonium uptake by species of freshwater algae common in blooms occurring in eutrophic waters are largely lacking. Half-saturation constants for nitrate have been determined for *Chlorella pyrenoidosa* and a mixed culture of green algae, green flagellates, and diatoms indigenous to the San Joaquin Delta in California (29). The genera represented in this mixture were *Scendesmus,*

Ankistrodesmus, Pandorina, Gonium, Melosira, Cyclotella, Tabellaria, Asterionella and *Navicula*. The $K_{T(NO_3^-)}$'s obtained for *C. pyrenoidosa* were 211.4 µM at 39.2°C, 98.6 µM at 35°C, 60.0 µM at 28.5°C, and 50.0 µM at 19°C. The mixed algae had a $K_{T(NO_3^-)}$ of 32.1 µM at 25°C.

An increase in the efficiency of nitrate uptake by *C. pyrenoidosa* occurred with increasing temperature up to 39.2°C. Also, the optimum temperature with respect to growth was not equal to the optimum temperature with respect to uptake of nitrate-N at relatively low nitrate concentrations.

The $K_{T(NO_3^-)}$ for the mixed culture of algae is substantially higher than those reported for marine algae. However, the mixed culture of algae did have a lower $K_{T(NO_3^-)}$ than did *C. pyrenoidosa*, indicating a higher specific growth rate at much lower nitrate-N concentrations. This is an indication of the adaption of the mixed algae to conditions of lower nitrate concentrations. The large $K_{T(NO_3^-)}$ for *C. pyrenoidosa* results from this species being originally isolated from soils rich in nitrate and its maintenance on nitrate-rich media over many years.

$K_{T(NO_3^- + NH_4^+)}$ of about 1 µM for *Selanostrum gracile* grown in Lake Tahoe water has been reported (30). Because of the interaction between nitrate and ammonium uptake mentioned above, and the lack of details concerning the addition of nitrogen in this study, a comparison of the K_T for this alga with the K_T values reported for other freshwater and marine species is difficult. However, it appears that the K_T reported for this species falls in the lower part of the range of K_T's determined for marine algae.

When ambient ammonium or nitrate are known to exist at concentrations large enough to permit maximum growth of phytoplankton (*i.e.*, ambient concentrations of NO_3^-—N or NH_4^+–N $\geqslant$ than those required for $V_{(NO_3^-)max}$ or $V_{(NH^+)max}$), the influence of other suspected limiting constituents, such as inorganic and organic nutrients, trace metals, and vitamins, on $V_{(NO_3^-)max}$ or $V_{(NH_4^+)max}$ can provide information useful in predicting which of these are possibly limiting growth. For example, the influence of Mo [nitrate reductase has Mo and Fe cofactors (31)] could be demonstrated if its addition resulted in altered nitrate kinetics such as an increase in $V_{(NO_3^-)max}$. The kinetic approach can thus be of assistance in determining which chemical species are limiting growth in natural situations. If enough information were available concerning the interactions between nutrients, irradiance, temperature, and the kinetics of nitrogen uptake, we could perhaps predict the growth success of individual species under a certain set of environmental conditions.

C.   Organic Nitrogen Uptake by Phytoplankton. Many types of organic nitrogen compounds are present in sewage and some can serve as sources of nitrogen for growth of marine and freshwater phytoplankton, but in general these are inferior to inorganic nitrogen as nitrogen sources. Urea and uric acid have been shown to be adequate nitrogen sources for both freshwater and neritic marine phytoplankton (32, 33). After urea, the amides (*e.g.*, acetamide, succinamide, asparagine, and glutamine) appear to be the organic nitrogen compounds most readily available for growth. Glycine is the most readily assimilable amino acid and the other amino acids vary in their ability to serve as nitrogen sources.

Qualitative and quantitative studies of the potential importance of dissolved amino acids in the nitrogen nutrition of the green flagellate *Platymonas* have been reported by North and Stephens (34, 35). Many amino acids were found to support growth as efficiently as nitrate in conventional batch cultures. *Platymonas* can assimilate ^{14}C-labeled amino acids in micromolar amounts, and the presence of these acids in protein fractions indicates that they rapidly enter metabolic pathways. The kinetics of amino acid uptake were found to be described by the Michaelis-Menten equation. A K_T of 5 μM for glycine was obtained. Uptake rates were found to be significantly higher in nitrogen-deficient cells and observed rates of glycine uptake from a 5×10^{-7} M solution could supply sufficient amino acid nitrogen to sustain growth.

Hellebust (36) reports that the marine diatom *Melosira nummuloides* takes up amino acids through active transport systems. Some amino acids are concentrated by a factor of more than two orders of magnitude when present in the medium at 10^{-4} M. Amino acids may contribute a substantial part of the cell's carbon and nitrogen at low rates of photosynthesis. However, amino acids do not support growth in the dark. It is suggested that this species is incapable of gluconeogenesis in the dark. The existence of active uptake mechanisms for amino acids suggests the possibility of interaction between this organic source of nitrogen and the primary inorganic sources, nitrate, nitrite, and ammonium.

The uptake capacities and specificities of organic substrates vary from species to species. Hellebust (36) conducted a general survey of the uptake capacities and specificities of some marine phytoplankton to find how widely distributed transport systems for organic substances are in marine photoautotrophs. Two members of the Chrysophyceae, *Coccolithus huxleyi* and *Isochyrsis galbana*, and two species of Chlorophyceae, *Dunaliella tertiolecta* and *Pyramimonas sp.*, did not take up any of the seven organic substrates offered. Two dinoflagellate species, *Gymnodinium nelsoni* and *Peridinium trochoideum*, took up glucose and the amino acids lysine and alanine. Most of the diatoms studied utilized acetate at low rates. Two diatoms from a salt water pond, *Cyclotella cryptica* and *Melosira nummuloides*, took up amino acids rapidly. The only open ocean diatom which took up amino acids to a measurable extent was *Cyclotella nana*.

The above information suggests that some organic nitrogen compounds can be used by marine phytoplankton as a direct nitrogen source, particularly when inorganic nitrogen is unavailable. Domestic sewage contains proteins of vegetable and animal origin, smaller organic molecules such as peptones, and minute amounts of free amino acids. During treatment and in the receiving waters the proteins are decomposed by bacteria to peptides, amino acids, and ammonia. The ammonium ion, in particular, and the amino acids can serve as sources of nitrogen for phytoplankton growth. Thus, the majority of the organic matter entering in sewage must first be mineralized by bacteria before the major portion of the primary nutrients are available for phytoplankton growth. However, it must be stressed that the capacity to use organic substrates varies widely in different species of phytoplankton, and organic nitrogen compounds probably are important sources of nitrogen when inorganic nitrogen compounds become low or exhausted.

D. Nitrogen Uptake by Macrophytes.   Studies of the nitrogen nutrition of fresh water and marine benthic macrophytes are limited. Data on the growth rates of these plants under different nitrogen enrichments is meager, and information is even sparse on the types of nitrogen compounds available to them as nitrogen sources.

Aquatic angiosperms can utilize nitrate as a nitrogen source (37). Those capable of growing either in the atmosphere or submerged in freshwater accumulate much more nitrate in the latter situation. Since soil angiosperms can use various organic nitrogen compounds as nitrogen sources, these nitrogen sources are also probably adequate for growth of aquatic angiosperms.

When other plant nutrients are present in sufficient amounts, there appears to be a direct relationship between the amount of available nitrogen and phosphorus in the water and the biomass of phytoplankton produced. Vascular aquatic plants apparently respond to nitrogen and phosphorus in a similar manner (38, 39).

Mulligan and Baranowski have studied the growth of freshwater vascular plants at different N and P levels (40). *Elodea canadensis* produced greatest yields over a range of nutrients from 2.6–12.29 µg-atom N and 0.65–2.1 µg-atom P. *Potamogeton crispus* and *Myriophyllum spicatum* var. *exalbescens* grew best over a narrow range of nutrients around 5.1 µg-atom N and 0.65 µg-atom P. *Elodea canadensis* produced about three times the biomass of the other two plants under similar nutrient levels.

These investigators examined the yields of *Myriophyllum* when using NH_4NO_3, $Ca(NO_3)_2$, and KNO_3 as nitrogen sources. KNO_3 was the best N source with $Ca(NO_3)_2$ next best and NH_4NO_3 the poorest.

The highest nutrient levels (3.57 mg-atom N, 0.161 mg-atom P) used in this study produced the greatest growth of phytoplankton but no vascular plant growth, while the lower concentrations resulted in increased yields of vascular plants. Filamentous algae grew best at intermediate nutrient concentrations. These data suggest that nutrient addition to lakes should favor changes in the benthic plant communities to communities dominated by *Elodea canadensis*, and if nutrient addition continues and nutrient concentrations increase, the filamentous algae and phytoplankton should become the dominate communities. An increase in phytoplankton results in decreased illumination; therefore, the bottom area with sufficient light to support benthic plants is diminished and this likewise promotes the extinction of bottom plants. In the end, the lack of adequate light and the inhibition of macrophytic growth in the presence of high nutrient concentrations will cause the lake to become choked with phytoplankton and devoid of benthic macrophytes.

E. Regeneration of Inorganic Nitrogen from Organic Matter

1. REGENERATION OF NITROGEN IN THE WATER COLUMN. The regeneration of nitrogen from organic matter present in aquatic ecosystems has, in general, been assumed to be similar to that occurring in terrestrial systems where regeneration results primarily from bacterial and fungicidal decomposition of organic substrates. However, it has recently been shown that zooplankton and protozoa can at times play a major role in the release of soluble nitrogen compounds in the marine environment (41). Aquatic animals, while metabolizing organic matter, release ammonium, free amino acids, and other organic nitrogen compounds. The importance of animals in nutrient regeneration in lakes and oceans can be explained by the predominance of grazing food chains in the aquatic environment compared to detritus food chains in terrestrial systems (42).

The regeneration of nitrogen from decomposing algae by bacterial activity has been extensively studied in the laboratory (43, 44), but few attempts have been made to assess *in situ* regeneration rates and in particular to establish the significance of zooplankton and other animals in regenerating nutrients. Mathematical descriptions of the extent of nutrient regeneration which occurs when algae decompose under oxic and anoxic conditions have been derived (45). The general agreement between predicted and measured values for both nitrogen and phosphorus regeneration in laboratory experiments under both oxic and anoxic decomposition conditions is sufficiently accurate to affirm the applicability of the hypothesized models to laboratory experiments. However, the applicability of such models to natural systems, in which zooplankton, protozoa, and other animals can apparently at times greatly accelerate nutrient regeneration, remains obscure.

The rates and even the mechanisms of inorganic nitrogen regeneration from organic matter in aquatic ecosystems in various states of eutrophication are not well understood. There is evidence that bacteria and/or other small organisms (*e.g.,* protozoa, fungi) play the major role in nitrogen regeneration in oligotrophic marine water (11), whereas in eutrophic situations the contribution from zooplankton and other animals appears to gain significance (41). Dugdale and Goering (11) in their study of the role played by nitrogen in marine primary productivity estimated that only about 10% of the observed ammonium uptake in the oligotrophic water surrounding Bermuda was supplied by zooplankton excretion. In certain regions of the northeastern subtropical Pacific Ocean, Wallen, Goering, and Baribault (unpublished data) estimated that the amount of ammonium released by net zooplankton (*i.e.,* zooplankton retained by a 0.5-mm mesh plankton net) was equivalent to only about 0.14% of the total microbial uptake of ammonium. Incidently, in this study, only about 4–8% of the ammonium assimilated at the 100, 50, 25, 10, and 1% light depths could be attributed to uptake by photosynthetic organisms. The amount of photosynthetically associated ammonium uptake was computed using a C–N uptake ratio of 6.1:1 (11). These results seem to imply that heterotrophs (*e.g.,* bacteria, fungi) are the major nutrient users in this area of the ocean.

The studies of Harris (46) and Martin (47) indicate that at certain times of the year zooplankton are important in nitrogen regeneration in productive marine coastal systems. The percentage of the annual phytoplankton nitrogen requirement supplied by release products from net zooplankton in Long Island Sound was estimated to be about 56% [based on ten determinations, range 11–113% (46)]. In Narragansett Bay in the winter, zooplankton production of nitrogen exceeds that assimilated by phytoplankton; during the spring bloom of phytoplankton, the contribution amounts to only 2.5% (47). The nitrogen requirements in these studies were computed from productivity data by assuming an uptake ratio of C–N similar to that found in phytoplankton rather than from instantaneous measurements of *in situ* nitrogen uptake. This approach cannot provide information on the total nitrogen uptake and, as mentioned previously, the uptake by microorganisms other than phytoplankton can at times be of major significance.

The importance of zooplankton as nitrogen regenerators in the lacustrine environment has not been extensively studied, but it seems appropriate to assume that they increase in significance in this respect with increasing eutrophication. It is documented that both plant and animal plankton standing stocks increase with eutrophication (48).

A limited number of turnover rates of inorganic nitrogen compounds in freshwater lakes and in the sea have been reported. Alexander (49), utilizing a ^{15}N isotope dilution technique, measured rates of ammonium supply in subarctic lakes. Turnover times for ammonium [turnover time $= (NH_4{}^+ -N$ ambient$)/(NH_4{}^+ -N$ supply rate$)$] of less than 24 hours were found during summer months in moderately productive subarctic lakes. During periods of high productivity the *in situ* regeneration rates for ammonium increased and were sufficient to account for the measured ammonium uptake.

Goering and Dugdale (unpublished data) using the ^{15}N technique have measured rates of ammonium supply in the euphotic zone of several different regions of the sea. Turnover times of 36–72 hours were obtained for the ammonium in tropical and subtropical seas. Rates of ammonium supply in the tropical ocean were uniform in the light and the dark. The data obtained from stations in the Arabian Sea suggest that ammonium supply and ammonium uptake are balanced quantitatively, the mean uptake of ammonium was 0.0335 μM $NH_4{}^+$/hour, while the mean ammonium supply was 0.0296 μM/hour. It must be remembered in interpreting these data that the balance indicated here holds only for the processes taking place inside an experimental flask. The effects of advection, predation, precipitation, and macrozooplankton excretion are excluded.

The above ammonium supply rates imply that the majority of the ammonium assimilated in productive surface waters of lakes and the sea is supplied by *in situ* regeneration from the particulate and dissolved organic matter. The sediments, therefore, appear to play a minor role in supplying nutrients to the surface water during the productive season. However, their significance may be greater on an annual basis.

2. REGENERATION OF NITROGEN IN SEDIMENTS. The accumulation of organic matter on the bottoms of lakes, rivers, estuaries, and other shallow areas in the sea is known to occur, but the rates of accumulation and processes altering the chemical make-up of this matter are not well defined. Settling probably accounts for the accumulation of organic matter in shallow water sediments, but evidence exists which strongly suggests that little organic matter is transported from the sea surface to the deep sea bottom by settling (50, 51). This conclusion is based on the fact that the vertical distribution of particulate carbon is homogeneous in time, space, and depth; and that the standing crop of phytoplankton has no measurable influence on the concentration of dissolved and particulate organic carbon occurring at depth. The majority of organic matter of recent origin in deep sea sediments, therefore, appears to be transported to depth by some mechanism other than settling, and perhaps an overlapping vertical series of different animal populations that feed on each other is responsible.

Lake and shallow marine sediments have been recognized as providing a source of nutrients for biological growth. The physical, chemical, and in particular, the biological processes that occur at the mud–water interface have been implicated in the release and removal of nutrients. All of these processes play a role in sediment–water nutrient interchange and the importance of each will depend on the individual sediment and basin in question.

The role of lake and ocean sediments in transforming organic nitrogen to inorganic nitrogen and its release into the overlying water has been assessed for three marine basins (52) and for numerous lakes (53, 54). Ammonium is generally the most abundant form of inorganic combined nitrogen present in the interstitial water of sediments. It results from excretion and from bacterial deamination of amino acids contained in peptides, proteins, and other nitrogenous compounds in both oxic and anoxic sediments. Its concentration has been shown to increase with depth in marine sediments down to 4 m (52) and in lake sediments down to 1.65 m (54). The increase with depth in marine sediments was found to be greater in a reducing basin and this was attributed to losses occurring primarily upward into the overlying water. Concentrations of ammonium in the interstitial water exceed those in the water overlying the sediment by more than 100-fold [$e.g.$, NH_4^+ in interstitial water of marine sediments ranges from about 0.014–1.14 mM (52); and in lake sediments, it ranges from about 0.429–10.71 mM (54)]. In oxidizing sediments, nitrate would be expected because of its production from ammonium by nitrifying bacteria, and nitrate has been found to exist down to considerable depths in marine and lake sediments which are or have been recently overlain by oxygenated water (52, 54). Nitrate like other soluble nutrients can be lost from sediments by diffusion into the overlying water. It can also undergo reduction to molecular nitrogen (denitrification by bacteria) when oxygen is absent or at least limiting microbial respiration. Denitrification in sediments appears to be an important sink in the nitrogen budget of aquatic systems (52, 55).

Mortimer (53) reported that ammonium, along with iron, silicate, phosphate, alkalinity, and conductivity, increased in water over reduced sediments. Rittenberg, Emery, and Orr (52) estimated from the difference in the decrease in organic nitrogen and the increase in ammonium with depth in the sediment, that in a marine basin containing reduced sediment a minimum of 1.6 μmoles of ammonium/cm^2 of sediment pass into the basin water annually. They pointed out that this amounts to only about one-third of the ammonium contained in the interstitial water, and is only about 0.4% of the annual amount of nitrogen used by phytoplankton in the euphotic zone in this region. Therefore, sediments are acting as an important sink for nitrogen, and must be considered of prime importance when constructing nitrogen balances. Brezonik and Lee (56) in their estimated sources and sinks for nitrogen in eutrophic Lake Mendota attributed 66.7% of the nitrogen lost from the lake to sediments and other minor sinks.

Little quantitative information is available on the types and amounts of nitrogenous organic compounds in lake and marine sediments. Apparently only small amounts of free amino acids are present in sediments overlain with oxygenated water, while large amounts of amino acid N are found in acid hydrolysates of sediments (57, 58). These probably exist in proteins, peptides, or in association with humic acids (59). Amino sugars (*e.g.*, hexosamine) and heterocyclic nitrogen compounds (*e.g.*, indole) also have been shown to occur in minute amounts in sediments (58).

Keeney, Konard, and Chesters studied the distribution of nitrogen in some Wisconsin lake sediments (58). The same forms of nitrogen were found to occur in sediments as in soils, but the depth distribution was somewhat different. The total organic nitrogen contents of deep- and shallow-water sediments were, in general, similar, and more than 98% of the total nitrogen occurred as organic nitrogen. Organic carbon–organic nitrogen ratios of the hard- and soft-water sediments were similar, averaging 11:1 and ranging from 9:1 to 15:1. Hard-water lake sediments exhibited no effect of trophic level on organic nitrogen content, but sediments from eutrophic soft-water lakes showed increasing organic nitrogen with increasing fertility.

No discernible effects of lake trophic levels on the fixed and exchangeable ammonium-N in sediments and nitrate-N in the interstitial water were evident. However, the ammonium-N in the interstitial water was considerably greater in eutrophic lake sediments than in oligotrophic lake sediments.

A decrease in hexosamine-N and an increase in amino acid-N occurred in sediments with increasing lake fertility. Hexosamines are major components of microbial cell walls and this would imply greater microbial turnover of organic-N in oligotrophic lakes. The authors suggest that the increased microbial activity may result from higher dissolved oxygen in the water overlying the sediments in oligotrophic lakes. Microbial activity in oligotrophic water likewise appears to play a more important role in nitrogen regeneration than it does in eutrophic waters.

Bacterial metabolism is probably the most important process causing nutrient transformations in sediments. Bacteria are known to be plentiful in surface sediments and their abundance decreases with depth. The environmental variables such as temperature, pH, redox potential, oxygen content, H_2S content, and other physical and chemical variables greatly influence the growth of these organisms and must be taken into account when assessing their activity in certain sediments. For example, the presence or absence of oxygen will drastically influence the growth of nitrogen transforming bacteria. Bacterial nitrification will occur mainly in sediment layers containing oxygen whereas denitrification is limited to anoxic layers.

The activities of animals in sediments also are significant in sediment–water interchange. Benthic organisms, particularly members of the infauna (*i.e.*, animals living buried or digging), physically alter the sediment by their activities and life processes. Their activities greatly enhance the exposure of the interstitial water to the water overlying the sediments. They also may be of importance in direct nutrient regeneration through excretion. Oysters have been shown to be important in the release of inorganic phosphate (60).

The information reviewed above indicates that soluble nutrients such as phosphate and ammonium abound in sediments and can become available for algal growth by diffusion into the overlying water. Various physical processes which cause water flow over the sediment or result in sediment being resuspended can greatly increase the rate of nutrient release. In shallow waters the resuspension of sediment can be brought about by wind-induced wave action. However, if sediments are not distributed by such turbulence, it then appears that nitrogen losses resulting from other processes, probably principally biological activities of various kinds (*e.g.*, stirring by benthic organisms, biogenic gas release from sediments) and physical desorption with changes in redox potential, amount to only about one-third of the total soluble nitrogen present in the interstitial water (52). Sediments appear, therefore, to be effective traps for removing nitrogen from aquatic ecosystems, and should be carefully studied before disturbing by dredging, which is often suggested as a method of reducing eutrophication in lake restoration projects. It should also be remembered that nutrients generally increase with depth in sediments down to considerable distances (2–4 m) and exposure to these deep sediments may in certain instances enhance the fertility of a body of water rather than reduce it.

F.   Volatilization and Absorption of Ammonia from Lake and Ocean Surfaces.   The significance of ammonia-N loss or absorption by surface waters of streams, lakes, and the ocean is generally ignored when the overall nitrogen cycle is studied. However, information is now available which suggests that volatilization and absorption can at times play an important role in the nitrogen balance of various surface waters.

In aqueous solution, ammonia-N is in equilibrium between the gaseous and hydroxyl forms:

$$NH_3 + H_2O \rightleftarrows NH_4^+ + OH^- \tag{5}$$

The reaction is very pH dependent; alkaline pH favoring the presence of the gaseous form of ammonia, NH_3, while at pH 7 or below the ammonia is almost all present in the ionic form NH_4^+.

The transfer of gaseous ammonia dissolved in water to the atmosphere is a function of the relative difference in partial pressures of the gas present in each medium. The net transfer of gas from liquid to atmosphere or atmosphere to liquid will occur in accordance with Henry's Law. The rate of ammonia transfer from an aqueous solution to the atmosphere obeys first-order reaction kinetics; that is, the rate of transfer is directly proportional to the concentration of ammonia nitrogen in solution (61). Many variables, such as pH, temperature, air movement over the water surface, and water turbulence, will drastically affect the rate of gas transfer.

The pH of many aquatic systems often rises to highly alkaline levels due to the photosynthetic activity of algae and benthic plants. pH values as high as 10 are not uncommon in warm eutrophic lakes. When such conditions prevail, the loss of ammonia by volatilization can be of importance.

Estimates of gaseous ammonia-N losses have been made for two small eutrophic impoundments in California (62). Loss rates of 2.4 mM NH_4^+ /($m^2 \times$ day) and 7.0 were obtained, respectively. Approximately 5.8% of the estimated total daily influent of ammonia-N in the first impoundment was lost by volatilization.

Absorption of ammonia gas from the atmosphere by simulated lake surfaces installed near cattle feedlots has also been demonstrated (63). The absorption rates ranged from about 14.3–17.4 μM NH_4^+/(hectare $\times$ week). Therefore, when the atmospheric concentration of ammonia is greater than its concentration in the water, as occurs near cattle feedlots and some industrial sites, the absorption of this excess ammonia can contribute significant amounts of nitrogen to surface water.

The above discussion indicates that warm eutrophic waters, exhibiting high pH values, are potentially able to liberate, through volatilization, significant quantities of gaseous ammonia-N to the atmosphere. Absorption of ammonia gas from the atmosphere can also occur if the pH of the surface waters is near neutrality and large amounts of ammonia gas are present in the atmosphere.

If bacterial nitrification which converts ammonium into a nonvolatile ion, nitrate, was inhibited naturally or artificially in lakes, streams, etc., then losses of ammonia by volatilization would be maximized, and perhaps could greatly diminish nitrogen, and thus assist in controlling fertility caused by nitrogen pollution. Inhibitors of nitrification are known (31), but all are probably too expensive or too toxic to beneficial microorganisms to be of value in lake restoration projects.

3-4.   Nitrification and Denitrification in Eutrophic Processes

A.   Nitrification.   The discharge of nitrogenous wastes from municipalities, industries and land runoff can not only influence the growth of aquatic plants, but can also stimulate growth of autotrophic bacteria that obtain energy by oxidizing compounds of nitrogen (*i.e.,* nitrifying bacteria). During oxidation of ammonium and nitrite, these bacteria use oxygen and if they are abundant and metabolically active, large declines in oxygen can occur. Several species of nitrifying bacteria are known but the genera *Nitrosomonas* and *Nitrobacter* are the most common in streams, lakes, and estuaries (5). In the open ocean *Nitrocystis oceanus* appears to be an important oxidizer of ammonium to nitrite (64).

Ammonium is the major form of nitrogenous waste released into streams, lakes, and estuaries. Organic nitrogen compounds can also be abundant, but these compounds undergo hydrolysis with ammonium as the major end product. The ammonium resulting from these sources can be oxidized under oxic conditions to nitrite by nitrifying bacteria of the genus *Nitrosomonas* (5, 31).

$$NH_4^+ + OH^- + 1.5O_2 \rightarrow H^+ + NO_2^- + 2H_2O + 54.9 \text{ kcal}   (6)$$

It is generally assumed, but has not been conclusively demonstrated, that nitrifying bacteria utilize ammonium ion as a substrate rather than gaseous ammonia. In aqueous solution, ammonia is in equilibrium between the gaseous and ionic forms (see Equation 5). The reaction is very pH dependent and at pH 8.5, which is the optimum pH for oxidation of ammonia by *Nitrosomonas* (31), about 90% of the ammonia occurs in the ionic form and about 10% in the gaseous form. Therefore, gaseous ammonia cannot be excluded as a possible substrate for *Nitrosomonas* growth. The further oxidation of nitrite by *Nitrobacter* proceeds according to the reaction

$$NO_2^- + 0.5O_2 \rightarrow NO_3^- + 18 \text{ kcal}   (7)$$

Stoichiometrically these reactions require 4.57 g oxygen/g of ammonia oxidized. These reactions therefore potentially use large quantities of oxygen, and nitrification can at times produce oxygen-depleted water in streams, lakes, and estuaries subjected to large amounts of high nitrogen wastes.

To describe quantitatively the influence of nitrifying bacteria on the conditions in streams, lakes, and estuaries it is necessary to know their growth-rate constants under the relevant environmental conditions and the half-saturation (Michaelis) constants of the substrates which support their growth. The growth kinetics of both *Nitrosomonas* and *Nitrobacter* have been studied by Knowles, Downing, and Barrett (65) and appear to conform to the equation of the Michaelis-Menton type. Growth-rate constants and half-saturation constants for both species were computed from experimental data on the course of nitrification in a mixed bacterial population obtained from the Thames Estuary.

The growth constants of both nitrifying organisms increase considerably with an increase in temperature in the range 8–30°C (*e.g.*, *Nitrosomonas* 0.2–1.8/day; *Nitrobacter* 0.5–1.3/day) (65). With *Nitrosomonas* the increase amounts to 9.5%/degree and with *Nitrobacter* the temperature coefficient is about 5.9%/degree. At a given temperature the value of the growth-rate constant for *Nitrobacter* is about 50% greater than that for *Nitrosomonas*.

The half-saturation constants of ammonium $[K_{T(NH_4^+)}]$ for *Nitrosomonas* and nitrite $[K_{T(NO_2^-)}]$ for *Nitrobacter* range from about 14–575 μM and increase with increasing temperature in the range 8–30°C (65). At temperatures below 20°C the constants for ammonium and nitrite are of the same order, but, whereas the value for both substrates increases with temperature, the rate of increase for $K_{T(NO_2^-)}$ appears to be larger than for $K_{T(NH_4^+)}$. The temperature coefficients are 11.8%/degree for $K_{T(NH_4^+)}$ (*Nitrosomonas*) and 14.5%/degree for $K_{T(NO_2^-)}$ (*Nitrobacter*).

Although the information on growth of nitrifying bacteria and phytoplankton under natural conditions is limited, a prediction on the competitive advantage of one over the other by estimating specific growth rates as a function of ammonium-N concentration appears warranted. The half-saturation constants for the uptake of ammonium-N by phytoplankton $[K_{T(NH_4^+)} = 0.1–10$ μM] are much lower than those for nitrifying bacteria $[K_{T(NH_4^+)} = 14–575$ μM]. The maximum specific growth rates of the two groups appear to be in the same general range, 1–5 doublings/day (27, 65). Thus in the euphotic zone where ammonium-N concentrations are generally much lower than 10 μM the phytoplankton would appear to be able to outcompete the nitrifying bacteria for the available ammonium (*i.e.*, grow faster) if other conditions, such as light and vitamins, are not limiting their growth. Verification of the competitive advantage of phytoplankton is indicated by the inability to demonstrate, using stable isotope tracer techniques, the oxidation of $^{15}NH_4^+$–N to NO_2^-–N or NO_3^-–N in water obtained from the non-light limited region of lakes and in the sea (Goering, unpublished data). Below the euphotic zone and in surface water during light-limited winter months the competition for ammonium by phytoplankton will be minimal and nitrification can occur if appropriate temperature, pH, oxygen tension, etc., prevail. In polluted situations, where ammonium-N reaches high concentrations (>10 µM), nitrification and phytoplankton growth can occur simultaneously. However, if ammonium-N concentrations rise to very high levels (*i.e.*, $\sim$75 µM), it becomes toxic and inhibits photosynthesis by phytoplankton (66).

B.   Denitrification.   Bacterial reduction of nitrate may be divided into two categories, assimilatory and dissimilatory. Assimilatory reduction occurs when nitrate is a source of nitrogen for synthesis of organic matter. Dissimilatory reduction is defined as reduction of nitrate in which nitrate serves as the essential hydrogen acceptor enabling bacterial growth. During dissimilatory nitrate reduction the gaseous end products NO, N_2O, and N_2, are formed. The term denitrification is used to describe this process which occurs when dissolved oxygen is depleted below the amount needed to saturate the oxygen using enzymes.

Denitrification has been shown to occur in both lacustrine and marine environments (67–69). The denitrification reaction in aquatic environments occurs in two separate steps. As nitrate disappears from anoxic seawater, nitrite accumulates in a 1:1 correspondence until nitrate concentrations are low or nitrate is depleted (70). Nitrite is then further reduced to gaseous nitrogen (N_2), since molecular nitrogen appears to be the only significant product of aquatic denitrifiers (67, 71).

Payne and Riley (72) have found in their studies of dissimilatory nitrate respiration ($NO_3^- \to NO_2^- \to NO \to N_2O \to N_2$) by *Pseudomonas perfectomarinus* that nitrate in very small quantities suppresses the activity of the reductase which converts NO to N_2O. Nitrite is also suppressive, but larger quantities of this ion are required to obtain the same amount of suppression as nitrate exerts. The suppression of NO reductase by nitrate would appear to explain the observed accumulation of nitrite in some low O_2 aquatic environments. Substantial rates of N_2 production, therefore, probably do not occur until nitrate is nearly depleted.

The effect of oxygen on aquatic denitrification is not well understood. Oxygen is known to be a potent inhibitor of the denitrifying process by virtue of its effective competition with nitrate as an electron acceptor in the energy metabolism of cells (73). The amount of dissolved oxygen which limits denitrification is uncertain, but it is thought to be a few tenths ml/liter and may be dependent on the concentration of nitrate (74). Skerman and MacRae (75) found no reduction of nitrate by *Pseudomonas dentrificans* at dissolved oxygen concentrations above 0.2 ml/liter. Perhaps with low O_2 concentrations (a few tenths ml/liter) localized anoxic zones exist which are the active sites of denitrification.

The significance of denitrification in aquatic environments as a sink in the nitrogen budget is still inadequately understood. Evidence presented in various studies suggests that under appropriate conditions the process is operative both in the water column and in sediments (55, 56, 69); and that large amounts of nitrogen may be lost from the system by this process.

3-5.   Dynamics of Nutrients and Organism Growth

There appears to be a delicate balance in aquatic ecosystems between nutrients and organismal growth. In Figures 3.1 and 3.2 a simplified diagram of the balanced flow of nitrogen through such a system is presented. If nutrients are limited, as in oligotrophic water, the balance is maintained (*i.e.*, the system is essentially in steady state). Also, as the water increases its nutrient content by natural or artificial fertilization the individual components (*i.e.*, phytoplankton, bacteria, zooplankton, ammonium, etc.) of the system increase in size, but remain in balance. Eventually, however, in late eutrophication, the balance is lost and large increases in the phytoplankton standing stock take place. McCoy and Sarles (76) suggest that the imbalance results from physical restraints of the habitat on the maximum size of the bacterial population. This would set an upper limit on the rate of nutrient assimilation by bacteria (*i.e.*, ρ_{13}, Figure 3.2), but not by phytoplankton (*i.e.*, ρ_{13}, Figure 3.1). Thus, when the bacterial population reaches its maximum, the phytoplankton could use any excess nutrient over that needed to maintain the maximum steady-state bacterial population and their own steady-state population to increase their standing stock, and the system would then become unbalanced.

Acknowledgments

Contribution No. 106 from the Institute of Marine Science, University of Alaska. This research was in part supported by the National Science Foundation Grants, NSF GB8274 and NSF GB8636, and by the Institute of Marine Science, University of Alaska.

REFERENCES

1. J. Chatt and G. J. Leigh, in *Recent Aspects of Nitrogen Metabolism in Plants* (Proceedings of the First Long Ashton Symposium), Academic Press, New York, 1968.
2. R. B. Heslop and P. L. Robinson, *Inorganic Chemistry,* Elsevier, Amsterdam, 1960.

3. R. F. Vaccaro, in *Chemical Oceanography* (J. P. Riley and G. Skirrow, eds.), Academic Press, New York, 1963.
4. L. H. N. Cooper, *J. Marine Biol. Assoc. U.K.* **22**, 183 (1937).
5. G. E. Hutchinson, *A Treatise on Limnology,* Vol. 1, Wiley, New York, 1957.
6. A. M. Hanson and T. F. Flynn, Jr., in *Proceedings of the 19th Industrial Waste Conference,* Purdue Univ., 1965.
7. Y. Knott and R. Ingerman, *Air Water Pollut.,* **10**, 603 (1966).
8. D. Rittenberg, A. S. Keston, F. Rosebury, and R. Schoenhiemer, *J. Biol. Chem.* **127**(1), 285 (1939).
9. J. C. Neess, R. Dugdale, J. Goering, and V. A. Dugdale, in *Radioecology* (Proceedings of the First National Symposium on Radioecology) (V. Schultz and A. W. Klement, Jr., Eds.), 1963, Rienhold Publishing Co., New York.
10. J. C. Neess, R. Dugdale, V. Dugdale, and J. Goering, *Limnol. Oceanog.* **7**, 163 (1962).
11. R. C. Dugdale and J. J. Goering, *Limnol. Oceanog.* **12**, 196 (1967).
12. C. W. Sheppard, *Basic Principles of the Tracer Method,* Wiley, New York, 1962.
13. R. C. Dugdale, *Limnol. Oceanog.* **12**, 685 (1967).
14. R. C. Dugdale and J. J. Goering, in *The Impingement of Man on the Sea* (D. W. Hood, Ed.), Wiley, New York (1971).
15. R. C. Dugdale and T. Whitledge, *Rev. Int. Oceanog. Med.,* Nice **17**, 859 (1970).
16. G. A. Riley, *Limnol. Oceanog. Suppl.* **10**, R202 (1965).
17. R. C. Dugdale, V. Dugdale, J. Neess, and J. Goering, *Science* **130**, 859 (1959).
18. V. A. Dugdale and R. C. Dugdale, *Limnol. Oceanog.* **7**, 170 (1962).
19. J. J. Goering and J. C. Neess, *Limnol. Oceanog.* **9**, 530 (1964).
20. P. L. Brezonik and C. L. Harper, *Science,* **164**, 1277 (1969).
21. L. Wilkinson, in *Advances in Water Pollution Research* (Proceedings of an International Conference, London, 1962), E. A. Pearson, Ed., Pergamon, Oxford.
22. J. Aguirre and E. F. Gloyna in *Nitrification and Denitrification in A Model Stabilization Pond,* Technical Report to FWPCA, U. S. Dept. of Interior, No. EHE-05-6701, Univ. Texas, 1967.
23. C. D. Parker, *J. Water Pollution Control Federation* **34**, 149 (1962).
24. R. C. Dugdale, *Limnol. Oceanog.* **12**, 685 (1967).
25. R. W. Eppley, J. L. Coatsworth, and L. Solorzano, *Limnol. Oceanog.* **14**, 194 (1969).
26. J. J. MacIsaac and R. C. Dugdale, *Deep-Sea Res.* **16**, 415 (1969).
27. R. W. Eppley, T. N. Rogers, and J. J. McCarthy, *Limnol. Oceanog.* **14**, 912 (1969).
28. R. W. Eppley and W. H. Thomas, *J. Phycol.* **5**, 365 (1969).
29. G. Shelef, W. S. Oswald, and C. C. Golueke, in *Proceedings of the 5th International Water Pollution Research Conference,* Pergamon, New York, 1971 (In Press).
30. E. A. Pearson, E. J. Middlebrooks, M. Tunzi, A. Adinorayana, P. H. McGoukey, and G. A. Rohlick, in *Proceedings of the Eutrophication Biostimulation Assessment Workshop* (Berkeley, Calif.) (E. J. Middlebrooks, T. E. Maloney, C. F. Powers, and L. M. Knack, Eds.), 1969.
31. K. V. Thimann, *Life of Bacteria,* Macmillan, New York, 1963.
32. P. S. Syrett, in *Physiology and Biochemistry of Algae* (R. A. Lewen, Ed.), Academic Press, New York, 1962.

33. R. R. L. Guillard, in *Symposium on Marine Microbiology* (C. H. Oppenheimer, Ed.), Thomas, Springfield, Ill., 1963.

34. B. B. North and G. C. Stephens, in *Proceedings of the 6th International Seaweed Symposium* (R. Margaleff, Ed.), Subsecretaria de la Marina Mercante, Madrid, Spain, 1968.

35. B. B. North and G. C. Stephens, *Biol. Bull.* **133**, 391 (1969).

36. J. A. Hellebust, in *Organic Matter in Natural Water* (Symposium) (D. W. Hood, Ed.), Univ. Alaska, College, Alaska, 1970.

37. H. S. McKee, *Nitrogen Metabolism in Plants,* Clarendon Press, Oxford, 1967.

38. G. C. Gerloff and P. H. Krombholz, *Limnol. Oceanog.* **10**, 529 (1965).

39. G. C. Gerloff, in *Eutrophication, Causes, Consequences, Correctives,* National Academy of Sciences, Washington, D. C., 1969.

40. H. F. Mulligan and H. Baranowski, *Verhandel. Int. Verein, Limnol.,* **17**, 802 (1969).

41. R. E. Johannes, in *Advances in Microbiology of the Sea,* Vol. 1 (M. Droop and E. Ferguson Woods, Eds.), Academic Press, New York, 1968.

42. A. MacFaden, in *Grazing in Terrestrial and Marine Environments* (D. J. Crisp, Ed.), Blackwell, Oxford, 1964.

43. T. Von Brand, *Biol. Bull.* **72**, 1–6, 165–175 (1937); **77**, 285 (1939).

44. E. V. Grill and F. A. Richards, *J. Marine Res.* **22**, 51 (1964).

45. E. G. Foree, W. J. Jewell, and P. L. McCarty, in *Proceedings of the 5th International Water Pollution Research Conference,* Pergamon, New York, 1971, (In Press).

46. F. Harris, *Bull. Bingham Oceanog. Coll.* **17**, 31 (1959).

47. J. H. Martin, *Limnol. Oceanog.* **13**, 63 (1968).

48. J. L. Brooks, in *Eutrophication, Causes, Consequences, Correctives,* National Academy of Sciences, Washington, D. C., 1969, pp. 236.

49. V. A. Alexander, in *Proceedings of the 25th Industrial Waste Conference,* Purdue Univ., Lafayette, Ind., 1970 (In Press).

50. D. W. Menzel and J. J. Goering, *Limnol. Oceanog.* **11**, 333 (1966).

51. D. W. Menzel, *Deep-Sea Res.,* **14**, 229 (1967).

52. S. C. Rittenberg, K. O. Emery, and W. L. Orr, *Deep-Sea Res.* **3**, 23 (1955).

53. C. H. Mortimer, *J. Ecol.* **29**, 230 (1941); **30**, 147 (1941).

54. A. R. Gahler, in *Proceedings of the Eutrophication Biostimulation Assessment Workshop* (Berkeley, Calif.) (E. J. Middlebrooks, T. E. Maloney, C. F. Powers, and L. M. Knack, Eds.), 1969.

55. J. J. Goering and M. M. Patmatmat, *Invest. Pesquera,* **35**, 233 (1971).

56. P. L. Brezonik and G. F. Lee, *Environ. Sci. Technol.* **2**, 120 (1968).

57. E. T. Degens, *Geochemistry of Sediments: A Brief Survey,* Prentice-Hall, Englewood Cliffs, N.J., 1965.

58. D. R. Keeney, J. G. Konard, and G. Chesters, *J. Water Pollution Control Federation* **42**, 411 (1970).

59. F. M. Swain, *Bull. Geol. Sur. Amer.* **72**, 519 (1961).

60. J. Foehrenbach, *J. Water Pollution Control Federation* **41**, 1456 (1969).

61. F. E. Stratton, *J. San. Eng. Div., Amer. Soc. Civil Eng.* **94**(SA 6), 1085 (1968).

62. F. E. Stratton, *J. San. Eng. Div., Amer. Soc. Civil Eng.,* **95**(SA 2), 223 (1969).

63. G. L. Hutchinson and F. G. Viets Jr., *Science* **166**, 514 (1969).

64. S. W. Watson, *Limnol. Oceanog., Suppl.* **10**, R274 (1965).

65. G. Knowles, A. L. Downing, and M. J. Barrett, *J. Gen. Microbiol.* **38**, 279 (1965).
66. K. V. Natarajan, *J. Water Pollution Control Federation* **42**, R184 (1970).
67. J. J. Goering and V. A. Dugdale, *Limnol. Oceanog.* **11**, 113 (1966).
68. J. J. Goering and R. C. Dugdale, *Science* **154**, 505 (1966).
69. J. J. Goering, *Deep-Sea Res.* **15**, 157 (1968).
70. J. J. Goering and J. D. Cline, *Limnol. Oceanog.* **15**, 306 (1970).
71. J. M. Barbaree and W. J. Payne, *Marine Biol.* **1**, 136 (1967).
72. W. J. Payne and P. S. Riley, *Proc. Exp. Biol. Med.* **132**, 258 (1969).
73. C. C. Delwiche, in *Inorganic Nitrogen Metabolism* (W. D. McElroy and B. Glass, Eds.), Johns Hopkins Press, Baltimore, 1956, pp. 233.
74. R. E. McKinney and R. A. Conway, *Sewage Ind. Wastes* **29**, 1097 (1957).
75. V. B. Skerman and I. C. MacRae, *Can. J. Microbiol.* **3**, 505 (1957).
76. E. McCoy and W. B. Sarles, in *Eutrophication, Causes, Consequences, Correctives,* National Academy of Sciences, Washington, D.C., 1969.

4 The Microbiology of Mine Drainage Pollution

D. G. Lundgren, J. R. Vestal,
and F. R. Tabita,
Department of Biology
Syracuse University
Syracuse, New York

4-1. Introduction

Acid waters originating from the exposure of sulfuritic materials to degradation processes constitute a water pollution problem of considerable magnitude. The seriousness of the problem is frequently stressed in nontechnical language in the news media and quasi-technical journals (1, 2). The United States Department of the Interior estimates that 10,000 miles of streams and 29,000 surface acres of impoundments and reservoirs are seriously affected by surface mining operations and that deep mining (and other types of mining) add to these figures significantly. It is estimated that acid mine drainage amounts to over 4 million tons per year of acidity from active and abandoned mines.

In nature both iron and sulfur oxidizing bacteria are associated with metal sulfide minerals which serve as rather unusual energy substrates for microorganisms. Metal sulfides such as arsenopyrite ($FeS_2 \cdot FeAs_2$), bornite (Cu_5FeS_4), calcocite (Cu_2S), calcopyrite ($CuFeS_2$), covellite (CuS), energite ($3Cu_2S \cdot As_2S_5$), galena (PbS), marcasite (FeS_2), millerite (NiS), molybdenite (MoS_2), orpiment (As_2S_3), pyrite (FeS_2), sphalerite (ZnS), and

69

tetrahedrite ($Cu_8Sb_2S_7$) are oxidized by microbes; this general subject has been reviewed (3).

The sulfide-bearing metals exist in different crystal structures and represent a challenge to the microorganisms as a utilizable energy source. Little is known of how bacteria actually attack a crystalline substrate and extract energy; however, the by-products of these metal sulfide oxidations are acid pollutants and metal precipitates such as "yellow boy."

Silverman (4) gives evidence that oxidation of pyrite can occur directly on the surface of the crystal as a result of the organism being absorbed to the surface (direct contact mechanism). A more satisfactory mechanism for oxidation however, may be the indirect one where the bacteria provide a high level of ferric ions in solution and these ions act as the oxidizing agent at the pyrite surface (see Reaction 4 below). An understanding of how metal sulfide oxidations, catalyzed by bacteria, occur will be important for any application of control measures designed to prevent acid formation and precipitate accumulation.

The abundance of iron and/or sulfur in natural sulfide metals makes the iron and sulfur oxidizing bacteria extremely important to man (3, 5). The organisms, microbial physiology, and biochemistry associated with iron and sulfur oxidations and the resulting acid formation will be the major theme of this chapter.

Metal sulfides when exposed to oxygen are oxidized to the corresponding metal sulfates and sulfuric acid. Reaction mechanisms are not fully understood but are generally indicated by the following equations, with pyrite used as an example.

$$2\,FeS_2 + 2H_2O + 7O_2 \longrightarrow 2FeSO_4 + 2H_2SO_4 \tag{1}$$
Pyrite   Ferrous sulfate   Sulfuric acid

The subsequent oxidation of ferrous sulfate yields ferric sulfate.

$$4FeSO_4 + O_2 + 2H_2SO_4 \rightarrow 2Fe_2(SO_4)_3 + 2H_2O \tag{2}$$
Ferrous sulfate   Ferric sulfate

The reaction can proceed to form ferric hydroxide and acid.

$$Fe_2(SO_4)_3 + 6H_2O \longrightarrow 2Fe(OH)_3 + 3H_2SO_4 \tag{3}$$
Ferric hydroxide

Pyrite oxidation can also occur in the presence of ferric iron at higher pH's.

$$FeS_2 + 14Fe^{3+} + 8H_2O \rightarrow 15Fe^{2+} + 2SO_4^{2-} + 16H^+ \tag{4}$$

Reaction 2 proceeds very slowly below pH 4 in the absence of a catalyst but is rapid in the presence of microbial catalysts such as iron-oxidizing bacteria. Although bacteria have been known for many years to be present in mine drainage (6, 7), the fact that they are major contributors to acid formation in nature has only recently been critically demonstrated (8). The rate-determining step in the oxidation of iron pyrite and the formation of acidity in streams associated with coal and copper mines has been shown to be the oxidation of ferrous iron. Bacteria exhibited the greatest effect of any catalyst in accelerating the oxidation of Fe^{2+} under sterile conditions after inoculations with untreated and with sterilized natural mine water showing that microbial mediation accelerates the reaction by a factor larger than 10^6.

4-2. Mine Drainage Classes

Although there is no "typical" mine drainage, waters discharging from coal mines which create the major problem can be divided into four general classes (Table 4.1).

Table 4.1 Mine Drainage Classes

	Class I, Acid Discharges	Class II, Partially Oxidized and/or Neutralized	Class III, Oxidized and Neutralized and/or Alkaline	Class IV, Neutralized and Not Oxidized
pH	2–4.5	3.5–6.6	6.5–8.5	6.5–8.5
Acidity, mg/l ($CaCO_3$)	1,000–15,000	0–1,000	0	0
Ferrous iron, mg/l	500–10,000	0–500	0	50–1,000
Ferric iron, mg/l	0	0–1,000	0	0
Aluminum, mg/l	0–2,000	0–20	0	0
Sulfate, mg/l	1,000–20,000	500–10,000	500–10,000	500–10,000

The type of drainage induced is dependent upon the kind of mine and the nature of the geological surroundings. For coal mines, drainage is dependent upon the amount of sulfides, their spatial distribution, crystallinity and size, the presence of bacteria, water level, and the presence or absence of calcium in the sulfide aggregates.

4-3. Microbiology of Drainage Waters

Acid mine waters contain a predominance of iron-oxidizing bacteria (*Thiobacillus ferrooxidans*) and sulfur-oxidizing bacteria (*Thiobaccillus thiooxidans*) whereas the autotrophs of alkaline waters are mostly non-acidophilic sulfur-oxidizing bacteria. Results of an ecological study (9) of synthetic mine drainage water report as many as 10^6 iron-oxidizing organisms per milliliter of acid water. Such studies as well as other investigations of acid waters from copper mines, report that among other microorganisms indigenous to acid drainage are Gram-positive and Gram-negative heterotrophic bacteria, algae, yeast, and protozoa (10, 11).

Hutchinson, Johnstone, and White (12), using data from numerical analysis, have produced an overall classification of the genus *Thiobacillus*. Adjunct tests (DNA base composition) to systematics have supported the observational classification arrived at by multivariate analysis (13). Unz and Lundgren (14) proposed a reclassification of the true acidophilic iron-oxidizing bacteria as listed in *Bergey's Manual of Determinative Bacteriology*, 7th ed., so that the bacteria would all be included in the family *Thiobacteriaceae*. It is well to remember that before one attempts to name a new isolate from mine drainage (or any other environment), it would behoove the investigator to consult the procedures for the identification of an unknown strain of the genus *Thiobacillus* (12).

Iron-oxidizing organisms can be cultivated on an artificial mine drainage medium such as the one shown below supplemented with a nitrogen source. The composition is contained in 1 liter of 0.01 N H_2SO_4:

$CaSO_4 \cdot 2H_2O$	0.344 g
$MnSO_4 \cdot H_2O$	0.024 g
$Al(SO_4)_3 \cdot 18H_2O$	0.186 g
$FeSO_4 \cdot 7H_2O$	0.997 g
$MgSO_4 \cdot 7H_2O$	0.246 g
$(NH_4)_2SO_4$	0.150 g

The final pH of this medium is approximately 3.2.

It was readily apparent from results of laboratory investigations of these bacteria that the simulated mine drainage medium would not support large numbers of bacteria because of limiting energy (iron) and nitrogen. The now much used 9K medium was developed to overcome this problem (15); this medium will support cell yields in excess of 3.0×10^8 cells/ml, an amount of cells required for laboratory studies. Harvesting procedures have also been improved to minimize loss of cells during the separation of bacteria from the precipitated mess (yellow boy) in the medium resulting during growth and iron oxidation.

Growth of *T. ferrooxidans* or *T. thiooxidans* on elemental sulfur (or thiosulfate) is easily achieved by replacing the ferrous iron in the above mentioned 9K medium with 0.5% elemental sulfur or sodium thiosulfate. Various media for the growth of thiobacilli on sulfur have been reviewed (16).

The general anatomy and fine structure of acidophilic iron and sulfur-oxidizing bacteria has been discussed (17–22). Figures 4.1 and 4.2 show

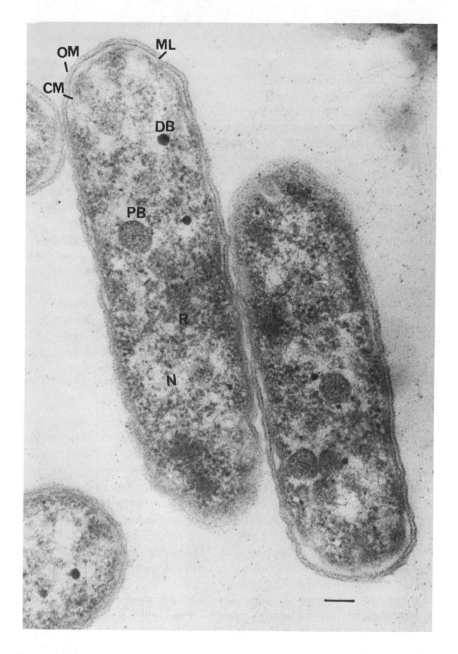

Figure 4.1 Thin section of an osmium-fixed cell of *T. ferrooxidans*. The cell envelope (wall plus cytoplasmic membrane) can be clearly seen and is labelled (OM), outer membrane or layer; (ML), middle layer or rigid layer; (CM), cytoplasmic membrane. CM is not shown in this micrograph surrounding the entire cytoplasm. The polyhedral body (PB), dense body (DB); ribosomes and nucleus (N) are identified in the cytoplasm. The bar marker represents 100 nm (22).

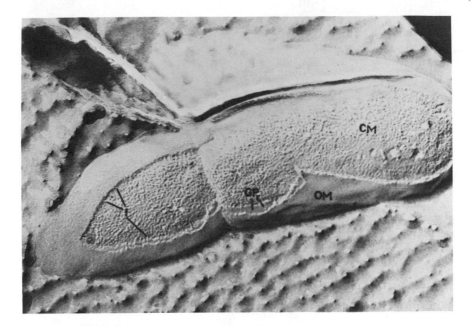

Figure 4.2 Frozen-etched preparation of a similar type of cell as shown in Figure 4.1 giving the reader a three dimensional-like view of the cell.

thin section and frozen etched profiles of *T. ferrooxidans*. Ultrastructural features of the envelope of *T. ferrooxidans* are similar to Gram-negative bacteria and the different layers of the cell envelope are shown accentuated in the frozen-etched preparation. The envelope outer layer is mostly lipopolysaccharide and lipoprotein and appears to have some function in Fe^{2+} binding. The cytoplasm of the cell contains ribosomes, nuclear material, and cell inclusions, the functions of which are not known. There is evidence that structural changes do occur in the thiobacilli when they are grown under different conditions; however, no physiological significance has as yet been attributed to these differences.

Alkaline mine drainage contains various kinds of thiobacilli that oxidize different reduced sulfur compounds (23). One thiobacillus isolated from alkaline drainage is similar to *Thiobacillus intermedius* whereas another isolate oxidizes thiosulfate incompletely and produces an alkaline byproduct; it appears to be similar to *Thiobacillus thioparus*. Preliminary work in the authors' laboratory indicates that the alkaline environment is suitable for supporting many types of bacteria capable of oxidizing reduced sulfur compounds. However, many more experiments need to be done before we begin to understand the microbiology of this type of mine drainage.

4-4. Growth of T. ferrooxidans on Iron

The conditions for growth of *T. ferrooxidans* on iron are rather unusual in the microbial world. The organism is acidophilic and grows poorly above pH 4.0. During growth, the organism produces sulfuric acid and thus decreases the pH. When the cells have reached maximum growth, under ideal laboratory growth conditions, the pH falls to about 1.5–2.0. Cells can oxidize iron at pH 1.0. The organism's ability to exist at a low pH has certain survival advantages, since above pH 5 reduced iron will oxidize chemically to Fe^{3+}, a form which possesses no potential energy for the bacterium.

The general reaction to express biological iron oxidation is:

$$4Fe^{2+} + O_2 + 4H^+ \rightarrow 4Fe^{3+} + 2H_2O \qquad (5)$$

In nature, the bacteria use a variety of reduced iron compounds such as $FeSO_4$ or FeS_2. The end product(s) formed during Fe^{2+} oxidation can occur as ferric oxides, hydroxides, sulfates, phosphates, etc. Under laboratory conditions the predominant precipitate is $Fe(OH)_3$ (24). During growth, the initial greenish-blue color of the medium due to the $FeSO_4$ is replaced by a dark burnt orange to red, the color of which is due to the ferric precipitates (yellow boy).

In the laboratory, these microorganisms grow under controlled conditions of temperature, pH, aeration, and nutrient composition. In nature, this is not possible and one must be cognizant of the differences in growth conditions. Such differences are becoming increasingly important, and the study of microbial ecology is attempting to wed laboratory results with what occurs in nature. Unfortunately, the quantitative measurements of growth of iron-oxidizing bacteria has mainly been in laboratory culture where conditions are defined.

The most widely used growth medium for *T. ferrooxidans* is the 9K medium (15). Cell yields in this medium are about 1 g (wet weight) of cells per 16 liters of media after 72 to 90 hours of incubation under forced aeration at 28°C. With a Fe^{2+} content of 9000 ug/ml of medium, it is easy to see that the utilization of Fe^{2+} as a primary energy source is not very efficient. Cell numbers, pH, and redox potential (*Eh*) when quantitated under these conditions (25) change in a characteristic manner

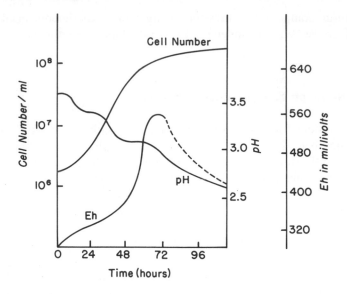

Figure 4.3 Typical growth curve of *T. ferrooxidans* grown in aerated 9K medium showing changes in pH and Eh.

(Figure 4.3). The redox potential of the growth medium rises gradually, followed by a rapid rise at about 48 hours; this rise is due to the biological oxidation of Fe^{2+}. The stepwise fall of pH can be correlated with the formation of $Fe_2(SO_4)_3$ between 24 to 30 hours of growth. The rapid drop between 30 and 40 hours was due to the increased rate of Fe^{2+} oxidation. A pH plateau is noted between 40 and 60 hours where the greatest rise in *Eh* is observed along with the formation of a reddish brown color to the medium. After 60 hours the pH continues to fall; this is associated with the formation of the $Fe(OH)_3$ precipitate. By 120 hours about 4×10^8 cells/ml are counted in the medium.

How efficient is this autotrophic growth using Fe^{2+} as the energy source and CO_2 as the sole source of carbon? Early estimates (26) of the efficiency of Fe^{2+} oxidation for fixing CO_2 are about 3.2%, based on the oxidation yield of 11.3 kcal of energy per gram atom of iron and assuming that 120 kcal of energy are required to fix 1 g of carbon. The conversion of CO_2 to cell material, as glucose, requires $+115$ kcal/mole of carbon synthesized. Experimental results showed a 182–1 weight ratio of Fe^{2+} oxidized to organic carbon synthesized; the value was based on O_2 uptake studies with an efficiency of Fe^{2+} oxidation of 20.5% (27).

The preferred nitrogen source for *T. ferroxidans* is $(NH_4)_2SO_4$ (17); inorganic nitrogen can be replaced by organic nitrogen sources such as alanine, glutamic acid, and lysine. However, there are sources of amine nitrogen that cannot substitute for $(NH_4)_2SO_4$.

The effects of some organic compounds on growth, with and without Fe^{2+} as the energy source, have been studied (17). Cells could not use methanol, formate, acetate, glyoxylate, pyruvate, lactate, oxalacetate, succinate, ribose, α-ketoglutarate, or yeast extract as a source of carbon and energy for growth. However, with Fe^{2+} present, detectible iron oxidation was found in the presence of methanol, acetate, glyoxylate, pyruvate, ribose, or α-ketoglutarate. Iron-grown cells can adapt to glucose as the sole source of carbon and energy. Heterotrophic growth will be discussed elsewhere in this chapter.

4-5. Iron Oxidation and Energy Production

Iron oxidation by resting cell suspensions of *T. ferrooxidans* has been studied using the conventional Warburg respirometer (27, 28). Maximum iron oxidation occurs between pH 3.0–3.7 at 37°C; growth of the bacterium, however, is maximal at 28°C. Concomitant with the oxidation of iron, CO_2, measured manometrically, is taken up by the cells. The efficiency of iron oxidation is greater than 92% of the theoretical value. Many workers have confirmed the fact that iron is oxidized by thiobacilli (29–31). Biological iron oxidation requires SO_4^{2-} and is partially inhibited by Cl^- at rather high concentrations of the anion (32). The anion phosphate partially inhibits iron oxidation, but citrate is a strong inhibitor. The development of a rapid and simple biological assay has made possible kinetic studies of iron oxidation (33). The assay procedure involves measuring a $FeCl_3$ complex at 410 nm in the spectrophotometer; the complex forms upon addition of HCl to the iron-containing reaction mixture. *T. ferrooxidans* has an apparent K_m for iron oxidation of $2.2 \times 10^{-3}\ M$ in a β-alanine–SO_4 buffer and $5.4 \times 10^{-3}\ M$ in an unbuffered acid–water system. The pH optimum for iron oxidation ranged from 2.5 to 3.8. The sulfate anion was necessary for the reaction but chloride, under these conditions, did not inhibit the oxidation. The sulfate requirement is partially replaced by HPO_4^{2-} and $HAsO_4^{2-}$, but not by BO_3^-, MoO_4^{2-}, NO_3^-, or Cl^-; $HCOO^-$ and MoO_4^{2-} inhibit the iron oxidation reaction.

When *T. ferrooxidans* is grown under conditions of "excessively vigorous aeration," the morphology of the bacteria is abnormal (30). Cells are normally seen as rods 1.0–1.7 μm long by 0.5 μm wide (Figures 4.1 and 4.2); the abnormal cells have diameters of about 1.0 μm and are coccoid in shape. Abnormal cells will not oxidize iron as efficiently as normal cells or cells which have reverted to normalcy after being coccoid. These observations suggest that the cell surface may be important in iron oxidation, for a decrease in the surface to volume ratio occurs when coccoidal cells are formed.

Where exactly does Fe^{2+} oxidation take place; is it on or in the bacterial cell? What role does the outer surface of the cell play in Fe^{2+} oxidation? How and where is the loss of the electron in Fe^{2+} oxidation coupled to energy production in the cell? These questions are fundamental to understanding biological iron oxidation and the search for these answers has motivated much of the research in the authors' laboratory.

The complexing of Fe^{2+} to cells has been demonstrated using polarographic and ^{59}Fe uptake studies (34); results of these investigations have helped to develop a working model to study biological Fe^{2+} oxidation (Figure 4.4 modified from Reference 34). Cell-free culture media give

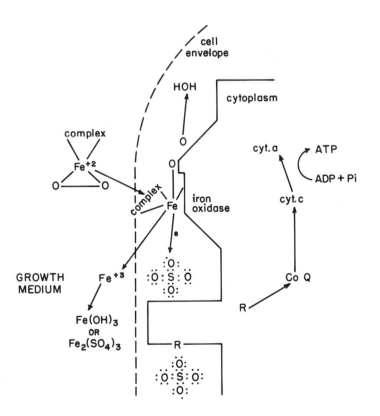

Figure 4.4 Schematic drawing of a possible model for iron oxidation by *T. ferrooxidans* (modified from 34).

two polaragraphic waves in the range of 0 to 0.9 V. These waves are not due to dissolved oxygen, Fe^{2+}, Fe^{3+}, or medium constituents. A "complex" between Fe^{2+} and oxygen is believed to form in the medium, but sulfate or phosphate may also be involved for they are known to complex with oxygen (35). Since SO_4^{2-} is essential to Fe^{2+} oxidation, and is bound to the cell (34), this anion could act as the binding link between the cell envelope and the Fe^{2+}–oxygen complex. After the "complex" is bound, the iron oxidase system (36–39) catalyzes the release of the Fe^{2+} electron. This electron could be transferred to an electron-deficient sulfate group which is possibly bound to the cell surface at an enzymatic site. The R group shown may be part of the electron transport system and thus the electron would pass down the transport chain and generate energy. The electron transport system (ETS) of *T. ferrooxidans* has been investigated and is abbreviated relative to aerobic heterotrophic organisms. One ATP is known to be produced during the oxidation and reduction of cytochrome *c* and cytochrome *a*, the only cytochromes present in the *T. ferrooxidans* ETS (39):

$$Fe^{2+} \begin{matrix} \rightarrow \\ Fe^{3+} \end{matrix} \quad \begin{matrix} \text{Oxidized cytochrome } c \\ \text{Reduced cytochrome } c \end{matrix} \quad \begin{matrix} \text{Reduced cytochrome } a \\ \text{Oxidized cytochrome } a \end{matrix} \quad \begin{matrix} O_2 \\ H_2O \end{matrix}$$

$$\text{ADP} + \text{Pi} \quad \longrightarrow \quad \text{ATP}$$

Coenzyme Q, known to be present in *T. ferrooxidans* (40), probably acts as an electron carrier in the electron transport system.

The oxygen atoms of the "complex," when split by the oxidase, may either be used as the terminal electron acceptor or may react chemically with the oxidized substrate (ferric ion) to form the oxides or hydroxides which precipitate as "yellow boy."

The components and mechanism of the aforementioned electron transport chain (or iron oxidase system) have been studied in detail at the biochemical level (39). Iron oxidation involves cytochromes *c* and *a*, and the enzymes iron–cytochrome *c* reductase and cytochrome oxidase, in that order. The iron oxidase system is closely associated with the cell membrane or particulate fraction of the cell, and is labile to heat and sonication. The iron–cytochrome *c* reductase has been purified to a high degree of purity and consists of two subunits, one protein (mol. wt. 27,000–30,000) and the other ribonucleic acid (mol. wt. 35,000–330,000). The exact mechanism of the iron-cytochrome *c* reductase reaction has been reported (41).

The binding of ^{59}Fe to the envelope of the cell and the elucidation of the electron transport system involving iron–cytochrome *c* reductase strongly suggest that the Fe^{2+} does not enter the cell prior to oxidation and that the oxidation occurs in or on the cell envelope. The involvement of the cell envelope in Fe^{2+} oxidation has been studied by trying to find constituents which will oxidize Fe^{2+}. The isolation of a rather large amount of phosphatidylserine (PS) and some phosphatidylethanolamine (PE) in the fraction of the cell believed to be the outer membrane of the cell envelope has been demonstrated (42). The large amount of PS suggests a special function for the phospholipid, such as providing a negative charge, which would facilitate the binding of Fe^{2+}. Recently this idea has been supported experimentally (42a). Purified PS isolated from *T. ferrooxidans* bound Fe^{2+} tightly, whereas PE, phosphatidylglycerol, phosphatidylcholine, and phosphatidylmonomethylethanolamine do not bind Fe^{2+} as readily. Other investigators (43) studying iron oxidation in a different organism report the phospholipid components to consist of large amounts of phosphatidylmonomethylethanolamine and phosphatidylglycerol; no PS was present and in this case no iron binding was noted.

The energy, generated as ATP during oxidative phosphorylation in the electron transport system of *T. ferrooxidans,* is used by the cell to support the necessary physiological reactions for growth. The reduction of CO_2 as the carbon source for growth requires energy and reduced forms of pyridine nucleotides. Since the redox potentials of $NAD^+/NADH$ and $NADP^+/NADPH$ are low ($E_0' = -0.320$ and -0.324, respectively) and the Fe^{2+}/Fe^{3+} redox couple is high ($E_0' = +0.77$), it is obvious that in the normal oxidation state, ferrous iron could not reduce the pyridine nucleotides. An ATP-dependent reduction of the pyridine nucleotides is known for *T. ferrooxidans* whereas in cell-free extracts NAD is reduced in the presence of ATP and reduced cytochrome *c* (44).

The utilization of carbon in these autotrophically grown microorganisms involves the reduction of CO_2 through the Calvin photosynthetic cycle (45). The cycle involves the carboxylation of ribulose-1,5-diphosphate to form two molecules of 3-phosphoglyceric acid, and this reaction is unique to autotrophic organisms. The function of the Calvin cycle in *T. ferrooxidans* as well as the likely involvement of phosphoenolpyruvate carboxylase, necessary for certain amino acid formation, has been shown (46, 47). The assimilation of $^{14}CO_2$ is dependent on the presence of Fe^{2+} (46).

4-6.   The Growth of T. ferrooxidans on Sulfur

In the laboratory, normal iron-oxidizing thiobacilli are capable of growth in the basal salts of the 9K medium containing sulfur (5 g/liter) substituted for iron as the energy source (15, 48, 49). Cell yields per liter of medium are about 150–180 mg of cells (wet weight) for sulfur-grown cells, compared to 60 to 70 mg (wet weight) for iron-grown cells.

In appearance the organisms grown on sulfur retain the typical Gram-negative features, but do, however, differ from iron-grown cells in that the cell ends are more pointed (Figure 4.5), called "nibs." Fine structure details do not appear to change. The ability to oxidize iron is retained by sulfur-grown cells which have gone through subsequent subculture.

When resting cell suspensions prepared from cells grown on either iron or sulfur are compared quantitatively with regard to their iron and sulfur oxidation ability, the $Q_{0_2}(N)$ for iron oxidation for both cell types is the same. This suggests a constitutive nature for the iron-oxidizing systems (48). This, however, is not the case for the sulfur-oxidizing system, since the rate of sulfur oxidation of iron-grown cells is about half that of sulfur-grown cells (ratio of 0.40). These results imply that the iron- and sulfur-oxidizing systems are different. Other workers have come to similar conclusions (50); heavy aged suspensions of *T. ferrooxidans* retain the ability to oxidize iron while losing sulfur-oxidizing ability. Other evidence in support of the difference in the two oxidative pathways is the differential sensitivity of iron and sulfur oxidation to *N*-ethylmaleimide and sodium azide (51). Thus, it seems that the ability to oxidize reduced iron and sulfur compounds is an inherent characteristic of these organisms, with the two mechanisms being located at different sites.

A most interesting facet of an organism's growth in a medium containing an insoluble substrate, such as sulfur, is its ability to attach and, by chemical means, extract the energy required for growth. It is well established (52, 53) that direct contact of the cell with the sulfur particle is necessary in growing cultures of *T. thiooxidans*. Direct contact between organism and sulfur is manifested by an erosion of the sulfur crystal in the area immediately adjacent to the cell (54); when large numbers of bacteria are present, the entire surface of the crystal becomes eroded.

Elemental "flowers of sulfur" (sp. gr. = 2.1) remains on the surface of the culture medium until the culture begins to grow. This suggests that a wetting agent is produced by the growing cells (55). Indeed, fresh sulfur added to spent culture filtrates becomes wet immediately. There is a certain optimum level of wetting agent required before adequate bacterial content is maintained in shaken culture (52). One of the wetting agents is known to be phosphatidylinositol (56). An extracellular complex forms in the culture filtrate of both iron- and sulfur-grown *T. ferrooxidans* (57) and the complex from sulfur-grown cells contains approximately 70% phospholipid, with phosphatidylserine the major component. This high percentage of phospholipid in filtrates from sulfur-grown cells, as in filtrates from cultures of *T. thiooxidans*, probably indicates a wetting agent role.

Bacterial membranes are composed of substantial quantities of phos-

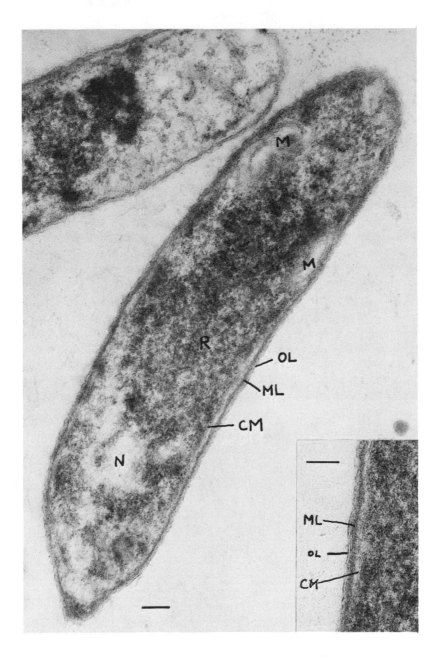

Figure 4.5 Thin section of *T. ferrooxidans* grown on sulfur showing the many polyhedral bodies. The cells have more pointed ends. The nucleus is labelled (N); cytoplasmic membrane (CM); middle layer (ML) and outer layer (OL).

pholipid and protein, and, from the results noted above, the cell envelope membrane could play an important role in the initial stages of oxidation and subsequent metabolism of sulfur. Cell envelope-membrane fragments, present in cell-free extracts of sulfur-grown *T. thiooxidans*, are able to oxidize elemental sulfur without any cofactors (58). Moreover, oxygen uptake during sulfur oxidation is markedly enhanced by the addition of wetting agents.

In *T. thiooxidans* the sulfur oxidizing system has been resolved into three fractions (59): a particulate (P_2) fraction (105,000 $\times$ g) and two soluble fractions, one being collodion-membrane permeable (S-P) and the other impermeable (S-IP). When reconstitution experiments are performed, sulfur oxidation takes place only in an assay system containing all three fractions. Furthermore, both reduced or oxidized forms of NAD or NADP at low concentrations could replace the S-P fraction. This fraction (S-P) does contain considerable amounts of NAD^+ and $NADP^+$. The reduced forms of these pyridine nucleotides are oxidized by the P_2 fraction. Cyanide inhibits both sulfur and NADPH oxidation, indicating that electrons from these donors are transferred to oxygen via common step(s) in the electron transport chain.

A particulate fraction (90,000 $\times$ g) isolated from the same organism is also able to oxidize thiosulfate ($S_2O_3^{2-}$) to tetrathionate ($S_4O_6^{2-}$) (60); trithionate ($S_3O_6^{2-}$) may also be formed during the reaction. Electron micrographs of this crude particulate preparation clearly show membrane fragments. Sulfur oxidation is associated with the "soluble" fraction of this preparation which oxidizes thiosulfate. The oxidation of sulfite has also been shown to be associated with a particle fraction (61). Thus, our most current information points to an increased importance of the envelope membrane in the mechanism of sulfur oxidation, yet structurally there does not appear to be significant difference in the cell envelope of sulfur-oxidizing thiobacilli (19, 20). Nothing can be said about differences at the molecular level at this time. Perhaps, as suggested by Trudinger (62), the soluble enzymes responsible for the oxidation of sulfur to sulfite and sulfite to sulfate, enzymes which normally require added cofactors (63, 64), may arise by a disaggregation of the membrane *in vitro*. Possibly the cell does not require added cofactors with active membrane preparations.

4-7. Metabolism of Inorganic Sulfur Compounds by Thiobacilli

The work relative to the enzymatic basis for sulfur oxidation has been adequately reviewed (62, 65, 66). However, many problems are not solved, and as Peck (67) points out, "the exact pathway for the oxidation of reduced sulfur compounds by these microorganisms remains in doubt, and it may be that different species of thiobacilli have modified pathways." Consequently, we will present an oxidation scheme (68, 69) for *T. ferrooxidans* based on published enzymatic data which will summarize the reactions known to occur in this organism:

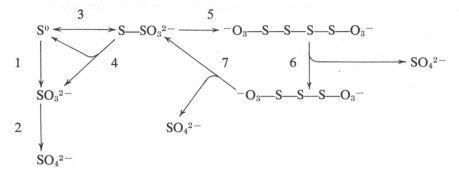

Reaction 1 is catalyzed by the sulfur-oxidizing enzyme where sulfite is the product of the reaction; sulfite is oxidized to sulfate via sulfite oxidase (Reaction 2). Thiosulfate, known to form nonenzymatically during the oxidation of elemental sulfur (Reaction 3), may then be cleaved by rhodanese (Reaction 4) to replenish sulfite and elemental sulfur. The thiosulfate molecule may also be oxidized to tetrathionate (Reaction 5), a reaction catalyzed by the thiosulfate-oxidizing enzyme. Reactions 6 and 7 are still unknown, but are postulated to occur based on data from other thiobacilli.

4-8.   Growth of T. ferrooxidans on Organic Substrates

In its normal mine water environment, *T. ferrooxidans* is in contact with relatively high concentrations of reduced iron and sulfur compounds, and their oxidation furnishes the energy needed for the reduction of CO_2 by reduced pyridine nucleotides and subsequent cell synthesis. It is also known that organic compounds are present in this environment, and these may have an effect on inorganic oxidation. Moreover, organic compounds can even be metabolized by these normally autotrophic organisms.

The oxidation of glucose by iron-oxidizing bacteria was initially studied in the authors' laboratory using glucose (0.5%) added to the 9K medium. The cells, in the presence of glucose, first oxidized iron and then glucose, following the cessation of iron oxidation (17). Cells removed from the medium containing both iron and glucose grow well in a glucose–salts medium free from substrate amounts of iron. However, when cells are taken directly from an iron–salts medium, they are not readily transferable to a medium containing only glucose as an energy source. Glucose-grown cells are at first slow in re-adapting to growth on iron, but once growth begins cells grow rapidly when transferred to fresh 9K medium.

Other strains of iron-oxidizing thiobacilli besides that of this laboratory also grow heterotrophically (70). However, this characteristic may not be common among the group (70). On glucose, cells have a generation time of about 4.5 hours as compared to about 10 hours on the iron containing medium. *T. ferrooxidans* thus resembles *T. novellus* and various *Hydrogenomonas* species in exhibiting a faster growth rate on organic matter. This expression is not necessarily true for all thiobacilli.

When grown solely on glucose, or in a medium composed of iron and glucose (mixotrophic culture), *T. ferrooxidans* shows some changes in fine structure as compared to iron-grown cells (21). In thin sections (Figure 4.6), glucose-grown cells exhibit many electron-transparent areas typical of storage compounds such as poly-β-hydroxybutyric acid (PHB) granules. Indeed, PHB has been chemically extracted from *T. ferrooxidans* (71). Another structural difference in glucose-grown cells is the absence of the electron-dense bodies seen in iron-grown cells (21).

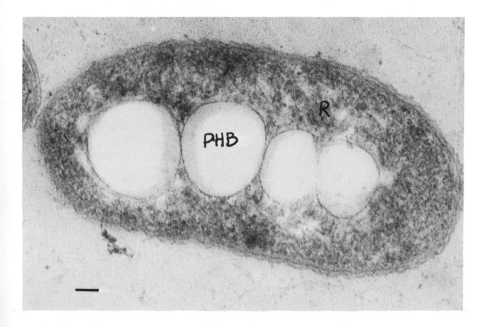

Figure 4.6 Thin section of *T. ferrooxidans* grown on glucose showing a shorter type of cell and the presence of the storage material, poly-β-hydroxybutyrate (PHB).

Cell-free extracts from *T. ferrooxidans* grown on glucose as an energy source contain the enzyme glucose-6-phosphate dehydrogenase and the enzymes of the Entner-Duodoroff pathway, namely 6-phosphogluconate dehydrase and 2-keto-3-deoxy-6-phosphogluconate aldolase, at high levels. The levels are dramatically enhanced over those of cell extracts from iron-grown cells (71a). This induction of the enzymes comprising the Entner-Duodoroff pathway was first reported in other chemoautotrophs such as *Hydrogenomonas H16* (72) and *T. intermedius* (73, 74).

Organic compounds do exert some effect on iron oxidation as shown by growing *T. ferrooxidans* mixotrophically and then testing these cells for iron-oxidizing ability, using the kinetic assay mentioned previously (33). Iron–glucose-grown cells harvested after both substrates were depleted show considerable loss in iron-oxidizing ability. However, iron–fructose-grown cells have nearly the same rate of iron oxidation as the rate for iron-grown cells. Iron–glutamate and iron–sucrose-grown cells are affected less in this regard than iron–glucose-grown cells. These differences in iron oxidation ability after cells are grown mixotrophically may be due to differing degrees of ribulose diphosphate carboxylase repression; this is a key enzyme concerned with CO_2 fixation. Such a precedent for enzyme repression exists in *Hydrogenomonas* (75). The interplay of organic substrates with oxidations of inorganic substrates in the natural environment may be a means for control of acid formation. It is the authors' belief that fundamental research on the physiology, biochemistry, and microbiology of bacteria that exist in mine drainage will contribute greatly to knowledge needed to control acids and precipitates now polluting these waters.

Acknowledgments

The authors wish to thank the following people for their contributions of material: Mr. Ron Hill, U.S. Department of the Interior, Cincinnati, Ohio, for the use of Table 4.1; Dr. Charles Remsen, Woods Hole Oceanographic Institution, Woods Hole, Mass., for Figure 4.2; Dr. Augustine Wang, Mississippi State University, State College, Miss., for Figures 4.5 and 4.6. This chapter was written while the authors were supported by Contract 14010 DAY of the FWQA, U.S. Department of the Interior.

REFERENCES

1. *Environ. Sci. Technol.* **3**, 1237 (1969).
2. *Chem. Eng. News* **48**, 33 (1970).

3. M. P. Silverman and H. L. Ehrlich, *Advan. Appl. Micro.* **6**, 153 (1964).
4. M. P. Silverman, *J. Bacteriol.* **94**, 1046 (1967).
5. J. E. Zajic, *Microbial Biogeochemistry,* Academic Press, New York, 1969.
6. A. R. Powell and S. W. Parr, Engineering Experiment Station Bulletin No. 111, Univ. Illinois (1919).
7. G. P. Hanna, Jr., J. R. Lucas, C. I. Randles, E. E. Smith, and R. A. Brant, *J. Water Pollution Control Federation* **35**, 275 (1963).
8. P. G. Singer and W. Stumm, *Science* **167**, 1121 (1970).
9. J. H. Tuttle, C. I. Randles, and P. R. Dugan, *J. Bacteriol.* **95**, 1495 (1968).
10. J. H. Tuttle, P. R. Dugan, C. B. Macmillan, and C. I. Randles, *J. Bacteriol.* **97**, 594 (1969).
11. H. L. Ehrlich, *J. Bacteriol.* **86**, 350 (1963).
12. M. Hutchinson, K. I. Johnstone, and D. White, *J. Gen. Microbiol.* **57**, 397 (1969).
13. J. F. Jackson, D. J. W. Moriarty, and D. J. D. Nicholas, *J. Gen. Microbiol.* **53**, 53 (1968).
14. R. Unz and D. G. Lundgren, *Soil Sci.* **92**, 302 (1961).
15. M. P. Silverman and D. G. Lundgren, *J. Bacteriol.* **77**, 642 (1959).
16. J. R. Postgate, *Lab. Pract.* **15**, 1239 (1967).
17. D. G. Lundgren, K. J. Anderson, C. C. Remsen, and R. P. Mahoney, *Develop. Ind. Microbiol.* **6**, 250 (1964).
18. C. C. Remsen and D. G. Lundgren, *J. Bacteriol.* **92**, 1765 (1966).
19. R. P. Mahoney and M. R. Edwards, *J. Bacteriol.* **92**, 487 (1966).
20. J. M. Shively, G. L. Decker, and J. W. Greenawalt, *J. Bacteriol.* **101**, 618 (1970).
21. W. S. Wang and D. G. Lundgren, *J. Bacteriol.* **97**, 947 (1969).
22. W. S. Wang, M. S. Korczynski, and D. G. Lundgren, *J. Bacteriol.* **104**, 556 (1970).
23. R. Tabita, M. Kaplan, and D. Lundgren, *Third Symposium on Coal Mine Drainage,* Bituminous Coal Research, Monroeville, Pa., p. 94 (1970).
24. H. Lees, S. C. Kwok, and I. Suzuki, *Can. J. Microbiol.* **15**, 43 (1969).
25. P. R. Dugan and D. G. Lundgren, *Develop. Ind. Microbiol.* **5**, 250 (1964).
26. K. L. Temple and A. R. Colmer, *J. Bacteriol.* **62**, 605 (1951).
27. M. P. Silverman and D. G. Lundgren, *J. Bacteriol.* **78**, 326 (1959).
28. J. V. Beck, *J. Bacteriol.* **79**, 502 (1960).
29. J. Landesman, D. W. Duncan and C. C. Walden, *Can. J. Microbiol.* **12**, 25 (1966).
30. M. P. Silverman and M. H. Rogoff, *Nature* **191**, 1221 (1961).
31. N. A. Kinsel and W. W. Umbreit, *J. Bacteriol.* **87**, 1243 (1964).
32. N. Lazaroff, *J. Bacteriol* **85**, 78 (1963).
33. C. A. Schnaitman, M. S. Korczynski and D. G. Lundgren, *J. Bacteriol.* **99**, 552 (1969).
34. P. R. Dugan and D. G. Lundgren, *J. Bacteriol.* **89**, 825 (1965).
35. W. Stumm and G. R. Lee, *Ind. Eng. Chem.* **53**, 143 (1961).
36. L. P. Vernon, J. H. Mangum, J. V. Beck and F. M. Shafia, *Arch. Biochem. Biophys.* **88**, 227 (1960).
37. B. A. Blaylock and A. Nason, *J. Biol. Chem.* **238**, 3453 (1963).
38. M. G. Yates and A. Nason, *J. Biol. Chem.* **241**, 4861 (1966).
39. G. A. Din, I. Suzuki, and H. Lees, *Can. J. Biochem.* **45**, 1523 (1967).
40. P. R. Dugan and D. G. Lundgren, *Anal. Biochem.* **8**, 312 (1964).
41. G. A. Din and I. Suzuki, *Can. J. Biochem.* **45**, 1547 (1967).

42. M. S. Korczynski, A. D. Agate, and D. G. Lundgren, *Biochem. Biophys. Res. Commun.* **29**, 457 (1967).

42a. A. D. Agate and W. Vishniac, *Bacteriol. Proc.* **70**, 50 (1970).

43. S. A. Short, D. C. White and M. I. H. Aleem, *J. Bacteriol.* **99**, 142 (1969).

44. M. I. H. Aleem, H. Lees, and D. J. D. Nicholas, *Nature* **200**, 759 (1963).

45. J. A. Bassham and M. Calvin, *The Path of Carbon in Photosynthesis,* Prentice-Hall, Englewood Cliffs, N.J., 1957.

46. G. A. Din, I. Suzuki, and H. Lees, *Can. J. Microbiol.* **13**, 1413 (1967).

47. W. J. Maciag and D. G. Lundgren, *Biochem. Biophys. Res. Commun.* **17**, 603 (1964).

48. P. Margalith, M. Silver and D. G. Lundgren, *J. Bacteriol.* **92**, 1706 (1966).

49. M. Silver, *Can. J. Microbiol.* **16**, 845 (1970).

50. J. V. Beck and D. G. Brown, *J. Bacteriol.* **96**, 1433 (1968).

51. D. W. Duncan, J. Landesman, and C. C. Walden, *Can. J. Microbiol.* **13**, 397 (1967).

52. T. M. Cook, *J. Bacteriol.* **88**, 620 (1964).

53. K. G. Vogler and W. W. Umbreit, *Soil Sci.* **51**, 331 (1941).

54. W. I. Schaeffer, P. E. Holbert, and W. W. Umbreit, *J. Bacteriol.* **85**, 137 (1962).

55. G. E. Jones and R. L. Starkey, *J. Bacteriol.* **82**, 788 (1961).

56. W. I. Schaeffer and W. W. Umbreit, *J. Bacteriol.* **85**, 492 (1962).

57. A. D. Agate, M. S. Korczynski, and D. G. Lundgren, *Can. J. Microbiol.* **15**, 259 (1969).

58. F. W. Adair, *J. Bacteriol.* **92**, 899 (1966).

59. A. Kodama, *Plant Cell Physiol.* (Tokyo) **10**, 645 (1969).

60. T. Tano, H. Asano, and K. Imai, *Agr. Biol. Chem.* (Tokyo) **32**, 140 (1968).

61. A. Kodama and T. Mori, *Plant Cell Physiol.* (Tokyo) **9**, 725 (1968).

62. P. A. Trudinger, *Advan. Microbiol. Physiol.* **3**, 111 (1969).

63. I. Suzuki, *Biochim. Biophys. Acta* **104**, 359 (1965).

64. H. D. Peck, Jr., *Biochim. Biophys. Acta* **49**, 621 (1961).

65. P. A. Trudinger, *Rev. Pure Appl. Chem.* **17**, 1 (1967).

66. D. P. Kelly, *Aust. J. Sci.* **31**, 165 (1968).

67. H. D. Peck, Jr., *Ann. Rev. Microbiol.* **22**, 489 (1968).

68. M. Silver and D. G. Lundgren, *Can. J. Biochem.* **46**, 1215 (1968).

69. D. B. Sinha and C. C. Walden, *Can. J. Microbiol.* **12**, 1041 (1966).

70. F. Shafia and R. F. Wilkinson, Jr., *J. Bacteriol.* **97**, 256 (1969).

71. D. G. Lundgren, R. Alper, C. Schnaitman, and R. F. Marchessault, *J. Bacteriol.* **89**, 245 (1965).

71a. F. R. Tabita and D. G. Lundgren, *Bacteriol. Proc.* **70**, 152 (1970).

72. H. G. Schlegel and H. G. Truper, *Antonie van Leeuwenhoek, J. Microbiol. Serol.* **32**, 277 (1966).

73. A. Matin and S. C. Rittenberg, *J. Bacteriol.* **104**, 234 (1970).

74. A. Matin and S. C. Rittenberg, *J. Bacteriol.* **104**, 239 (1970).

75. B. A. McFadden and C. L. Tu, *J. Bacteriol.* **93**, 886 (1967).

Part II
Microbial Changes Induced by Organic Pollutants

5 Energetics of Organic Matter Degradation

Perry L. McCarty, Civil Engineering Department, Stanford University, Palo Alto, California.

Natural aquatic systems are dynamic in nature and have energy flowing through them. The flow of energy gives rise to and maintains some order in the system. When the energy flux through an aquatic system remains nearly constant with time, the system will approach a steady-state condition which is kept away from a state of equilibrium by the flux of energy and the ordering which results. For this reason the concentrations of materials present cannot be determined from equilibrium calculations alone. The dynamics of the biological and physical processes occurring must also be considered.

In this chapter the energetics and kinetics of bacterial growth and substrate utilization in natural aquatic systems are discussed. Bacterial growth occurs at the expense of energy released by the flow of electrons from donors to acceptors mediated by bacteria. However, bacteria are open systems in which irreversible processes are occurring and only a portion of the free energy released can be captured for useful work. The remainder escapes as heat. The extent to which bacterial growth occurs is a function of the energy released by the electron transfer and the efficiency of energy utilization by the organism mediating the transfer. Those organisms which can bring about the transfer most rapidly and can capture released energy most efficiently in a given environment will tend to dominate as their rate of growth will be greatest. The concentrations of electron donors and acceptors resulting under steady-state conditions will be governed to a large extent by the dynamics of the processes of biological growth and decay.

5-1.   Thermodynamic Relationships in Biological Systems

A.   Energy Relationships.   Free energy values for various half reactions of interest in natural water systems are listed in Table 5.1. The unusual fractional coefficients result from normalizing each reaction to a single electron. $\Delta G^\circ(W)$ represents the Gibbs free energy for the reaction in which one mole of electrons is transferred, assuming unit activity for all species except the hydrogen and hydroxide ions which have activities for neutral water of 10^{-7}. The free energy of the electron at unit activity is zero by convention.

Free energy values for reactions of interest in natural waters can be obtained by subtracting one-half reaction from another as illustrated in Table 5.2. Typical electron acceptors for bacterially mediated reactions are oxygen, nitrate, sulfate, and carbon dioxide. When oxygen is used as electron acceptor, the energy released is a maximum, while when carbon dioxide is used, the energy released is a minimum. In addition, nitrite formed during denitrification can be used as an electron acceptor, and when it is the energy release is greater than from oxygen usage.

A common observation in mixed bacterial systems is that denitrification, or the reduction of nitrate or nitrites to nitrogen gas, does not occur in the presence of oxygen, and that sulfate and carbon dioxide are not used as electron acceptors in the presence of either oxygen or nitrate. While it is appealing to use a thermodynamic argument to explain this phenomenon, such arguments are not valid. It is thermodynamically possible, for example, for acetate oxidation to be coupled with sulfate reduction even in the presence of oxygen. The energy released, of course, would be much smaller than if oxygen were used, but this of itself does not exclude sulfate reduction. Theoretically, oxidation of acetate and reduction of sulfate in the presence of oxygen could lead to the formation of a metastable intermediate, such as sulfide, which could subsequently be oxidized by autotrophic organisms to sulfate using oxygen as the terminal electron acceptor. This combination of reactions does not occur in natural systems because the sulfate-reducing organisms find an oxygen atmosphere toxic. Thus rather than a thermodynamic argument for the absence of sulfate, nitrate, or carbon dioxide reduction in the presence of oxygen, the proper argument appears to be that the presence of oxygen is inhibitory to the organisms which can mediate these reactions. Nitrate is also inhibitory to the organisms using sulfate and carbon dioxide as electron acceptors if it is present in significant quantities.

The formation of metastable intermediates is common with bacteria. A good example is the anaerobic fermentation of glucose to methane. The complete conversion under natural conditions requires the intervention of several different bacterial species. Table 5.3 lists a typical series of reactions for the methane fermentation of glucose, each being mediated by at least one different species. The initial conversion of glucose to acetate, propionate, butyrate, and hydrogen is typical of fermentations occurring in the rumen of animals (2, 3) and in anaerobic sludge digestion of municipal wastes (4). This reaction probably results from the action of several species operating together in various combinations. The subsequent four reactions in which methane is formed are, from all indications, brought about by separate individual species of methane forming bacteria. The overall reaction yields more energy than any combination of reactions in the series; however, no organism has been found which can mediate it. The fact that this reaction is the most favorable one from a thermodynamic viewpoint has in no way precluded the occurrence of the less favorable reactions nor has it prohibited the production of metastable intermediates. The evolution of an organism capable of mediating the overall reaction from glucose to methane could have a highly undesirable effect on ruminating animals as the metastable intermediates are their major source of energy.

Another series of reactions which leads to metastable intermediates is anaerobic denitrification in which nitrate is used as an electron acceptor. In this case the electron acceptor is reduced to nitrogen gas in two steps, nitrite being formed as the metastable intermediate. The ability to bring about denitrification is characteristic of a wide variety of common facultative bacteria including the genera *Pseudomonas, Archromobacter,* and *Bacillus.* Some species, however, can reduce nitrates to nitrites only, some can reduce nitrites to molecular nitrogen only, and some can bring about the reduction of both nitrates and nitrites to molecular nitrogen. Under natural conditions, nitrites may or may not appear as intermediates. Again, from a thermodynamic viewpoint, the overall reaction is the most favorable. The initial reaction in which nitrate is reduced to nitrite is far less rewarding and results in only a small portion of the overall energy resulting from denitrification to nitrogen gas. This is another example to illustrate that use of thermodynamic arguments to exclude those reactions leading to the formation of metastable intermediates are not valid.

Table 5.1 Free Energies for Various Half Reactions[a]
(Reactants and products at unit activity except $[H^+] = 10^{-7}$)

Half Reaction	$\Delta G^o(W)$, kcal/mole electrons	$E_H{}^o(W)$, V	$pE^o(W)$
1. $\frac{1}{3} NO_2^- + \frac{4}{3} H^+ + e^- = \frac{1}{6} N_2 + \frac{2}{3} H_2O$	-22.263	0.966	16.35
2. $\frac{1}{4} O_2 + H^+ + e^- = \frac{1}{2} H_2O$	-18.675	0.808	13.7
3. $Fe^{3+} + e^- = Fe^{2+}$	-17.780	0.770	13.0
4. $\frac{1}{5} NO_3^- + \frac{6}{5} H^+ + e^- = \frac{1}{10} N_2 + \frac{3}{5} H_2O$	-17.128	0.743	12.0
5. $\frac{1}{2} NO_3^- + H^+ + e^- = \frac{1}{2} NO_2^- + \frac{1}{2} H_2O$	-9.425	0.408	6.91
6. $\frac{1}{8} NO_3^- + \frac{5}{4} H^+ + e^- = \frac{1}{8} NH_4^+ + \frac{3}{8} H_2O$	-8.245	0.357	6.05
7. $\frac{1}{8} SO_4^{2-} + \frac{19}{16} H^+ + e^- = \frac{1}{16} H_2S + \frac{1}{16} HS^- + \frac{1}{2} H_2O$	5.085	-0.220	-3.75
8. $\frac{1}{8} CO_2 + H^+ + e^- = \frac{1}{8} CH_4 + \frac{1}{4} H_2O$	5.763	-0.250	-4.25
9. $\frac{1}{8} CO_2 + \frac{1}{8} HCO_3^- + H^+ + e^- = \frac{1}{8} CH_3COO^- + \frac{3}{8} H_2O$	6.609	-0.286	-4.85
10. $\frac{15}{92} CO_2 + \frac{1}{92} HCO_3^- + H^+ + e^- = \frac{1}{92} CH_3(CH_2)_{14} COO^- + \frac{31}{92} H_2O$	6.657	-0.289	-4.85
11. $\frac{1}{7} CO_2 + \frac{1}{14} HCO_3^- + H^+ + e^- = \frac{1}{14} CH_3CH_2COO^- + \frac{5}{14} H_2O$	6.664	-0.289	-4.89

Table 5.1 (Continued)

	Half Reaction	$\Delta G^0(W)$, kcal/mole electrons	$E_H{}^0(W)$, V	$pE^0(W)$
12.	$\dfrac{1}{8}CO_2 + H^+ + e^- = \dfrac{1}{12}CH_3CH_2OH + \dfrac{1}{4}H_2O$	7.592	−0.329	−5.57
13.	$\dfrac{1}{6}CO_2 + \dfrac{1}{12}HCO_3^- + \dfrac{1}{12}NH_4^+ + H^+ + e^- = \dfrac{1}{12}CH_3CHNH_2COOH + \dfrac{5}{12}H_2O$	7.639	−0.331	−5.60
14.	$\dfrac{1}{6}CO_2 + \dfrac{1}{12}HCO_3^- + H^+ + e^- = \dfrac{1}{12}CH_3CHOHCOO^- + \dfrac{1}{3}H_2O$	7.873	−0.341	−5.78
15.	$\dfrac{1}{5}CO_2 + \dfrac{1}{10}HCO_3^- + H^+ + e^- = \dfrac{1}{10}CH_3COCOO^- + \dfrac{2}{5}H_2O$	8.545	−0.370	−6.26
16.	$H^+ + e^- = \dfrac{1}{2}H_2$	9.670	−0.419	−7.10
17.	$\dfrac{1}{4}CO_2 + H^+ + e^- = \dfrac{1}{24}C_6H_{12}O_6 + \dfrac{1}{4}H_2O$	10.020	−0.432	−7.35
18.	$\dfrac{1}{2}HCO_3^- + H^+ + e^- = \dfrac{1}{2}HCOO^- + \dfrac{1}{2}H_2O$	11.480	−0.497	−8.40

[a] From McCarty (1).

Table 5.2 Examples of Calculations for ΔG_r

Electron Donor	Electron Acceptor	Half Reactions Combined (from Table 5.1)	Complete Reaction for One Electron Equivalent of Change	ΔG_r, kcal/ electron equivalent
Acetate	O_2	(14)–(2)	$\frac{1}{8}$ CH_3COO- + $\frac{1}{4}$ O_2 = $\frac{1}{8}$ CO_2 + $\frac{1}{8}$ HCO_3^- + $\frac{1}{8}$ H_2O	-25.28
Acetate	NO_3^-	(14)–(5)	$\frac{1}{8}$ CH_3COO- + $\frac{1}{2}$ NO_3^- = $\frac{1}{8}$ CO_2 + $\frac{1}{8}$ HCO_3^- + $\frac{1}{8}$ H_2O + $\frac{1}{2}$ NO_2^-	-16.03
Acetate	SO_4^{2-}	(14)–(10)	$\frac{1}{8}$ CH_3COO- + $\frac{1}{8}$ SO_4^{2-} + $\frac{3}{16}$ H^+ = $\frac{1}{8}$ CO_2 + $\frac{1}{8}$ HCO_3^- + $\frac{1}{8}$ H_2O + $\frac{1}{16}$ H_2S + $\frac{1}{16}$ HS^-	-1.52
Acetate	CO_2	(14)–(12)	$\frac{1}{8}$ CH_3COO- + $\frac{1}{8}$ H_2O = $\frac{1}{8}$ CH_4 + $\frac{1}{8}$ HCO_3^-	-0.85

Table 5.2 (*Continued*)

Electron Donor	Electron Acceptor	Half Reactions Combined (from Table 5.1)	Complete Reaction for One Electron Equivalent of Change	ΔG_r, kcal/electron equivalent
Glucose	O_2	(29)–(2)	$\dfrac{1}{24}C_6H_{12}O_6 + \dfrac{1}{4}O_2 = \dfrac{1}{4}CO_2 + \dfrac{1}{4}H_2O$	-28.70
Glucose	NO_3^-	(29)–(5)	$\dfrac{1}{24}C_6H_{12}O_6 + \dfrac{1}{2}NO_3^- = \dfrac{1}{4}CO_2 + \dfrac{1}{4}H_2O + \dfrac{1}{2}NO_2^-$	-19.45
Glucose	SO_4^{2-}	(29)–(10)	$\dfrac{1}{24}C_6H_{12}O_6 + \dfrac{1}{8}SO_4^{2-} + \dfrac{3}{16}H^+ = \dfrac{1}{4}CO_2 + \dfrac{1}{4}H_2O + \dfrac{1}{16}H_2S + \dfrac{1}{16}HS^-$	-4.94
Glucose	CO_2	(29)–(12)	$\dfrac{1}{24}C_6H_{12}O_6 = \dfrac{1}{8}CO_2 + \dfrac{1}{8}CH_4$	-4.26
Glucose	Glucose[a]	(29)–(20)	$\dfrac{1}{24}C_6H_{12}O_6 = \dfrac{1}{12}CH_3CH_2OH + \dfrac{1}{12}CO_2$	-2.43

[a] Fermentation to ethanol.

Table 5.3 Examples of Biological Reactions Occurring in Series and Resulting in Metastable Intermediates

Step in Series	Reaction	ΔG kcal	Percent of Overall Total
	Methane Fermentation with Glucose:		
1	$0.042C_6H_{12}O_6 + 0.072HCO_3^- = 0.044CH_3COO^- + 0.016CH_3CH_2COO^- + 0.012CH_3CH_2CH_2COO^- + 0.096H_2 + 0.139CO_2 + 0.043H_2O$	-2.74	64.3
2	$0.012CH_3CH_2CH_2COO^- + 0.012HCO_3^- = 0.024CH_3COO^- + 0.006CH_4 + 0.006CO_2$	-0.075	1.8
3	$0.016CH_3CH_2COO^- + 0.008H_2O = 0.016CH_3COO^- + 0.012CH_4 + 0.004CO_2$	-0.093	2.2
4	$0.084CH_3COO^- + 0.084H_2O = 0.084CH_4 + 0.084HCO_3^-$	-0.602	14.1
5	$0.096H_2 + 0.024CO_2 = 0.024CH_4 + 0.048H_2O$	-0.75	17.6
Overall:	$0.042C_6H_{12}O_6 = 0.126CH_4 + 0.126CO_2$	-4.26	100.0
	Dentrification with Methanol:		
1	$0.067CH_3OH + 0.2NO_3^- = 0.067CO_2 + 0.133H_2O + 0.2NO_2^-$	-7.35	28.2
2	$0.100CH_3OH + 0.2NO_2^- + 0.2H^+ = 0.1CO_2 + 0.3H_2O + 0.1N_2$	-18.74	71.8
Overall:	$0.167CH_3OH + 0.2NO_3^- + 0.2H^+ = 0.167CO_2 + 0.433H_2O + 0.1N_2$	-26.09	100.0

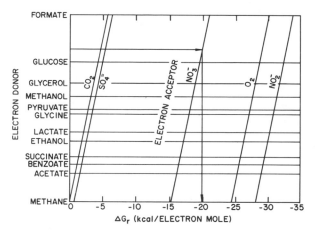

Figure 5.1 Free energy release per electron mole for various organic electron donors and inorganic electron acceptors. Reactants and products at unit activity except $[H^+] = [OH^-] = 10^{-7}$.

The energy released by various oxidation-reduction reactions using values from Table 5.1 is illustrated in Figure 5.1. The spread in energy availability from oxidation of organic matter is represented by the difference between methane, with the least available, and formate, with the most available, among the common naturally occurring organic materials. When oxygen is used as electron acceptor, the energy released per electron mole for methane oxidation is about -24.4 kcal and for formate about -30.2, not a very large relative difference. For this reason the yield of organisms per electron equivalent of organic matter oxidized is relatively constant. However, when carbon dioxide is used as the electron acceptor, the energy released ranges from zero with methane oxidation to -5.72 kcal with formate oxidation. The free energy yield is low and the relative spread in free energy values is great. For this reason bacterial yields per electron equivalent from methane forming reactions are low and vary considerably with the organic compound being oxidized. In general, fatty acids produce the lowest yields and carbohydrates the highest.

B. Redox Potential and Electron Activity. An individual oxidation-reduction reaction occurring in a natural water system can be written in the following generalized way:

$$0 = \sum_i v_i A_i \tag{1}$$

The stoichiometric coefficients v_i are positive for products, negative for reactants, and equal to zero for components that do not take part in the reaction. The Gibbs free energy change (ΔG) for the reaction can then be written as

$$\Delta G = \Delta G^0 + RT \sum_i \nu_i \ln a_i \tag{2}$$

where ΔG^0 is the standard free energy for the reaction and a_i is the activity of component A_i.

Electrode or redox potentials are related to the Gibbs free energy change per mole of electrons associated with a given reduction reaction as follows:

$$E^0 = \frac{\Delta G^0}{nF} \tag{3}$$

F equals the Faraday and n is the number of electrons involved in the reaction as written. Potentials are usually written with respect to the standard hydrogen electrode (E_H). Values of $E_H^0(W)$ determined in this way are listed in Table 5.1 and are for all constituents at unit activity except the hydrogen and hydroxyl ions which are assumed to have activities associated with neutral water. The redox potential for other than unit activity can be determined for a given reaction through the Peters-Nernst equation:

$$E = E^0 + \frac{RT}{nF} \sum_i \nu_i \ln a_i \tag{4}$$

The electron activity ($pE = -\log [e^-]$) is conceptually related to free energy and redox relationships. As indicated by Stumm (5) and Sillen (6), pE is a convenient measure of the oxidizing intensity of a system at equilibrium and is related to the reversible redox potential (E_H) and Gibbs free energy as follows:

$$pE = E \left(\frac{F}{2.3RT} \right) \tag{5}$$

$$pE = \frac{-\Delta G}{2.3nRT}$$

Since at 25°C, $2.3RT/F$ equals 0.059 V/equivalent; and $2.3RT$ equals 1362 cal/equivalent:

$$pE = \frac{E}{0.059} = \frac{-\Delta G}{1362n} \tag{6}$$

pE is an intensity factor and is conceptually similar to pH. It represents the electron free energy level per mole of electrons and has a computational advantage over other free energy parameters by eliminating cumbersome conversion factors:

$$pE = pE^0 + \sum_i \nu_i \log a_i \tag{7}$$

Values for $pE^0(W)$ are listed in Table 5.1 and are representative for all constituents at unit activity except the hydrogen and hydroxyl ions which have their activities in neutral water.

The electron activity for a given redox reaction can be defined if the activities of the reactants and products are known. In a system with many constituents, different redox reactions are usually possible and unless the system is in equilibrium a conceptually meaningful pE for the system as a whole cannot be defined. The pE concept, however, can be useful in viewing a nonequilibrium system.

First, consider an aqueous redox system containing various possible inorganic electron acceptors and their reduction products, all in equilibrium at a given electron activity. If the concentration of a key component in a given redox reaction is given, then the individual concentrations of reduced and oxidized forms can be specified for the given electron activity. In Figure 5.2 the relative concentrations of the electron acceptors, oxygen,

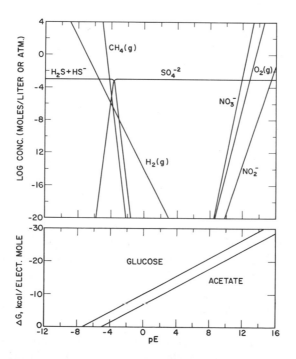

Figure 5.2 Relation between pE, concentration of various electron acceptors and their reduced end products, and free energy availability from oxidation of an electron equivalent of acetate or glucose introduced at 10^{-3} moles/liter.

$$[CO_2] = 10^{-3}, \qquad [N_2] = 0.78, \qquad [H_2S] + [HS^-] + [SO_4^{-2}] = 10^{-3}$$

nitrate, sulfate, and carbon dioxide, and their reduction products are illustrated for different electron activities for typical conditions in natural waters as listed in the figure legend.

Now, if a perturbation were caused by the introduction of organic material into such a system at a given electron activity, the system would be displaced from its former equilibrium state. Since the concentration of each electron acceptor and its reduced end product in the system is defined, the Gibbs free energy for substrate oxidation can be determined. Free energy values for acetate and glucose, when introduced at 10^{-3} M concentration into equilibrium systems at various electron activities as defined by the upper portion of Figure 5.2, are illustrated in the lower portion of the figure. For this particular set of conditions, the lower the original pE of the system, the lower is the free energy available from oxidation of organic matter introduced into the system. Free energy values associated with use of oxygen as an electron acceptor occur at pE values greater than 13, with use of nitrate at pE values greater than 12, with use of sulfate at pE values greater than -4, and with use of carbon dioxide in methane fermentation at pE values greater than -5.

Consider the introduction of acetate and a heterogeneous population of bacteria to mediate its oxidation into an isolated aerobic aquatic system initially at equilibrium and with a pE of 13.5. Organisms using oxygen as electron acceptor would first come into action and would consume dissolved oxygen while oxidizing the acetate. As the oxygen concentration fell, pE_c the calculated equilibrium pE of the system (neglecting the presence of the acetate), and the free energy available from subsequent acetate oxidation would decrease. When the oxygen concentration and pE_c reached a sufficiently low point, organisms using nitrate as electron acceptor would come into play, reducing the nitrate concentration and further reducing pE_c. Finally, nitrates would be exhausted, and then sulfates would come into play as electron acceptor and finally carbon dioxide.

Thus the pE or redox potential of an equilibrium system gives an indication of the nature of the reactions likely to occur and free energy availability resulting from the introduction of organic matter into that system. Aerobic systems are characterized by high pE values and high energy availability following the introduction of organic matter, while anaerobic systems are characterized by low pE and redox values and low energy availability.

Use of electrodes in an attempt to measure the redox potential of natural systems has frequently been made and certainly gives positive values in aerobic systems and negative values in anaerobic systems. However, as discussed in detail by Stumm (5), such measurements fail to give unambiguous interpretation with respect to solution composition. Reversible electrode potentials cannot be established for most of the important redox components in natural aqueous systems and under the nonequilibrium conditions that normally exist, so that E_H measurements usually represent mixed potentials that are not amenable to quantitative interpretation. However, conceptually defined redox potentials can be calculated from analytical information of a few important components such as the electron acceptors for biological reactions and their reduction products. Meaningful interpretations of the nature of the system with respect to redox reaction and energy potentials can be made from such calculations.

5-2. Steady-State Phenomena

A. Equilibrium and Steady State. Significant concentrations of organic matter exist in all fresh waters and the oceans. Calculations of the equilibrium concentrations of organic matter which would exist at any electron activity likely to be encountered in natural waters indicate concentrations many orders of magnitude less than normally found. The concentration of acetate which would be present in a water body in equilibrium with the atmosphere at $pE = 13.5$ can be determined from Equation 7 to equal about 10^{-800} moles/liter ($[O_2] = 0.21$, $[CO_2] = 3 \times 10^{-4}$, $[HCO_3^-] = 10^{-3}$). Similarly, under anaerobic conditions at $pE = -4.2$, in equilibrium with an atmosphere containing $[CH_4] = 0.65$ and $[CO_2] = 0.35$, the acetate molar concentration would be about 10^{-29}. While the equilibrium concentration of acetate is much larger under anaerobic than under aerobic conditions, in both cases the concentration is so small as to be immeasurable. Calculations for other organic compounds give similar results.

At equilibrium under expected natural aqueous conditions, the concentration of organic matter present would be negligible. It is apparent that equilibrium models give a poor indication of the concentration of organic materials occurring in natural systems. Such systems have relatively large quantities of organic matter introduced into them from natural land drainage as well as from waste water discharges. More important, organic matter is continually being produced in natural systems through photosynthetic processes. Some bodies such as the ocean approach a steady state in which the organic matter formed through photosynthetic activity and introduced from land drainage is just balanced by consumption through plant respiration, biological decay, predation, and by loss through incorporation into bottom sediments. Freshwater bodies such as lakes, on the other hand, are characterized by annual cycles of growth and decay, and stratification and destratification, so that steady-state conditions seldom result. In order to describe steady-state or time-variant systems, dynamic models are required which incorporate the effect of a continual influx of organic materials and loss through decay and other processes. While equilibrium energy considerations cannot be used to describe such rate processes, they can be used to gain some insights into the factors which tend to govern the near steady-state conditions often found.

B. Bacterial Growth Rate. One of the simplest steady-state models is that of a completely mixed body of water with water and nutrients continuously flowing in and out at fixed rate and concentration. The growth rate of a given species of bacteria growing on a single limiting nutrient introduced into this system can be estimated from the empirical equation presented by Monod (7) as modified by van Uden (8) to consider the influence of specific maintenance rate, maximum yield factor, and transport of substrate into the cell:

$$\mu = Y_m k_m \frac{S}{K_s + S} - b \tag{8}$$

This equation is similar to the Michaelis-Menton equation for enzyme kinetics. The fractional increase in bacterial mass per unit time is given by μ, Y_m represents the maximum yield of organisms per unit of substrate consumed for energy, k_m is the rate at which substrate is consumed for energy per unit mass of bacteria, S represents the concentration of the growth rate-limiting substrate (usually taken to be the electron donor for the energy reaction), and K_s is a coefficient which represents the substrate concentration at which the growth rate is one-half the maximum rate (considering b to equal zero), and b is a coefficient expressing the cell decay rate.

Van Uden termed b the specific maintenance rate to account for the decrease in microbial mass through endogenous respiration, death, and lysis. In mixed populations this term can also be used to include decrease in mass through predation, and might better be termed the organism decay rate. Typical values for b have ranged from 0.01 day^{-1} to 0.05 day^{-1} in mixed cultures. In pure culture, values as high as 0.43 to 0.67 day^{-1} have been noted (2). Organism decay rates under anaerobic conditions tend to be lower in general than under aerobic conditions (9). Values for K_s, the substrate concentration at which the growth rate is half the maximum value, are generally less than 10^{-4} mole/liter for aerobic decomposition (7, 10). In methane fermentation of acetate under anaerobic conditions, on the other hand, values of K_s equal to 10^{-3} mole/liter seem typical (9).

C. Energetics and Bacterial Growth. Yield coefficients for bacterial growth vary widely for autotrophic versus heterotrophic organisms and for aerobic versus anaerobic conditions. McCarty presented a relationship between free energy of reaction and maximum cell yield which was felt to be applicable for both heterotrophic and chemosynthetic autotrophic bacteria (1, 11). In heterotrophic growth an organic substrate is used, a portion is converted to end products for energy and another portion is synthesized into cellular material. For such growth a method using electron equivalents was devised to evaluate the proportion converted to each use (1). The electron equivalents of a given mass of cells can be determined by starting with an empirical cell formulation such as $C_5H_7O_2N$ and writing a half reaction for its oxidation:

$$\frac{1}{20} C_5H_7O_2N + \frac{9}{20} H_2O = \frac{1}{5} CO_2 + \frac{1}{20} HCO_3^-$$
$$+ \frac{1}{20} NH_4^+ + H^+ + e^- \tag{9}$$

The weight, c, in grams of cells per electron equivalent, can be determined from this half equation and equals 113/20 or 5.65.

Using the above relationship along with the proper conversion factors, a relationship for cellular yield is obtained:

$$Y_m = \frac{c}{hA} \tag{10}$$

Here, A represents the electron equivalents of electron donor converted for energy per electron equivalent of cells synthesized, c is as determined above, and h equals the number of electron moles actually transferred from a donor molecule divided by the electron equivalents per mole as given by the half reaction as listed in Table 5.1. Normally h equals 1.0, but it may be less than unity in certain fermentations. A knowledge of this reaction mechanism is needed in order to choose a correct value for e when organic substrates are not oxidized completely.

In order to solve Equation 10 for Y_m, a value for A is needed. McCarty (1) estimated A from a model developed from energy considerations. He postulated that energy was transferred in two steps, first from the energy source to an energy carrier such as ADP, and then from the energy carrier to a synthesis reaction. For each transfer a certain loss of energy was assumed to result and this was related to an energy transfer efficiency for the reaction. The net free energy lost during the synthesis of an electron equivalent of cells must equal the net free energy change from the overall synthesis reaction

$$A\Delta G_r + \Delta G_s = \text{energy lost} \tag{11}$$

Here ΔGr represents the Gibbs free energy change per electron equivalent of substrate oxidized and ΔG_s represents the free energy required for synthesis of an electron equivalent of cells. The energy lost is equal to the portion $(1 - k)$ of the substrate energy which is not transferred to the energy carrier. Thus, energy loss equals $(1 - k)A\Delta G_r$. With this substitution Equation 11 can be solved for A:

$$A = - \frac{\Delta G_s}{k\Delta G_r} \tag{12}$$

The free energy required for synthesis of an electron equivalent of cells was postulated to consist of three energy requirements which, when substituted for ΔG_s in Equation 12, resulted in the following equation for A:

$$A = - \frac{\Delta G_p/k^m + \Delta G_c + \Delta G_n/k}{k\Delta G_r} \tag{13}$$

The term ΔG_p represents the free energy required to convert the carbon source used for cell synthesis to an intermediate level, ΔG_n represents the free energy required to convert the inorganic nitrogen source into ammonia, the oxidation state of nitrogen in cellular material, and ΔG_c represents the free energy required to convert both the intermediate-level carbon and the ammonia into cellular material. The k terms are introduced to account for losses in transfer of energy from energy carrier to the synthesis reactions. A k term is not used with ΔG_c since the value for this of 7.5 kcal/electron equivalent of cells was estimated from empirical observations which embodied all transfer inefficiencies. The value for ΔG_p is obtained by subtracting $\Delta G^0(W)$ for the pyruvate half reaction (Equation 15, Table 5.1) from $\Delta G^0(W)$ for the half reaction involving the cell carbon source. The ΔG_n is obtained from values in Table 5.1 and by assuming that an electron equivalent of cells contains 1/20 mole of nitrogen (Equation 9). The ΔG_n equals zero if the nitrogen source is ammonia, 4.17 kcal for nitrate, 3.25 kcal for nitrite, and 3.78 kcal for atmospheric nitrogen.

The value for ΔG_r in the denominator of Equation 11 can be obtained by subtracting the $\Delta G^0(W)$ value in Table 5.1 for the electron acceptor from $\Delta G^0(W)$ for the electron donor as illustrated in Table 5.2. More precise values can be obtained through use of Equation 4 and actual activities of reactants and products, although such corrections are usually small.

The value of k represents the average efficiency of all energy transfers, and for autotrophic and heterotrophic bacteria growing under optimum conditions generally varies from about 0.4 to 0.8 with an average of about 0.6. Typical yield values using this average efficiency and calculated for oxidation of different organics with various electron acceptors are listed in Table 5.4 along with energy values. The electron donor in all cases was assumed to serve as carbon source for cell synthesis.

Table 5.4 Cell Yield Coefficients Estimated from Energetics of Substrate Oxidation
(Assuming $k = 0.6$ and ammonia used as nitrogen source)

Electron Donor	Electron Acceptor	kcal		A	Y_m, g/electron equivalent
		$\Delta G_r{}^a$	ΔG_p		
Acetate	O_2	−25.28	1.94	0.71	7.96
Acetate	$NO_3{}^{-b}$	−23.74	1.94	0.76	7.43
Acetate	$SO_4{}^{2-}$	− 1.52	1.94	11.8	0.48
Acetate	CO_2	− 0.85	1.94	21.1	0.27
Glucose	O_2	−28.70	−1.48	0.38	14.90
Glucose	CO_2	− 4.26	−1.48	2.58	2.19
Ethanol	O_2	−26.27	0.95	0.58	9.76
Ethanol	CO_2	− 1.83	0.95	8.3	0.67

a Assuming products and reactants at unit activity and pH = 7.
b Reduction to N_2.

Comparisons between measured values of A and calculated values assuming a k value of 0.6 are given in Tables 5.5 and 5.6. A wide range of anaerobic and aerobic heterotrophic conditions are represented by the data listed. In general the calculated values for A are within $\pm 50\%$ of the measured values with calculated values ranging from 0.38 for aerobic growth to 21.1 for some anaerobic growth conditions. This is a reasonably close comparison when the difficulty in accurately measuring bacterial growth yields is considered.

While the comparison between energetics of reaction and growth yield is considered good, there are many factors which would tend to result in growth yields much less than calculated. Less than maximum efficiency can result when bacteria are growing under adverse conditions. Lower efficiency might result from what Gunsalus and Stanier (12) termed "nutrient limitation," a condition sometimes occurring during rapid rates of growth where a high percentage of substrate carbon is converted to cells. Such diversion may result in (a) the accumulation of polymeric products either in storage form or as unusable waste, (b) dissipation as heat by "ATPase mechanisms," and (c) activation of shunt mechanisms by-passing energy-yielding reactions. In addition, organisms living under adverse conditions such as in the presence of inhibiting materials, unbal-

Table 5.5 Comparison between Calculated Values for A and Measured Values for Aerobic Heterotrophic Bacterial Growth

Energy Reaction	Organism	ΔG_r, kcal/ Electron Mole	A		Ref.
			Calc.[a]	Meas.	
Glucose + $6O_2$ = $6CO_2$ + $6H_2O$	Enrichment	-28.7	0.38	0.27	28
Fructose + $6O_2$ = $6CO_2$ + $6H_2O$	Enrichment	-28.7	0.38	0.35	28
Lactose + $12O_2$ = $12CO_2$ + $11H_2O$	Enrichment	-28.7	0.38	0.35	28
Sucrose + $12O_2$ = $12CO_2$ + $11H_2O$	Enrichment	-28.7	0.38	0.34	28
2 Glycine + $3O_2$ = $2CO_2$ + $2HCO_3^-$ + $2NH_4^+$	Enrichment	-27.1	0.47	0.92	28
Alanine + $3O_2$ = $2CO_2$ + HCO_3^- + NH_4^+ + H_2O	Enrichment	-26.3	0.57	0.92	28
2 Glutamate + $9O_2$ = $6CO_2$ + $4HCO_3^-$ + $2NH_4^+$ + $2H_2O$	Enrichment	-26.3	0.57	0.81	28
Butanol + $6O_2$ = $4CO_2$ + $5H_2O$	Enrichment	-25.8	0.66	1.04	28
2 Benzoate + $15O_2$ = $12CO_2$ + $2HCO_3^-$ + $4H_2O$	Enrichment	-25.6	0.67	1.19	28
Butyrate + $5O_2$ = $3CO_2$ + HCO_3^- + $3H_2O$	Enrichment	-25.5	0.67	0.50	28
2 Propionate + $7O_2$ = $4CO_2$ + $2HCO_3^-$ + $4H_2O$	Enrichment	-25.3	0.68	0.74	28
Acetate + $2O_2$ = CO_2 + HCO_3^- + H_2O	Enrichment	-25.3	0.68	0.74	28

[a] Assuming $k = 0.60$.

Table 5.6 Comparison between Calculated Values for A and Measured Values for Anaerobic Heterotrophic Bacterial Growth

Energy Reaction	Organism	ΔG_r, kcal/ Electron Mole	A Calc.[a]	A Meas.	Ref.
Glucose + 2HCO$_3^-$ = 2 Lactate + 2H$_2$O + 2CO$_2$	S. faecalis	−2.15	5.2	5.5	16
Glucose + 2HCO$_3^-$ = 2 Lactate + 2H$_2$O + 2CO$_2$	S. faecalis	−2.15	5.2	6.5	27
1.5 Glucose + 3H$_2$O = 2 Propionate + 2 Acetate + 4H$_2$O + 4CO$_2$	P. pentosaceum	−3.37	3.3	3.0	16
Glucose = 2 Ethanol + 2CO$_2$	S. cerevisiae	−2.43	4.5	6.1	16
8 Pyruvate + 2SO$_4^{2-}$ + 3H$^+$ = 8 Acetate + H$_2$S + HS$^-$ + 8CO$_2$	D. desulfuricans	−2.24	5.6	5.7	29
6 Pyruvate + 2SO$_3^{2-}$ + 3H$_2$O = 6 Acetate + H$_2$S + HS$^-$ + 3HCO$_3^-$ + 3CO$_2$	D. desulfuricans	−2.61	4.8	6.6	29
4 Lactate + 2SO$_4^{2-}$ + H$^+$ = 4H$_2$O + 4 Acetate + H$_2$S + HS$^-$ + 4CO$_2$	D. desulfuricans	−1.77	8.1	5.5	29
Glucose = 1.4 Acetate + 1.4H$^+$ + 1.6CO$_2$ + 1.6CH$_4$	Enrichment	−3.86	2.8	2.7	30
2 Octanoate + 8H$_2$O = 11CH$_4$ + 3CO$_2$ + 2HCO$_3^-$	Enrichment	−0.894	19.8	13.5	30
Acetate + H$_2$O = CH$_4$ + CO$_2$ + HCO$_3^-$	Enrichment	−0.846	21.1	15.7	30
4 Methanol = 3CH$_4$ + CO$_2$ + H$_2$O	Enrichment	−3.20	3.5	5.5	14
4 Propionate + 6H$_2$O = 7CH$_4$ + CO$_2$ + 4HCO$_3^-$	Enrichment	−0.90	19.6	13.5	14
4 Benzoate + 22H$_2$O = 15CH$_4$ + 9CO$_2$ + 4HCO$_3^-$	Enrichment	−1.13	19.3	8.1	14
Succinate + 1.7H$_2$O = 0.84CH$_4$ + 0.19 Acetate + 0.41 Propionate + 0.15CO$_2$ + 1.4HCO$_3^-$	Enrichment	−0.875	18.9	16.2	14
Lactate + 0.7H$_2$O = 1.2CH$_4$ + 0.07 Acetate + 0.14 Propionate + 0.5CO$_2$ + 0.8HCO$_3^-$	Enrichment	−1.92	7.5	11.7	14

[a] Assuming k = 0.60.

anced ionic concentrations, or other than optimum pH may have smaller cell yields because of higher energy expenditure for maintenance of favorable balances within the cell. In addition, pure culture studies have shown that some bacteria do not have enzyme systems required to obtain energy efficiently and so exhibit low growth yields (13).

Returning to other variables in the growth rate Equation 8, the electron transfer rate, k_m, is a function of temperature, but appears to equal about 1 to 2 electron moles/g-day at 25°C for a variety of heterotrophic and autotrophic bacteria (1). Values estimated from growth data found in the literature are summarized in Table 5.7. The spread of values found is

Table 5.7   Electron Transfer Rates

Electron Donor	Electron Acceptor	k_m, electron moles/g-day			E_a, kcal/ mole	Ref.
		6–15°C	16–25°C	26–35°C		
Glucose	O_2	0.12	0.35	1.0	18.2	19
Glucose	O_2	—	1.1	2.8	13.5	7
Maltose	O_2	—	0.9	1.9	11.7	7
Acetate	CO_2	—	0.6	1.1	9.9	12
Propionate	CO_2	—	1.0	1.0	—	12
Butyrate	CO_2	—	—	1.0	—	12
NH_4^+	O_2	—	—	2.9	—	20
NH_4^+	O_2	0.42	0.6	1.3	12.9	21
NH_4^+	O_2	0.7	1.1	2.8	16.8	22
NO_2^-	O_2	—	—	2.0	—	20
NO_2^-	O_2	0.53	0.7	1.2	9.8	21
NO_2^-	O_2	1.3	1.8	3.2	10.3	22
Fe^{2+}	O_2	—	—	0.6	—	23
Fe^{2+}	O_2	—	—	2.0	—	24
H_2	CO_2	—	—	3.1	—	25
H_2	SO_4^{2-}	—	—	2.4	—	26

unexpectedly small. The close agreement between such widely varying energy reactions suggests that electron transport is the rate-limiting step in most bacterially mediated reactions. This seems an important phenomenon requiring further substantiation.

The values for k_m tend to increase with increase in temperature as expected. Using the Arrhenius equation:

$$k_m = Be^{-E_a/RT} \qquad (14)$$

values for E_a were determined from the observed rates and are listed in Table 5.7. The values for the temperature range from 10 to 40°C vary between about 9 and 18 kcal/mole. This spread is quite large and perhaps reflects the limitations of using the Arrhenius equation for biological data, especially over a very large range of temperatures. Bacteria have strict temperature limitations and inclusion of data for temperature conditions outside of their normal range of growth can lead to extreme calculated values for E_a.

D. Steady-State Substrate Concentration. Consider a completely mixed steady-state system with volume V containing a concentration of organisms, X, and substrate, S. Steady-state flow rate through the system is Q and the influent contains a constant concentration of substrate S^0. The growth rate of the microorganisms in the system $(\mu X V)$ must equal the loss of microorganisms from the system at steady state (XQ) so that:

$$\mu = \frac{XQ}{XV} = D \qquad (15)$$

The growth rate and dilution rate (D, time^{-1}) for the steady-state system are equal (14). Substitution into Equation 8 and rearranging results in the following relationship for the effluent concentration of substrate:

$$S = \frac{(D+b)K_s}{Y_m k_m - (D+b)} \qquad (16)$$

The resulting steady-state concentration of an organic substrate continuously fed into the body of water from land runoff, or treatment plant effluent discharge, or formed within the body of water from degradation of algae and other aquatic plants can be obtained from Equation 16. An essential assumption is that an individual organic substrate, such as acetate or glucose, is consumed by a single species of bacteria under steady-state conditions.

Herbert, Elsworth, and Telling (14) presented experimental verification for the relationships described by Equations 15 and 16 using pure cultures of aerobic bacteria. Lawrence and McCarty (9) presented similar verification for the opposite spectrum of organic matter degradation, the anaerobic methane fermentation using enrichment cultures. Figure 5.3 represents a summary of some of the data collected for the methane fermentation of acetate. The relatively good fit between the experimental data and the model can be seen.

Estimates of the resulting concentration of glucose and acetate from degradation either under aerobic or anaerobic conditions for various dilution rates and using Equation 16 are contained in Table 5.8. As can be seen from these data and from Figure 5.3 for dilution rates of less than about 0.1 (corresponding to a detention time of 10 days), the resulting concentration changes very little. As the dilution rate approaches zero (detention time approaches infinity), the substrate concentration does not

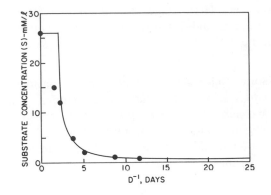

Figure 5.3 Steady-state relationship between dilution rate and substrate concentration for methane fermentation of acetate at 35°C (9). Curve based upon $K_s = 3.45 \times 10^{-3}$ moles/liter, $k_m = 1.54$ electron moles/g-day, $Y_m = 0.32$ g/electron-mole, $b = 0.015$ day^{-1}, and $S_0 = 2.62 \times 10^{-2}$ moles/liter.

decrease to zero, but approaches a finite value, S_1, which can be obtained from Equation 16 by setting D equal to zero:

$$S_1 = \frac{bK_s}{Y_m k_m - b}, \qquad D = 0 \qquad\qquad (17)$$

The influence of b in the denominator is negligible. The resulting substrate concentration at zero dilution rate is a direct function of b and K_s,

Table 5.8   Estimated Steady-State Concentrations of Acetate and Glucose under Aerobic and Anaerobic Conditions and at Different Dilution Rates
($K_s = 10^{-3}$ moles/liter, $b = 0.03$ day^{-1}, $k_m = 1.5$, Y_m as determined from Equation 10)

Dilution Rate, days^{-1}	Steady-State Conc., (moles/liter) $\times 10^8$			
	Acetate		Glucose	
	Aerobic[a]	Anaerobic[b]	Aerobic[a]	Anaerobic[b]
0	25	800	13	91
10^{-3}	26	830	14	95
10^{-2}	34	1100	18	122
10^{-1}	110	4700	58	410
1	940	∞	480	4500
10	5300	∞	8100	∞

[a] Using oxygen as electron acceptor.
[b] For methane fermentation, assuming the metabolism was carried out by a single species.

parameters reflecting the rate of energy consumption required for maintaining cellular integrity, and the influence of substrate concentration in the bulk liquid on its rate of diffusion into the cell. The substrate concentration is also inversely proportional to $Y_m k_m$, a product which gives the maximum growth rate under unlimiting nutrient conditions. Thus, the resulting substrate concentration at zero dilution rate can be interpreted as that concentration which just permits a sufficient rate of diffusion of energy-yielding substrate into the cell so that the growth rate from substrate consumption just balances the rate of organism decay. In other words, it is the concentration which permits a balance between energy-yielding and energy-consuming reactions.

In anaerobic methane fermentation, K_s tends to be much larger, at least with acetate fermentation, and $Y_m k_m$ much smaller than for aerobic oxidations (1). These factors tend to make the resulting substrate concentration at zero dilution rate much larger than under aerobic conditions as indicated in Table 5.8.

A zero dilution rate may be approached in lakes with long retention time and in the ocean. Vaccaro (15, 16) and coworkers working with seawater and Hobbie and Wright (17) with freshwater reported glucose concentrations varying from about 10^{-8} up to 10^{-6} moles/liter, which is within the range for aerobic values listed in Table 5.8. For most of the ocean areas studied, the glucose concentration was found to decrease with depth and the former group of researchers hypothesized that glucose was excreted by phytoplankton and the quantity of material excreted decreased with depth as did the rate of photosynthesis. The rate of glucose uptake by the consuming microorganisms was generally 10^{-11} moles/hour or less, although in cases where glucose was most prevalent, the uptake rates were almost two orders of magnitude higher. From this datum, the turnover time of glucose in the ocean can be estimated to be about 40 to 400 days or longer. This would correspond to the reciprocal of the dilution rate in Equation 15. Substituting these low dilution rates into Equation 16 would lead to values very closely approximating those for a zero dilution rate.

The close agreement between calculated concentrations assuming steady-state conditions and measured concentrations of glucose in the ocean is encouraging. One of the most critical and uncertain values in Equation 16 is that for K_s, the saturation coefficient. Vaccaro and Jannasch (15) have indicated this value varies considerably with different organisms. Values as low as 4×10^{-8} were found for certain marine organisms using glucose, well below typical reported values of 10^{-5} to 10^{-3} moles/liter. The marine organisms appear to have adapted to the very low concentrations of substrates which are available to them in the oceans. Perhaps organisms with similar characteristics would be found in freshwaters having long retention times.

E.   Multiple Substrates.   Results from the study of steady-state systems employing a multiple of substrates and mixed organisms as would occur in natural systems are limited. Enrichment culture studies of this type for anaerobic conditions have been evaluated by Lawrence and McCarty (9) for a mixture of butyrate and acetate, and a mixture of propionate and acetate. O'Rourke (18) evaluated the steady-state anaerobic treatment of sewage sludge, which is a highly complex mixture of carbohydrates, fats, and fatty acids, and protein, plus a great variety of undefined materials. The changes in concentration of some of the major components with dilution rate are depicted in Figure 5.4. The concentrations of each component at

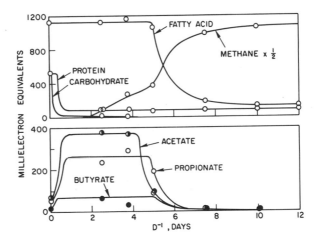

Figure 5.4   Steady-state concentrations of various components (measured in electron equivalents) *vs.* the reciprocal of dilution rate during anaerobic methane fermentation of sewage sludge at 35°C (18).

each dilution rate studied are the result of a multitude of different organisms operating in series and in parallel in a complex chain of reactions. Embodied in these data are the net results of organisms operating competitively, symbiotically, and antagonistically in as complex a fashion as might be found in any natural system. Nevertheless, the data indicate that for individual substrates, the predictions from the model of Equation 16 tend to be obeyed. O'Rourke (18) found that the kinetics of reaction determined by studying the individual substrates could be applied with success to the complex case.

A great variety of organic metabolites, including carbohydrates, proteins, and fatty acids, are released or excreted from growing and decaying phytoplankton and discharged to natural waters from municipal and industrial treatment facilities. If each soluble metabolite were decomposed by a single bacterial species, and if each species were capable of metabolizing only a single organic substrate, then the total quantity of organic matter present in the water at steady state would equal the sum total of all the steady-state concentrations of each of the metabolites as perhaps calculated by Equation 16. Considering all the amino acids, the nucleotides, the various fatty acids, and carbohydrates which would be included, the result would be a total organic concentration of perhaps 2 to 4 orders of magnitude greater than that listed in Table 5.8 for individual metabolites. Organic concentrations an order of magnitude higher than calculated for individual metabolites are, in fact, typical. The ocean contains approximately 1 mg/liter of dissolved organic material, and dissolved organic concentrations as high as 10 to 20 mg/liter are typical in many freshwaters. Many of the components have yet to be identified. However, it appears that the mixture consists of complex polymeric materials highly resistant to microbial degradation, as well as simple metabolites present in concentrations too low for microbial metabolism.

The ability to consume a variety of organic substrates is characteristic of many organisms. Preferential metabolism of one substrate over another, various types of substrate inhibition, and release of intermediate metabolites during organic decomposition are typical characteristics among microorganisms. Many of the preformed simple intermediates are used, not as an energy source, but for direct synthesis by photosynthetic and nonphotosynthetic organisms. All of these complex mechanisms play a role in contributing to and decreasing the available supply of simple organic materials present in natural waters. Thus, while steady-state calculations assuming single substrate–single organism interactions appear to give a reasonable estimate of the concentrations of organic materials present in natural waters, the actual mechanisms of organic material turnover are no doubt considerably more complicated than these simple calculations would tend to indicate. Nevertheless, such calculations can serve as first approximation in ascertaining some of the factors influencing the nature and concentration of organic materials found present in natural waters.

Notation

The following symbols have been adopted for use in this chapter.

A = electron equivalents of substrate converted for energy per electron equivalent of cells synthesized.

a_i = activity of component A_i.

A_i = a chemical component in a reaction.

b = specific maintenance rate, units equivalent biomass per unit biomass present per unit time.

B = constant, time^{-1}.

c = grams of cells formed per electron mole of carbon synthesized.

D = dilution rate, time^{-1}.

E = electrode potential.

E_a = energy of activation, kcal/mole.

E_H = electrode potential with respect to hydrogen electrode.

F = the Faraday, 23,060 cal/eV.

ΔG = Gibbs free energy for a given reaction.

ΔG_c = free energy of conversion of one electron equivalent of intermediate to one electron equivalent of cells.

ΔG_n = free energy per electron equivalent of cells for reduction of nitrogen source to ammonia.

ΔG_p = free energy of conversion of one electron equivalent of cell carbon source to intermediate.

ΔG_r = free energy per electron equivalent of substrate converted for energy.

ΔG_s = carrier free energy required for synthesis of one electron equivalent of cells.

$\Delta G^0(W)$ = redox potential, free energy change per mole electron when all species are at unit activity except H^+ and OH^- which are at activity of neutral water.

h = electrons transferred in energy reaction per molecule of substrate used divided by the electron equivalents per molecule as given in Table 5.2.

k = efficiency of energy transfer.

k_m = electron transport rate, electron moles transferred for energy per gram of bacteria per day.

K_s = saturation coefficient, or growth limiting substrate concentration at which specific growth rate is one half μ_m.

m = constant, equal to $+1$ when ΔG_p is positive, and -1 when ΔG_p is negative.

n = number of electron moles transferred per mole of substrate utilized for energy.

pE = electron activity $(-\log [e^-])$.

Q = flow rate, volume/time.

R = universal gas constant, 1.99 cal/(mole)(deg).

S = concentration of growth limiting substrate.

S_1 = concentration of growth limiting substrate at zero dilution rate.

T = temperature, degrees Kelvin.

μ = specific growth rate, biomass formed per unit biomass present per unit time.

μ_m = maximum specific growth rate occurring at high substrate concentration.

v_i = stoichiometric coefficient.

X = microorganism concentration, mass/volume.

Y_m = maximum yield factor, units biomass formed per unit mass of energy source consumed, if no energy required for maintenance; expressed in grams of bacteria synthesized (dry volatile weight) per mole electrons transferred for energy.

REFERENCES

1. P. L. McCarty, "Energetics and Bacterial Growth," in *Proceedings of the Fifth Rudolf Research Conference*, Rutgers, The State University, New Brunswick, N.J., in press.

2. A. G. Marr, E. H. Nilson, and D. J. Clark, *Ann. N.Y. Acad. Sci.* **102**, 536–548 (1963).

3. R. E. Hungate, *The Rumen and Its Microbes*, Academic Press, New York, 1966.

4. P. L. McCarty, "Natural Succession of Microbial Processes Constituting the Anaerobic Decomposition of Organic Compounds, A Discussion," in *Proceedings of the Fourth International Conference on Water Pollution Research*, Prague, in press.

5. W. Stumm, *Proceedings of the Third International Conference on Water Pollution Research,* Munich (1966).

6. L. G. Sillen and A. E. Martell, *Stability Constants of Metal-Ion Complexes*, Spec. Pub. No. 17, The Chemical Society, London, 1964.

7. Monod, *Recherches sur la Croissance des Cultures Bacteriennes,* Hermann and Cie, Paris, 1942.

8. N. van Uden, *Arch. Mikrobiol.* **58**, 145–154 (1967).

9. A. W. Lawrence and P. L. McCarty, *J. Water Pollution Control Federation* **41**, R1–R17 (1969).

10. E. Stumm-Zollinger, *Appl. Microbiol.* **14**, 654–664 (1966).

11. P. L. McCarty, "Thermodynamics of Biological Synthesis and Growth," in *Second International Conference on Water Pollution Research,* Pergamon, New York, 1965, pp. 169–199.

12. I. C. Gunsalus and R. Y. Stanier, "Energy-Yielding Metabolism in Bacteria," in *The Bacteria,* Vol. **2**, Academic Press, New York, 1961.

13. T. Bauchop and S. R. Elsden, *J. Gen. Microbiol.* **23**, 457–469 (1960).

14. D. Herbert, R. Elsworth, and R. C. Telling, *J. Gen. Microbiol.* **14**, 601–622 (1956).

15. R. F. Vaccaro and H. W. Jannasch, *Limnol. Oceanog.* **11**, 596–607 (1966).

16. R. F. Vaccaro, S. E. Hicks, H. W. Jannasch, and F. G. Carey, *Limnol. Oceanog.* **13**, 356–360 (1968).

17. J. E. Hobbie and R. T. Wright, *Limnol. Oceanog.* **10**, 471–474 (1965).

18. J. T. O'Rourke, *Kinetics of Anaerobic Waste Treatment at Reduced Temperatures,*" Ph.D. Thesis, Stanford University (March 1968).
19. H. Ng., *J. Bacteriol.* **98**, 232–237 (1969).
20. H. Loudelout, P. C. Simonart, and R. van Droogenbroeck, *Arch. Mikrobiol.* **63**, 256–277 (1968).
21. F. E. Stratton and P. L. McCarty, *Environ. Sci. Technol.* **1**, 405–410 (1967).
22. G. Knowles, A. L. Downing, and M. J. Barrett, *J. Gen. Microbiol.* **38**, 263–278 (1965).
23. G. A. Din, I. Suzuki, and H. Lees, *Can. J. Biochem.* **45**, 1523–1546 (1967).
24. J. V. Beck, *J. Bacteriol.* **79**, 502–509 (1960).
25. T. G. Shea, W. A. Pretorius, R. D. Cole, and E. A. Pearson, *Water Res.* **2**, 833–848 (1968).
26. K. R. Butlin and J. R. Postgate, "Microbial Formation of Sulphide and Sulfur," in *Symposium on Microbial Metabolism, VIth International Congress of Microbiology,* Fondazione Enanvele Paterno, Rome, 1953, pp. 126–143.
27. J. T. Sokatch and I. C. Gunsalus, *J. Bacteriol.* **73**, 452–460 (1957).
28. C. E. Burkhead and R. E. McKinney, *Proc. Amer. Soc. Civil Eng.* **95**, SA2, 253–268 (1969).
29. J. C. Senez, *Bacteriol. Rev.* **26**, 95–107 (1962).
30. R. E. Speece and P. L. McCarty, "Nutrient Requirements and Biological Solids Accumulation in Anaerobic Digestion," in *First International Conference on Water Pollution Research,* Pergamon, London, 1964, pp. 305–333.

6 Stream Purification

K. Wuhrmann, Federal Institute for Water Resources and Water Pollution Control, Associated with the Swiss Federal Institutes of Technology, Zurich Switzerland

6-1. Introduction

Self-purification deals with the chemical composition of the water and of the sediments in a river. The term "purification" obviously means the *elimination* (used here as a neutral expression irrespective of the mechanisms involved) of dissolved or particulate matter with "polluting" properties in a river reach. The chemical compounds in the water and deposits are pertinent ecological factors for the organism associations growing at any location in a water course. Self-purification, therefore, induces *secondary effects* such as biological gradients and gradients of oxygen concentration. The basic cause for these effects is a moving ratio between photosynthetic oxygen production and community respiration in the length profile of a polluted water course.

In theory a river represents a *continuous fermentation system* of the

119

plug flow type in which part of the biomass is in free suspension and part of it is fixed to the bottom (epibenthic biomass) or lives in the sediment (benthic biomass). A river is, of course, far from being an ideal plug flow fermenter because of irregularities of flow, longitudinal mixing, variance in substrate feeding, etc. The dominant properties of the system are nevertheless clearly recognizable: (*a*) there is no feedback of information in the system, that is reactions occurring at any location have no influence on future events upstream. (*b*) The system may be considered as a single batch reaction and the actual state of affairs in subsequent river reaches represents stages of development of one and the same situation, observed at the uppermost reach as a function of the residence time (flow time) from one point of observation to the next. (*c*) Every river reach of a suitable length may be considered as an individual batch in which the water quality is constant and the biocenosis is in a state of "equilibrium" with the ecological conditions, characteristic for the reach. (*d*) In large rivers where true potamoplankton represents the major part of the biomass in the system, continuous inoculation of a reach from an upstream location is essential for maintaining the level of organism concentration in the water mass.

Items (*b*) and (*c*) presuppose strictly constant feeding conditions qualitatively and quantitatively as well as a constant dilution rate of the system. In nature, none of these conditions is fulfilled and consequently "batches" with certain characteristics of water quality and biology are "moving" up and down in a river, according to the change of external conditions such as flow rate, change of pollution intensity, and seasonal influences.

Polluting matter subject to "self"-purification may include compounds completely inert against organisms or products which are influencing the metabolism (as substrates, inhibitors, toxicants) of the photo- or zoocenoses in a stream. The majority of work in the field of self-purification deals with pollutants which are utilizable substrates, and the biocenological effects produced by "saprobic" conditions have been extensively studied. In contrast to clean rivers, however, the aspects of the *trophic structure* and the *energy relationships* in situations of organic pollution have not yet found an adequate emphasis. Biological self-purification representing an activity of community metabolism may, of course, be included in a concept of ecosystem development. It must be admitted, however, that essential data for quantitative considerations are lacking to a large extent (for instance, growth rates and yield coefficients of essential heterotrophic phytocenoses, or consumption rates and yield coefficients of macroinvertebrates).

This chapter summarizes facts which have been chosen from the literature or from experimental work in the author's laboratory. It is aimed at a more precise meaning of the diffuse expression of self-purification and at encouraging the investigation of community metabolism in more or less polluted streams.

6-2. Definitions

An adequate general definition of the term "self-purification" was given by Steinmann and Surbeck (1) in 1918: "Self-purification is the sum of all those processes which bring a polluted water body back into its normal original state." This definition includes the two terms "polluted water" and "normal original state," the exact meanings of which are open in practice to subjective judgment or legal regulations. A more general definition is desirable, therefore, from a scientific point of view. Adopting, for instance, the concept of maturity of ecosystems (2), we may say that self-purification comprises all those reactions affecting the transfer from an immature to a mature ecosystem. This definition makes it clear that self-purification refers to physical or chemical activities whereby the quality of the water or of the deposits changes and new ecological conditions are created. "New" in this context means a unidirectional development of an ecosystem and, therefore, of its biocenosis toward less specificity. The less precise definition by Steinmann and Surbeck permits, however, the inclusion of self-purification processes regarding ecologically inactive pollutants which may be of great practical importance, such as parasitic organisms or radioactive material. From an entirely *practical* point of view, self-purification in a river is understood as the decrease of mass transport of water contents within a given flow distance. The material involved may be dissolved inorganic or organic substances and living or inert solid particles.

All the above definitions apply to *products* of biological, chemical, or physical reactions. They do not include the *effects* of these reactions, that is, the manifold ecological consequences of self-purification. A frequent mistake in technical literature is that causes and effects are not clearly separated.

6-3. Mechanisms of Self-Purification

Starting, for the moment, from the restrictive definition of self-purification as a diminution of mass transport of "pollutants" (whatever may be included in this expression), we first find that *dilution* (*e.g.*, by the effluent of a clean river into a polluted mainstream) cannot be considered a self-purification reaction, irrespective of the beneficial secondary effects the dilution may produce. True self-purification reactions in rivers and their mechanisms may be summarized into four groups as follows:

Overall Reactions	*Main Mechanisms*
1. Transport and incorporation of compounds from the water mass into deposits.	Sedimentation (eventually after flocculation reactions) of inorganic or organic particulate matter.
2. Reactions within the water mass and on surfaces of suspended matter (eventually transport of reaction products into the sediment).	*Chemical:* Acid-base reactions, oxidation-reduction processes, adsorption, precipitation, etc. *Biochemical:* Dissimilation of inorganic or organic metabolizable compounds by organisms. Kill or other inactivation of parasitic organisms.
3. Exchange reactions of volatile compounds in the water mass with the atmosphere.	Loss of volatile matter to the atmosphere. Equilibration reactions of dissolved gases with the atmosphere (loss or uptake of O_2, CO_2, N_2, etc.).
4. Transformation of sediments from reduced to more oxidized states, including destruction of organic material.	Chemical and biochemical oxidations within the sediment.

The variety of substances subject to chemical or biochemical transformations in a water body or its sediments necessarily requires a large number of reactions which may interact by mutual inhibition or acceleration.

We have already shown in the introduction that a river can be understood as a series of ecosystems in which self-purification is represented by the change of environmental conditions occurring from one ecosystem to the next in a downstream direction. These changes may be interpreted and measured in various ways:

1. *Chemically.* Concentration or mass transport, increase (pollution) or decrease (self-purification) of specific compounds.

2. *Thermodynamically.* Gain (pollution) or loss (self-purification) of the amount of total free energy of the system by import, photosynthesis and respiratory dissimilation, respectively.

3. *Physically.* Degree of equilibration of dissolved gases with the atmosphere; increase or decrease of mass transport of suspended solids.

4. *Hygienically.* Numerical change of specific parasitic species in the water body (bacterial self-purification).

Units and *methods* for measuring self-purification will vary according to the intentions and aims of the investigation. It is obvious, however, that in any case the basic units will have the dimensions of either mass or energy per unit volume or mass or energy transported per unit time. These units may include functional parameters or coefficients linking self-purification with environmental factors such as temperature, weight of biomass, and hydraulic radius of a river. The mechanisms leading to self-purification need not to be known in detail. They might be of interest, however, when they are rate-limiting.

6-4.   Units for Measuring Self-Purification

From the previous remarks two units for measuring self-purification are self-evident:

1. *Amount of self-purification* (decrease of mass transport within a river reach) assuming no additional load between points of observation:

$$S_m = Q(c_0 - c_u) \qquad [\text{Mol s}^{-1}] \qquad (1)$$

2. *Rate of self-purification* (concentration decrease of pollution per unit time at constant flow):

$$S_r = \frac{dc}{dt} = \frac{c_0 - c_u}{t} \qquad [\text{Mol m}^{-3}\,\text{s}^{-1}] \qquad (2)$$

The S_r may also be expressed in units of travel distance. The precision of its determination depends largely on the definition and the measurement of flow time between points of observation.

These units gain much more significance when related to ecological parameters. We may express, for instance, the amount of self-purification as a function of biomass in the river:

$$S_e = \frac{S_m}{G} = \frac{Q(c_0 - c_u)}{t(g'Pv + g''Q)} \qquad [\text{Mol g}^{-1}\,\text{s}^{-1}] \qquad (3)$$

or the rate of self-purification as a function of biomass and the settled wetted surface:

$$S_r = S_e \left(\frac{g'}{R} + g'' \right) \qquad [\text{Mol m}^{-3}\,\text{s}^{-1}] \qquad (4)$$

S_e is equivalent to the specific elimination rate exerted by the total biomass which is in direct contact with the flowing water (i.e., the suspended organisms plus the sessile epibenthic flora and fauna). Introducing into

Equation 2 the hydraulic radius as a geometrical parameter and the value of S_e, we emphasize in S_r the *space distribution* of the biomass within a cross section (4).

The symbols appearing in Equations 1–4 are defined as follows:

Q flow [m^3 s^{-1}]

c_0, c_u concentration of compounds at the upstream (c_0) and the downstream (c_u) end of the self-purification reach Mol m^{-3}

t flow time within self-purification reach [seconds]

G total biomass [grams]

g' attached biomass [g m^{-2}]

g'' suspended biomass [g m^{-3}]

P length of wetted cross profile [m]

R hydraulic radius [m]

v flow velocity [m s^{-1}]

The term "self-purification capacity" is frequently used in technical literature and will be briefly discussed here. The word "capacity" is normally understood to mean an ability to remove a certain amount of pollution. What this amount will be, however, is open to subjective judgment. The normal assumption is that the capacity will be exhausted when, for instance, an assumed minimum of oxygen concentration in the water appears or an undesirable association of organisms starts to develop (*e.g.*, sewage bacteria or fungi). It is obvious, therefore, that the term "self-purification capacity" is applicable only when some well-defined boundary conditions (either chemical or biological) are stated. Under these premises the term then becomes identical with the above "amount of self-purification."

The four equations may be applied for any quantitatively measurable material or substance considered to be a polluting ingredient in the water. In most cases, weight units or molarities will be applied. Other discrete units are not excluded, however, as for instance numbers of organisms (bacteria) or energy units.

It has to be realized that descriptive units for *secondary* self-purification *effects*, such as the biochemical oxygen demand of river water, or the concentration or mass transport of oxygen, or the type of organism associations, are by definition unsuitable parameters for the quantification of self-purification.

6-5.   Biological Self-Purification

A.   Concepts.   Biological self-purification may be defined as the entity of reactions due to organismic activity which cause a concentration change (and hence a change in mass transport) of water compounds considered as pollution. The majority of reactions are of a biochemical nature. Physicochemical processes such as adsorption on organism surfaces or flocculation by biochemically produced polymers may help, however, to remove inert materials such as mineral turbidity.

Biochemical elimination of a compound from the environment invariably involves its entrance into the food chain. Two mechanisms are then responsible for a concentration decrease.

Entrance of pollutants into the food chain.

Permanent loss from the system due to oxidation in the energy metabolism.

Change of distribution and chemical composition within the system caused by concentration in progressively fewer but larger organisms (biomass production at various trophic levels).

Except for organic particles which may enter the food chain at some higher level (consumption by detritus feeders), self-purification processes for dissolved or suspended organic pollutants start at the microbial (predominantly bacterial) level.

It is tempting to discuss biological self-purification on the basis of a mass (or energy) balance in a river reach. The dominant gain factors affecting this balance are: Formation of organic matter by heterotrophic organisms according to their yield coefficient in the consumption of dissolved polluting material; phototrophic growth (import of light energy); import of residual organic pollutants from upstream reaches; and import of biomass detached upstream. The loss factors are: Respiration of organic substrates by both heterotrophic and phototrophic organisms; loss of residual organic pollutants; loss of organisms by detachment and displacement; and loss of biomass by consumers leaving the habitat actively or passively (insects, fishes, etc.).

At any location in the river, the quality of the ecosystem and, more specifically, the water quality and the quality of the sediment are characterized by the equilibrium conditions between mass (energy) gains and losses. Self-purification is observed when the losses prevail over the gains. If this negative mass balance is repeating downstream from reach to reach, a final state of low mass and energy turnover due to the impoverishment of the medium is achieved, that is, a situation which is typical for a clean stream of low nutrient content, or what Steinmann and Surbeck had in mind with their expression of "original state."

B.   Pertinent Factors Influencing Biological Self-Purification in Rivers.
It is appropriate to discuss separately the complex of biological self-
purification reactions in the water mass of a river and the processes
occurring in the sediments. A limited interaction between the two phases
exists which will, however, be neglected for the moment.

From the previous general considerations numerous factors have to be
emphasized which might influence the rate of concentration change and
the final fate of polluting compounds entering a river reach: The type and
quantity of biomass actively metabolizing pollutants in intimate and continuous contact with the flowing water; the geometry of the biotope, that is,
the space distribution of the biomass in the river cross section; the contact
time between biomass and water; the type and concentration of the compounds to be removed; the temperature, pH, and oxygen tension in the
medium; and additional factors such as light incidence, inorganic composition of the water, and flow velocity which might affect the growth and
metabolic rate of river biocenoses.

Innumerable interactions among these factors have to be expected, some
of which will be indicated later in this section. It is conceivable that the
complexity of a natural river as a purification system poses very difficult
problems in view of the evaluation of the relative and absolute magnitude
of action of the various parameters. Consequently, very few observations
exist which could be interpreted on the basis of the previously defined self-
purification units. Much more information has been gained in model
experiments with artificial channels although a direct transfer of such
results to natural conditions is not yet possible.

The specific elimination rate of the biomass S_e is the only adequate
parameter for activity comparisons of various organism associations.
Table 6.1 contains S_e values from measurements in artificial channels in
which a preset biocenosis had been developed prior to the self-purification
determinations. The table demonstrates that so-called "pure water" biocenoses, formed exclusively by phototrophic microphytes, are largely inactive in removing the organic pollutants applied in the experiment to the
river. As expected, however, a heterotrophic "Aufwuchs," such as
Sphaerotilus natans, exhibits a considerable self-purification effect.

There was still a measurable substrate removal by the seemingly completely autotrophic biocenoses mentioned in Table 6.1. Direct microscopic
and cultural methods demonstrate that phototrophic submersed micro- and
macrophytes carry a considerable epiphytic population of heterotrophic
bacteria. Their density allows only for a small organic substrate consumption rate as shown in Table 6.1, and it would be quite ineffective for the
removal of sizable organic pollution in a river. It is, however, sufficient
to decompose organic compounds excreted by chemo- and photoauto-

Table 6.1 Self-Purification Activity of Various Phytocenoses

Dominant Organisms[a]	P/H Index	Temp., °C	Concentration in the Flowing Water, μM/liter					S_e, 10^{-3} μM/(g)(sec)				
			Glucose	Fructose	Saccharose	Galactose	Total Sugars	Glucose	Fructose	Saccharose	Galactose	Total Sugars
Diatoms[a] (D. hiemale)	36	7	2.22	2.12	2.20	2.04	8.58	0.22	0.24	0.08	0.04	0.58
Cladophora[a]	30	15	2.96	2.85	3.08	2.79	11.68	0.12	0.38	0.88	0.09	1.47
Sphaerotilus[a]	0.7	15	3.25	2.96	3.36	3.41	12.97	2.09	1.98	1.14	4.55	9.76
Sphaerotilus + Fungus (Fusarium?)[b]	1.3	7	2.56	2.38	2.57	2.44	9.95	0.34	0.13	0.60	0.12	1.19

[a] Masses of floating strands and filaments, largely occupying total space of channel cross section.
[b] Dense, compact mat on channel bottom.

trophic phytocenoses in clean streams and to prevent the accumulation of secondary pollutants (3–5).

The density of the heterotrophic component of the biocenosis, that is, its growth rate and standing crop, is strictly related to the amount of substrate supplied. In a river a pronounced gradient of growth intensity for heterotrophs must occur, therefore, below an outfall importing suitable organic pollutants. This gradient is independent of photoautotrophic organisms and only a function of the substrate concentration gradient produced by consumption. We shall discuss this point more extensively later.

Biomass geometry: The term geometry is understood as the spacial distribution of the biomass in the cross section and the length profile of a river. This parameter determines the organismic surface exposed to the flowing water mass and thus greatly influences the transport of polluting particles to the individual consumer organisms. There are two points to be considered; (*a*) The proportion of attached and suspended biomass in a river profile, and (*b*) the substrate supply into dense macrocolonies of attached organism growths.

1. Current data suggest that in shallow rivers the heterotrophic biomass attached to the bottom or other solid surfaces (including algae filaments or macrophytes) dominates by far the quantity of heterotrophic microphytes traveling as suspended particles in the water (Table 6.2). It is obvious, however, that with increasing depth (increasing hydraulic radius) of a stream, self-purification depends more and more on the suspended organisms. *Self-purification rates* will therefore decrease from the shallow headwater regions of a river to its downstream reaches with comparable pollution situations. This, of course, does not necessarily apply to the *displacement distance* of a pollutant because of the decreasing flow velocity downstream.

2. Substrate transport velocity to the individual organism may easily be rate limiting for self-purification where dense heterotrophic growths cover a river bottom. The disentanglement and suspension of attached colonies of filamentous green algae (*Cladophora, Ulothrix, Hormidium*, or diatoms such as *Diatoma* and *Melosira*) or of heterotrophs (such as *Sphaerotilus, Leptomitus* and *Fusarium*) are fairly sensitive to the flow velocity. At 20 cm/sec or more such colonies may easily be lifted up to 10 or 15 cm above the bottom, exposing the majority of cells to a continuous stream of medium. At 10 cm/sec or less, they collapse to more or less dense mats, offering only a fraction of their total surface to the flowing water. Figure 6.1 clearly demonstrates this effect with two types of biocenoses. The data originate from experiments where the flow quantity and the gradient were

Table 6.2 Distribution of the Heterotrophic Biomass in Rivers

		Heterotrophic Biomass above 1 m² of River Bottom		
			Free Suspended	
River	Depth of River, m	Sessile, g	g	% of Total
Experimental channel ground water[a]	0.1	1.5×10^{-2}	5×10^{-7}	0.003
Experimental channel ground water + 12% sewage	0.1	2.8	6×10^{-2}	2
Silver springs[b]	1.5	0.59	7.5×10^{-5}	0.01
Nile River	0.1	0.55[c]	1×10^{-4}	0.02
Hypothetical river with 0.1% sewage[d]		0.5	5×10^{-4}	0.1
	1	0.5	5×10^{-3}	1
	10	0.5	5×10^{-2}	10

[a] From Wuhrmann (5).
[b] Assuming 100 bacteria/ml in spring water. From Odum (3).
[c] Fixed on suspended mineral particles, bottom flora neglected, assuming unstable (moving) sediment. From Jannasch (30).
[d] Assuming mixed phototrophic/heterotrophic Aufwuchs of 50 g/m² with 1% heterotrophs. Sewage with 10^7 bacteria/ml and river water with 10^3 bacteria/ml.

changed in model rivers which were already occupied by well-developed biocenoses. The graph shows that the self-purification rate exerted by one and the same biocenosis may increase from 0.9×10^{-3} μM total sugars per second at ca. 4 cm/sec flow velocity to more than the tenfold value, that is, 11.5×10^{-3} μM/sec at *ca.* 24 cm/sec. It is also evident that a biocenosis with a high metabolic rate for the organic pollutant is much more affected by the limiting factor of transport rate than less active phytocenoses as is shown by the curves for the green algae and the bacterial associations. Since the limitation of the nutrient transport automatically reduces the growth rate of the consuming organisms, it necessarily also profoundly affects their competitive situation within the river community. In channels identically polluted with diluted sewage, the formation of macrocolonies by organisms like *Sphaerotilus* or protozoans (*e.g.*, *Carchesium polypinum*), occurs only at flow velocities above 0.05 m/sec (6). There is also a consistent correlation between *Sphaerotilus* growth and flow velocity at constant composition of the river water (7).

Temperature: Self-purification, being a function of the substrate consumption rate and the growth rate of heterotrophic river microorganisms, is expected to depend on the temperature like any other biochemical reaction. It is not easy, however to prove temperature effects on self-purifica-

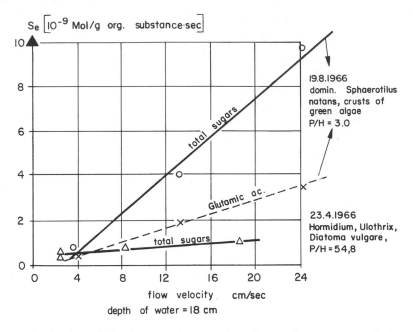

Figure 6.1 Specific elimination rate for organic pollutants by two types of phytocenoses as a function of the flow velocity. Experiments in model rivers fed with ground water at identical hydraulic conditions. Pollution: mixture of saccharose, glucose, fructose, and glutamic acid. Length of self-purification reach = 100 m (25).

tion rates. Other factors, such as extremely low substrate concentrations or high competition for ubiquitous nutrient compounds between the species of a diversified heterotrophic microbial flora, exert rate limitations which may obscure temperature effects. However, by emphasizing specific compounds such as phenols or NH_4^+, which can be oxidized only by a small group of organisms, the influence of temperature can easily be demonstrated.

As an example, the water temperature and the amount of ammonia and phenols carried by the Rhine River at Lobith, NL, has been plotted in Figure 6.2 for 1963–1968 (biweekly observations). It can be assumed that the main source of the two pollutants found at Lobith is the industrial complex of the Ruhr Valley about 90 km upstream. From this region a fairly constant quantity of phenols and ammonia, together with other products, is discharged into the Rhine. The graph clearly indicates the strong inverse relationship between temperature and the mass transport of the two compounds at Lobith. The highly significant correlation (see Table 6.3) is easily explained by the microbial oxidation of phenols and ammonia,

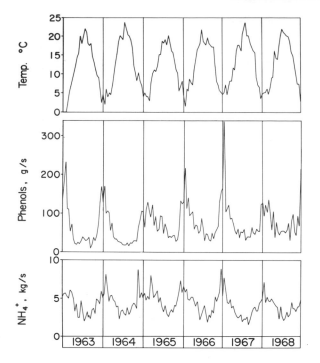

Figure 6.2 Temperature and transport of total phenols and ammonia in the Rhine River at Lobith (NL). Biweekly observations from 1963–1968 (from the annual data sheets of the International Commission for the Protection of the Rhine against Pollution).

Table 6.3   Influence of Temperature on the Phenol, Ammonia, and B.O.D. Removal in the Rhine River, 1963–1968

	NH_4^+	Phenols[a]	B.O.D.
Number of observations, N	156	155	155
Mean transport $Q\,c$, kg/sec, Lobith, NL	4.01	0.060	15.5
Regression of $\ln Q\,c$ on temperature	−0.042	−0.072	10^{-4}
Correlation coefficient	0.606[b]	0.733[b]	$<10^{-6}$

[a] Phenols: compounds reacting with 4 amino-antipyrin.
[b] Significance: $P < 0.001$.

occurring at temperature dependent rates, in the self-purification reach between the Ruhr and Lobith. Temperature coefficients between 1.7 and 2.0/10°C can be estimated for the two self-purification reactions. In Table 6.3 the B.O.D. values, observed at the same station, are included. It is not surprising to see that this analytically untenable "pollution" parameter is not significantly correlated with temperature.

The increase of metabolic reaction rates with increasing temperature does not necessarily mean that self-purification will be larger in summer than in winter. In cold climates river flow is at a minimum in the cold season and higher water temperatures may be combined with high water flows. The geometrical conditions for self-purification are then more favorable in winter than in summer and consequently self-purification rates may be much higher in winter even though the metabolic reaction rates of the organisms involved will be low.

Oxygen: The dissolved oxygen concentration in a river has to be emphasized in two directions: (*a*) its effect on the rate of the respiratory metabolism of organisms responsible for primary self-purification reactions, and (*b*) its influence on the species composition of epibenthic river communities. Oxygen concentrations required in the external medium of a cell of aerobic microphytes are well below 1 ppm for unlimited respiration rates, and a small fraction of this value satisfies half-rate requirements. Dispersed cells in the epibenthic space of a river will probably always find these small concentrations. Sufficient oxygen supply to the cells in the sediments and in the dense colonies of potent heterotrophic self-purifiers (filamentous bacteria, fungi, macrocolonies of protozoans) assumes, however, much higher D.O. values in the water. The quantity of oxygen available for an organism in the interior of an agglomeration is determined by the ratio of the oxygen-consumption rate of the organism and the oxygen-transport rate from the external medium. The driving force for this transport is the concentration gradient. A close analogy exists in this respect to the oxygen supply of organisms in activated sludge or in trickling filters. Methods for calculating oxygen transport rates in biomass agglomerations of sewage purification systems have been summarized by Wuhrmann (8). After all, there is no fundamental difference in the mechanisms determining the rate of oxygen supply or of other dissolved metabolites, and the findings of the previous paragraph concerning the role of flow velocity.

The main effect of the oxygen concentration in the river water is undoubtedly felt in the *sociology of the zoocenoses* at higher trophic levels. Although protozoans and higher invertebrate animals have only a very slight direct influence on the removal of dissolved substrates from the water, their activity as consumers of primary biomass and imported detritus represents an essential part of self-purification in rivers. In addition to this scouring of river bottoms, their continuous breaking up of deposits greatly enhances the microbial oxidation of organic matter in sediments. It is certainly misleading, therefore, to assume a relatively small influence of benthic decomposition on the river water as is suggested by the classical experiments by Fair, Moore, and Thomas (9). In this investigation no higher invertebrates were present in the mud samples. In nature, however, enormous numbers of mud dwellers are normally found, the metabolism of which has an immediate and large influence on the epibenthic water mass (10–12). Because of the large variance in the oxygen requirements of invertebrates and in their anatomical equipment for oxygen uptake (13), fairly strong gradients of the zoocenoses as a function of the oxygen concentration and/or the flow velocity are found in rivers. Unfortunately, knowledge of the "self-purification capacity," that is, the substrate consumption rate and the metabolic efficiency of these consumers is very restricted. Their obligate aerobiosis suggests, however, that the self-purification accomplished by the *totality* of a river biocenosis (*i.e.*, invertebrates included) will depend much more on the oxygen concentration in the water than does the metabolic activity of its individual members.

6-6. Bacterial Self-Purification

The expression of bacterial self-purification has been introduced by Phelps in his book on stream sanitation (14). He defines it, in general terms, as the decrease of bacteria of all types, and especially those of fecal origin, as a function of flow distance or flow time in a river. Suitable parameters for bacterial self-purification are, in analogy to previous definitions, decrease of cell mass or number per unit water volume and decrease of cell mass or number transported per unit time or flow distance. Both parameters are to be based on defined groups or single species of organisms.

Common saprophytes or *chemoautotrophic bacteria* are omnipresent in a river. We think, therefore, that the ups and downs of this indigenous bacterial population due to the natural interrelationship between substrate concentration, bacterial growth, and the continuous inoculation by soil washings cannot be readily assessed in a quantitative manner and expressed in terms of self-purification parameters. It is ecologically sound to concentrate in this context exclusively on *enteric bacteria, viruses, or protozoans* found in rivers which originate primarily in domestic sewage and stormwater overflows of sewerage systems, agricultural wastes, and washings from pastures or manured land. The species to be considered epidemiologically are almost exclusively of fecal origin, predominantly salmonellae and other enterobacteriae, enterococci, and some enteroviruses.

As with chemical pollutants, the behavior of populations of micro-organisms in a river can only be clearly observed below a more or less defined *point source* and in a river reach which is long enough for the manifestation of self-purification effects. The investigations of Streeter and his coworkers in Cincinnati on the Ohio River (15) are still classic, although numerous other water courses have been observed bacteriologically since then.

Due to the obvious difficulty of investigations in nature, many authors have preferred to investigate the dynamics of bacterial kill in flask experiments in the laboratory, suspending well-defined test organisms in samples of natural river water and establishing death rate curves. This procedure reduces the number of variables which might affect the killing rate of bacteria considerably. The simplicity of laboratory experiments permits the quantitative investigation of many parameters of self-purification mechanisms which, at least qualitatively, might be useful in the interpretation of observations in nature. Many such laboratory investigations produced results corresponding within reasonable limits to measurements in natural environments. Some arbitrarily chosen examples assembled in Table 6.4 represent values of T_{90} of laboratory and river observations from various sources.

The main result of Streeter's Ohio River investigation on the self-purification of enteric bacteria can be summarized at follows: the maximum density of coliforms occurs at some distance downstream of a source of pollution; and based upon the maximum numbers of organisms, the percentage of surviving cells downstream follows a curve with a slightly decreasing slope when a log linear plot is applied. The variance of single observations rapidly increases with a diminishing of the absolute numbers of organisms (as is well known from experiments on killing rates).

The Cincinnati group interpreted the course of events as the resultant of two different first-order reactions of bacterial kill adhering to two groups of coliform organisms. A satisfactory correlation of the cell counts was achieved with this mathematical construction. Over 90% of the coliforms in the Ohio belonged to the group with the higher death rate constant. It is highly improbable that this empirical mathematical best fit to the numerical values of the bacterial counts has any bearing on the actual elimination mechanisms of the bacteria. Kittrell and Furfari (16) have demonstrated, however, that the death rate curves for coliforms in deep, slow flowing rivers is sufficiently regular to allow mathematical predictions, irrespective of whether they are biologically interpretable or not. Some examples from various sources (Figure 6.3) sustain this pragmatic approach.

In strict analogy to the elimination rates of degradable chemical com-

Table 6.4 Killing Times for the First 90% (T_{90}) of Enteric Bacteria in River Water

[Flask experiments (F) and nature (N)]

Organism	Medium	Temp., °C	T_{90}, hours	Ref.
E. coli	(F) Tap water + 0.1% sewage	20	60	Pike et al. (26)
	(N) Ohio River	Summer	47	Streeter (14)
		Winter	51	
	(N) "Stream"	22	1.6[a]	Kittrell et al. (27)
	(N) Missouri River	Winter	115	
	(N) Tennessee River	Summer	53	
	(N) Sacramento River	Summer	32	Kittrell et al. (16)
	(N) Cumberland River	Summer	10	
	(N) Glatt River	Summer/winter	2.1[a]	Waser et al. (28)
	(N) Ground water stream	10	110	Wuhrmann (unpubl.)
S. typhi	(F) Thames River	0	172	
		5	108	Houston (14)
		10	77	
	(F) Surface (bog) water	4	24	
		22	~12	Kraus and
S. paratyphi B	(F) Tap water	4	120	Weber (29)

[a] Small shallow streams.

pounds, *killing* rates of bacteria in natural streams depend largely on the *hydraulic* and *biological characteristics* of the stream. In shallow rivers, where a large ratio of wetted surface to water flow exists, and a considerable amount of fixed biomass is normally present, rate constants for the initial part of the death rate curves (*e.g.*, until T_{90}) of coliforms might be found which are 20 or more times higher than reported from the Ohio River (see examples in Table 6.4.)

Flask experiments demonstrate further that increasing *temperature* enhances bacterial disappearance considerably (θ_{10} values around 2). Similar temperature effects might exist in rivers, although they normally will be obscured by the large variance of bacteriological river data. Where statistically sufficient observations have been available, such as in the Ohio or Potomac River studies, death rates in summer were consistently higher than in winter.

The influence of the *organic pollution* load on the persistence of coliforms or other fecal bacteria in a river is controversial. The multiplication of

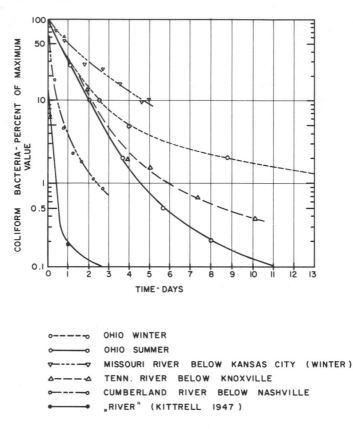

Figure 6.3 Death rate curves for coliform organisms in various rivers (14, 16).

pathogenic organisms in natural waters is certainly negligible (sediments very rich in fecal matter may be an exception). Dissolved organic pollutants probably indirectly increase the death rate by favoring the growth of secondary consumers (bacteria feeders).

Systematic studies on the *removal of viruses* in rivers are practically nonexistent. The numerous point observations on the presence or absence of virus particles do not allow for any conclusions as to the kinetics of virus inactivation. Preliminary information may be found, however, in a model experiment published by Rische *et al.* (17). They seeded a small river with Vi Phage Ll and observed particle frequency in a reach about 35 km long. The river was rather polluted in some places; oxygen concentrations never dropped below 7.5 ppm, however. During the main experiment the temperature varied from 11 to 17.8°C and the pH was between 6.5 and 7.5. Flow velocities from 0.3 to 0.5 m/sec were found and the total flow time within the reach was about 23 hours. No significant losses of virus particles nor a change of their lysis properties could be detected. The persistence of the phage was evidently at least as high in the river as with the previously cited viruses in the flask experiments.

The presently available information clearly shows that the kinetics of virus inactivation in streams needs further investigation *in situ*. The mechanism of destruction of bacteria and viruses in natural waters is discussed by Mitchell in Chapter 11.

It has to be recognized, however, that in spite of many classical investigations on microbial self-purification, the kinetics of inactivation of enteric bacteria and viruses in rivers still deserve our full interest.

6-7. Effects of Self-Purification in Rivers

The removal of biodegradable material, suspended solids and certain cations or anions from the water mass by self-purification reactions has two main consequences:

1. It promotes a shift in the ratio of available free chemical and light energy within the system. The steady change of this ratio downstream of a source of organic pollution creates a *gradient of ecological conditions* which is superimposed on gradients of external factors such as slope, water depth and current velocity, bottom configuration, and temperature.

2. As a result of the above consequence, a *biocenological gradient* is formed which is essentially characterized by an increasing ratio of photo-autotrophic to heterotrophic microphytes. This phytosociological shift implies automatically a change of the ratio of photosynthetic O_2-production (P) and community respiration (R) with increasing travel distance of the river; a gradual decrease of the absolute amount of oxygen respired per volume of water takes place and will profoundly influence the oxygen balance and the *absolute oxygen concentration* in the river water.

A. Biocenological Effects of Self-Purification. It is a legitimate question how far the biocenological gradients induced by self-purification can be used to assess self-purification reactions and rates quantitatively and qualitatively, and whether definite assertions as to the degree of self-purification, achieved at a given location below an outfall, can be made from biological investigations. Posing the question immediately leads into the complex problem of *indicator organisms* and so-called *saprobiological systems*. We shall not enter into this controversial subject, however, but restrain the discussion to a few considerations of general nature.

In essence, two mechanisms are responsible for the modeling of river biocenoses by pollution:

1. *Inhibition of growth and propagation*, or even kill of typical species by nonmetabolites as, for example, heavy metals, extreme pH, and undegradable organic biocides ("toxic" effects). A decrease of the species diversity and of the total biomass will occur. Resistant organisms may eventually grow, however, into the emptied niches and the standing crop of the biomass may recover to a limited extent, though with a distorted composition of the original association. Self-purification is functionally not bound to biological activity under these premises since the dominant self-purification reactions are of abiotic nature [predominantly adsorption on inorganic or organic suspended matter; see, for example, Herrig (18) for some German rivers], and *no interaction between the self-purification process and the biocenosis* in the river exists.

2. *Competitive displacement* of the original members of the associations and overgrowth by new species, specifically favored by the pollution (substrate effect). Again a decrease of species diversity occurs. Biomass production and standing crop will mostly be greatly increased due to the additional energy input into the system. *Self-purification* is strictly bound to the growth of the new organisms and is, therefore, *inseparable from the biocenological shift.*

The two mechanisms cannot always be neatly separated. Biological self-purification reactions in highly polluted waters or in copious organic sediments may produce, for example, fermentation products which are quite toxic for many species. They will then aggravate the competitive situation for active organisms living in an already unfavorable environment.

Both pollution situations converge in one common effect: the *reduction of species diversity*, although the reasons are entirely at variance in the two cases. A nice illustration of this difference has been reported by King and Ball (19) from a river polluted with plating wastes at an upstream location and with domestic sewage further downstream. The plating wastes caused an abrupt decrease of the aquatic insect/oligocheate (tubificides) ratio within 1 mile, obviously by heavy metal toxicity. This ratio gradually was restored in a self-purification reach about 20 miles long down to the point where domestic sewage started to pollute the river. This new impact produced almost the same drop in the insect population as the plating wastes and the insect–tubificide ratio was again very low, although the type of pollution and the ecological mechanism for this effect was certainly different from the upstream reach.

An evaluation of the progress of self-purification by its biological effects (indicator organisms or associations) has to emphasize the *natural bio-*

logical gradient already existing in a river from the spring region to its mouth. Depending on the topography of the drainage basin, drastic changes of the ecological conditions along the length profile occur, accompanied by a respective variance of the phyto- and zoocenoses. The organisms at higher trophic levels are, of course, predominantly involved in these changes. The summarizing essay by Illies and Botosaneau (20) on the problem of biological river zonation indicates clearly that in case of the extension of a self-purification reach over several physiographic regions, a fundamental change of the baseline for biocenological evaluation has to be emphasized.

In the present context the biocenological effects of *organic pollutants* serving as microbial substrates are of prime interest, and the next paragraphs are devoted to this aspect. Theoretically, a strict relationship exists between the type and concentration of biodegradable substances at any point in a river and the quantity of heterotrophic microorganisms utilizing the compounds as nutrients, thereby inducing biological self-purification. The biomass production of these organisms (classified as "destruents" in other ecological concepts) represents—just as photosynthetic biomass production—the substrate for consumer communities at higher trophic levels, building up the food chain in a river. At self-purification rates that cause very low oxygen tensions or even anaerobiosis in the benthic and epibenthic space, the colonization of the biotope by consumers will be largely inhibited, and an intensive accumulation of biomass in the system occurs. The *first biocenological effect* of self-purification is felt at the second and higher trophic levels, therefore, and may consist in a more or less pronounced *truncation of food chains*. Their restoration is only possible when self-purification rates have sufficiently diminished to permit primary consumer populations to exist.

The *second biocenological* effect of self-purification—occurring at the lowest trophic level—is a direct consequence of substrate consumption by the heterotrophic phytocenoses. It consists in the *re-establishment of the phototrophic communities*, shifting primary biomass production from heterotrophic growth to photosynthesis. Theoretically it should be possible to assess quantitatively the gradient of the self-purification activity in a river reach on the basis of the growth rate and the standing crop of the heterotrophic biomass producers, provided their yield factors are known. Figure 6.4 indicates schematically the trophic shift due to self-purification.

The difficulties in verifying these obvious relations in a natural river are evident. Experiments under simplified ecological conditions and on a reduced scale can, however, fill the gap to a certain extent. Some of the above raised questions can indeed be answered on the basis of model assays (21). An artificial channel, 600 m long, was fed with clean ground

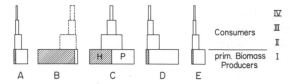

Figure 6.4 Scheme of biomass production and consumption along the length profile of a polluted river. Locations A = E as in Figure 6.7. H = heterotrophic, P = photosynthetic biomass formation, I = IV = successive trophic levels; areas indicating relative biomass production per unit time per river reach. Biomass production in the consumer levels is always heterotrophic.

water and a solution of sugar beet molasses, fortified with amino acids, was added as pollutant. The development of phytobiocenoses along the channel and the elimination of the individual compounds of the pollutant were measured after various flow periods. Figure 6.5 indicates the rapid concentration decrease of the pollutants as soon as a measurable quantity of biomass had built up in the channel, the P/H index in Figure 6.5 represents the extinction of the methanolic chlorophyll extract of 1 g fresh biomass at 663 μm in proportion to the loss by ignition of the same material. The strict parallelism of the growth rate of the heterotrophic microphytes with the rate of self-purification of individual compounds is remarkable. In the first reach, 135 m in length, a rate of biomass increase in the river model of roughly 2 g volatile solids $m^{-2}day^{-1}$ or a total growth of 108 g/day or *ca.* 43 g organic C/day was observed. Assuming an average value for the yield coefficient of about $Y = 0.3$ as is found for many saprophytic bacteria in aerobic, favorable media, the respective carbon uptake of the river organisms from the water would be about 130 g organic C/day. In the experiment a flow of 2.5 liter/sec with an added concentration of 1.2 mg organic C/liter was maintained, amounting to a total supply of 260 g organic C/day to the 135 m reach. Consumption by this microflora (predominantly *Sphaerotilus* in this experiment) was, therefore, about 50% of the pollution which corresponds roughly to the observed concentration decrease from 0 to 135 m of flow distance. This large self-purification cannot be generalized, of course, because in large rivers a much smaller weight of biomass is in continuous contact with the flowing water, and natural pollutants normally have less favorable substrate properties than the chemicals applied in this experiment.

The example merely served to illustrate the fact that *biological self-purification produces biomass,* the quantity of which amounts to 30% or more of the pollution removed from the water (as organic C). The experiment also yields information on the limitation of the removal rates exerted by the absolute concentration of the polluting compounds. Figure 6.6 summarizes the specific self-purification rates S_e of saccharose

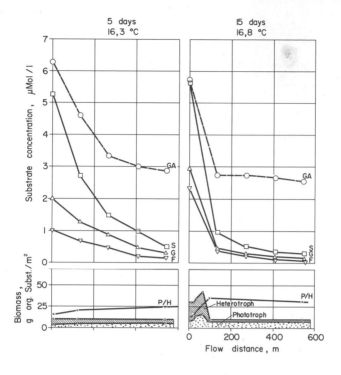

Figure 6.5 Self-purification at the fifth and fifteenth day after start of pollution in a model river. $Q = 2.5$ liter/sec ground water. Pollution: Saccharose (S), glucose (G), fructose (F), glutamic acid (GA). Above: Concentration in μM/liter of pollutants at various flow distances. Below: Length profile of the amount of heterotrophic and phototrophic biomass (g/m^2) and of the P/H index (21).

at various concentrations in the water by the epibenthic biomass at various ratios of heterotrophic and phototrophic microphytes. These data are compiled from several self-purification experiments with the same original substrate mixture. The conclusions from this graph are straightforward: the phototrophic microphytes are almost inactive "self-purifiers" in regard to the organic compounds tested. The self-purification rate in a river decreases rapidly, therefore, with increasing dominance of phototrophs. Independent of this factor the *absolute concentration of individual polluting compounds* limits the self-purification rate at concentrations below a certain level. This is, of course, a consequence of the well-known Monod relationship between growth rate and concentration of limiting substrates. Since rivers usually contain very low absolute concentrations of individual organic compounds (even under grossly polluted conditions), maximum

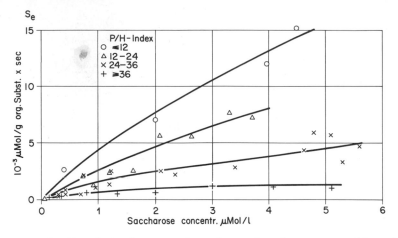

Figure 6.6 Specific elimination rate of saccharose by phytocenoses with various P/H indices as a function of saccharose concentration in the flowing water. Basis: Experiment of Fig. 6.5 (21).

growth rates of the organisms responsible for self-purification are probably never met in nature.

B.   Relationship of Oxygen Balance and Self-Purification in Rivers. The oxygen tension in water is one of the most potent ecological factors. Its variance may trigger a chain of sociological reactions in the river, the extent of which depend on the duration and the intensity of the amplitudes of the oxygen concentration changes (especially their minima). Since, on the other hand, the organism associations in a river determine the rates and types of most of the self-purification processes, a complicated system of interaction exists which is still unexplored in its details.

The dissolved oxygen concentration (D.O.) and the oxygen saturation in a river reach depend at any time on:

Oxygen Gains

Uptake rate from the atmosphere (reaeration rate at undersaturation conditions), $k_2(c_s - c)$ when $c < c_s$.

Photosynthetic oxygen production rate at daytime (P).

Oxygen Losses

Respiration rate of allochthonous substrates in water and sediments (R_{all}).

Respiration rate of autochthonous substrates from photo- or chemosynthetic production (R_{aut}).

Oxygen consumption rate by abiotic oxidations (Ox).

Oxygen desorption rate to the atmosphere at supersaturation conditions $k_2(c_s - c)$ when $c > c_s$.

The net result of the momentary balance situation in subsequent river reaches ("batches" in a plug flow concept of the river), is then ($S =$ surface of the reach, $h =$ mean water depth, $\bar{c} =$ mean D.O. in reach):

Rate of change of oxygen quantity below 1 surface unit	$=$	Gains	$-$	Losses

$$\frac{c_0 - c_u}{S\,h}\,Q \qquad = k_2(c_s - \bar{c}) + (P) - R_{\text{aut}} - R_{\text{all}} - \text{Ox}$$

$$\text{(Day only)} \qquad \text{Self-purification}$$

In the classic concept of Streeter and Phelps (31) the balance equation is written as

$$\frac{dD}{dt} = k_1 L - k_2 D$$

where all rates of losses and gains are lumped together in the rate constants k_1 and k_2. Innumerable investigations have been published aiming at numerical values for the two constants and to emphasize the influence of the local pollution conditions and of other ecological factors on their magnitude (22).

In the present context the question of interest is the following: Which information can be gained from the oxygen sag curve in regard to self-purification rates or the actual stage of self-purification achieved at a given point of observation downstream of a source of pollution?

It is obvious that to answer this question, information is needed which relates the *biocenological effects* caused by self-purification with the oxygen content of the river water. The most convenient parameter for this purpose is the *diurnal variance* of oxygen concentration at any one locality in the length profile.

Because of the phytocenological gradient caused by progessing self-purification, the oxygen sag curve will show an increasing *variance between dark and light conditions* as is schematically indicated in Figure 6.7. The size of this diurnal difference naturally depends to a large extent on the type, the absolute amount, and the geometry of the biomass in a river reach. In a deep turbid water course with a moving silty bottom, unfit for intensive algal colonization, photosynthesis will influence the oxygen balance much less than in a shallow, densely settled river. It is remarkable, however, that considerable day–night excursions of the D.O. have even been reported from the Ohio River (23). The relationship of the diurnal

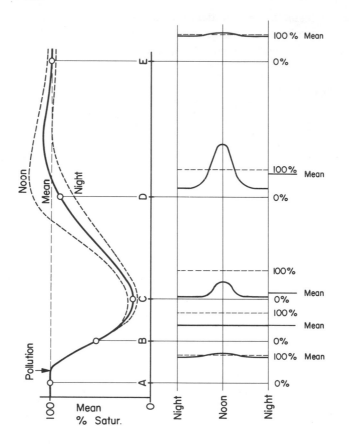

Figure 6.7 Theoretical day–night shift of the oxygen concentration at various locations in a polluted river. Left: Scheme of D.O.-sag curve. Right: Scheme of 24 hours of D.O.-saturation profiles.

shift of oxygen consumption and production rates with the ratio of epibenthic phototrophic and heterotrophic microphytes (P/H index) is exemplified in Figure 6.8. The observations were made in model rivers in the course of several independent experiments. They show a considerable scatter due to variations in the external experimental conditions; the general trend, however, is unmistakable:

1. A completely heterotrophic phytocenose (P/H < 10) with a high oxygen consumption rate due to a rich substrate supply expectedly demonstrates a negligible daily shift in respiration rate. It is rarely nil because there are always some individuals of phototrophic species present.

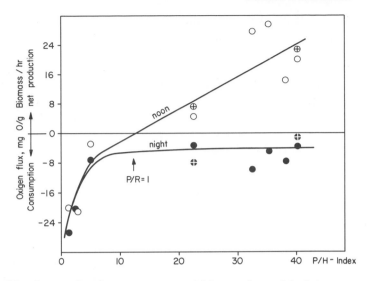

Figure 6.8 Oxygen flux between water and biomass in model rivers at noon and in the dark as a function of the P/H index of the biomass. Pairs of points on the same abscissa correspond to observations of the same biocenosis. Difference between points = gross oxygen production. Ordinate above the zero line = net oxygen production (data from Eichenberger, unpublished).

2. This situation is maintained with decreasing respiration rate (decreasing self-purification rate) until phototrophs successively recolonize the biotope (P/H > 10–15).

3. Minimal growth of phototrophs in a community, still dominated by heterotrophs, produces remarkable day–night differences in the direction of net oxygen flux.

Various points (A–E) of the oxygen sag curve may be characterized by the correlation of self-purification with the D.O. dynamics, always keeping in mind that there is a *continuous transition* in the sequence of events.

1. A refers to a clean head water, poor in nutrients, and with low productivity. P/R is around 1, and the photosynthetic oxygen production by the small autotrophic biomass in daytime or the nocturnal community respiration are nearly compensated by the immediate oxygen exchange with the atmosphere. The average D.O. corresponds to saturation.

2. Assuming that the organic pollution imported below station A is sufficient to replace the phototrophs completely by bacteria and fungi, the P/R ratio at station B will drop to zero, no light influence on the

daily D.O. profile will be visible, and community respiration rate may be far more rapid than reaeration. Self-purification is at its highest intensity.

3. When the organic substrate concentration can no longer sustain a maximum growth rate of its heterotrophic consumers, the self-purification rate decreases rapidly, and phototrophic organisms start to reappear and will finally dominate the phytobiomass. A diurnal D.O. wave with increasing amplitude appears, although the average saturation value may still be very low or already at the minimum of the sag curve, due to the cumulative oxygen consumption in the upstream reaches C. The P/R ratio will again approach 1, but at biocenological associations which are very different from those at stage A.

4. At some further flow distance the imported organic substrate will be exhausted and the self-purification rate for *dissolved* compounds drops near zero. Sediments may still be very rich in organic substrates and maintain large invertebrate consumer populations. There may remain, however, considerable quantities of *inorganic* fertilizing elements in the water which will increase phototrophic growth and promote large D.O. supersaturation by day and heavy nocturnal deficits, D. P/R is larger than 1 and average dissolved oxygen concentration may approach the saturation level.

5. Only after long further travel may the fertilizing effects disappear because of nutrient losses into the atmosphere or permanent incorporation into the sediment. The original situation is restored to a P/R ratio near 1. It depends on the definition of self-purification whether the reactions in the transition from stage C to D or from D to E are considered as terminal self-purification processes.

In nature the idealized picture of Figure 6.7 will normally be much less clear due to distortions caused by fluctuations in the pollution load, variations in light intensity or temperature, and longitudinal mixing. Furthermore, the possibility of very drastic benthic respiration, which in the length profile will in general lag far behind the decrease of the epibenthic oxygen consumption rate, may obliterate the seemingly simple situation. A beautiful example of the consequences of the delay of benthic self-purification in relation to the epibenthic self-purification has been reported by Owens and Edwards (11).

Schmassmann (24) proposed the grouping of rivers on the basis of trophic levels, in analogy to lake classification. As a parameter he used the diurnal oxygen balance, approximately according to stages A to E above. This interpretation of the natural situation is not appropriate, however, since a river is a continuum with transients ever moving in location as well as in time.

In conclusion, it is evident from the preceding discussion that

1. The debit side of the oxygen balance of a river informs only qualitatively and in a rather general way on self-purification and, furthermore, only in regard to self-purification reactions involving aerobic biochemical oxidations,

2. It is impossible, therefore, to derive from oxygen profiles (sag curves) any quantitative data concerning the amount of pollution removed from the water or to predict self-purification rates. The main reason is the fact that an unknown and variable proportion of the resorbed substrate is respired by organisms. This proportion depends on the type of organisms, the kind and combination of substrates, the oxygen tension, and possibly on other environmental factors.

3. The amplitudes of the diurnal oxygen concentration shifts at points below a pollution discharge will in general reveal more information on the progress of self-purification than the oxygen sag curve based on *mean* D.O. values.

6-8.   Energy Relationships in Self-Purification Reaches

In a river polluted with biodegradable organic compounds, the production of biomass in the lowest trophic level is promoted by two energy sources (Figure 6.4):

1. Free energy of the organic pollutants E_s. The quantities available at a given location below the point of discharge correspond to the respective mass transport of substrates.

2. Light energy E_L. Its influx is equal at all points downstream. It depends, however, on the season and the day length.

A common basis for the two imports is possible when the daily calorie flux into the water colum below a unit water surface is used as a unit.

The ratio of the quantities of the two energy forms E_s/E_L varies steadily in a self-purification reach since self-purification consists essentially of processes consuming the imported organic energy. Correlating the actual ratio E_s/E_L at any location with the type of biocenosis, for example, in terms of the ratio of heterotrophic and phototrophic biomass (P/H index), an insight can be gained into the conditions required for heterotroph or phototroph to dominate its competitor for growth space and attachment area. The ratio E_s/E_L strongly depends on the transparency, the biomass geometry, and the absolute dimensions of a river. Comparable values can only be gained, therefore, from observations in the same river

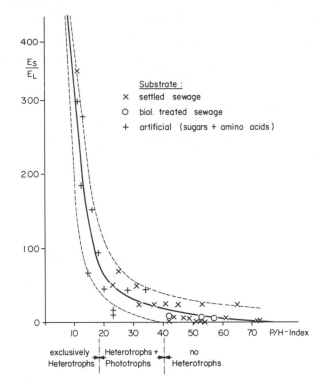

Figure 6.9 P/H index of phytocenoses in model rivers at comparable hydraulic conditions as a function of the ratio E_s/E_L. River water: Ground water with the addition of various pollutants. E_s and E_L in kcal/(dm²)(day). Data compiled from numerous independent model river studies from 1965–1970.

under various pollution conditions or from model experiments. An evaluation of numerous earlier investigations in river models, emphasizing the present point of view, permits the construction of Figure 6.9. For these calculations the light incidence E_L was uniformly assumed to be 7.5 kcal/dm²/day (9 hours at 0.14 gcal/cm²/min). E_s was calculated from the substrate import within 1 day into the water column below 1 dm² of the river surface. With artificial substrates the sum of the free energies of the individual compounds at their measured concentration in the river was used. In experiments with domestic sewage a value of 100 kcal/gMol of dissolved organic carbon was assumed. An amazingly strong interdependence between the ratio of the two energy forms flowing into the biotope and the dominance of the physiological groups of primary energy consumers results. The large variance of the observations is mainly due to the very crude estimate of the energy supply, the disregard of the efficiency of energy utilization, and the lumping together of observations from experiments rather different in organization and execution.

Under the geometrical conditions of the model rivers (water depth of 10 to 15 cm and flow velocities around 20 cm/sec) the observations indicate that at a supply of organic energy sources exceeding 50 to 100 times or more the influx of light energy, heterotrophic microphytes completely overgrow phototrophs. When the flux of organic pollution has diminished by self-purification to less than 10 to 20 times the value of the daily irradiation, the heterotrophic biomass reduces to only microscopically visible epiphytic growth on the phototrophic thallus.

It must be recalled that the above findings—as far as numerical values are concerned—strictly apply to the conditions of the experiments. Similar evaluations of measurements in natural rivers are presently not available. Published energy balances concern "clean" streams only, where phototrophic productivity and the subsequent transformation of photosynthetic primary biomass in the ascending trophic levels dominate by far the biocenological aspect of chemical energy imports. In view of the fact that the almost exclusive emphasis on the oxygen balance in polluted streams has amply demonstrated its inadequacies, an approach on the grounds of energy relationships merits at least some thought.

The above discussion explains why seemingly large seasonal variations in the biocenological effects of pollution and self-purification occur at otherwise constant conditions: the continuous change of the value of E_L over the year causes—at unchanged pollution import—a shift of the ratio E_S/E_L which automatically acts on the ratio of heterotrophic to phototrophic microphytes. The development cycles of many primary biomass predators (water insects) indirectly adds to variations in the phytobiocenoses. The resulting seasonal change in the aspect of the river associations does not imply a corresponding change in self-purification, however, because the growth of the heterotrophic consumers of pollution is exclusively a function of the available substrate. Therefore, high P/H indices are by no means an unequivocal measure for self-purification, they may indeed camouflage the presence of considerable heterotrophic phytocenoses.

6-9. Final Remarks

The essential point which has to be kept in mind is the fact that self-purification in a river—irrespective of the details of its definition—is not some mysterious force inherent to flowing water, but represents a number of definable reactions induced by pollution. It is a matter of discussion, therefore, whether the term "self-purification" should be retained. Nothing occurs by "itself," and one might as well replace the diffuse word by the use of the biological, chemical, or physical terminology necessary for nominating the actually occurring reactions. It must be admitted, however, that the large complex of diversified events in a polluted river is conveniently summarized by a single expression like self-purification.

It is appropriate to come back to a few practical questions concerning the use of self-purification in water management. As already indicated, a river "gains" the property of self-purification only after pollution. The widespread opinion that a clean river will have an *a priori* self-purification "capacity" has to be firmly rejected because it infers the idea that there will be a discrete amount of pollution which will be accepted and eliminated by self-purification without any repercussion on the river's ecosystem. A stream undisputably can acquire a capacity to "assimilate" pollution. This capacity will increase with the amount discharged and according to the specific growth conditions of the pollution assimilating organisms. Only in the case of preset boundary conditions of the water quality or the sediments, as, for example, minimum oxygen concentrations, can the self-purification "capacity" of the river be predetermined.

The apparent ease of mathematical treatment of the self-purification of rivers in terms of oxygen balances has magically attracted innumerable investigators for decades. Although the oxygen concentration is indeed a pertinent ecological factor, there is no justification for neglecting almost all other aquatic contents or their respective ecological effects and to look at self-purification with only the "oxygen eye." In fact, river biocenoses and the metabolism of their members (*i.e.,* self-purification) are much more affected by organic compounds (with or without B.O.D.!) or metal ions than by oxygen. "Loading plans" based on so-called self-purification capacities, admitting some arbitrarily chosen minimum of the oxygen sag curve, are certainly not adequate. On the contrary, oxygen values may easily hide pollutants instead of indicating them. The practice of river protection should aim at the prevention of the entrance of ecologically detrimental compounds into rivers and not to impose on them a self-purification capacity they never have as clean streams.

Acknowledgment

I wish to thank Dr. E. Eichenberger for the data of Figure 6.8 and the other members of our team for their careful work in the series of model river experiments.

REFERENCES

1. P. Steinmann and G. Surbeck, Die Wirkung organischer Verunreinigungen auf die Fauna schweizerischer fliessender Gewässer. Schweiz. Dept. d. Innern, Bern, 1918.
2. R. Margalef, *Perspectives in Ecological Theory*, Univ. Chicago Press, Chicago, 1968.
3. H. T. Odum, *Ecol. Monographs* **27**, 55 (1957).
4. J. M. Teal, *Ecol. Monographs* **27**, 283 (1957).
5. K. Wuhrmann, in *Principles and Applications in Aquatic Microbiology* (H. Heukelekian and N. C. Dondero, Eds.), Wiley, New York, 1964.
6. P. Zimmermann, *Schweiz. Z. Hydrol.* **23**, 1 (1961).
7. J. D. Phaup and J. Gannon, Wat. Res. **1**, 523 (1967).
8. K. Wuhrmann, in *Advances in Biological Waste Treatment* (W. W. Eckenfelder and J. McCabe, Eds.), Pergamon, New York, 1963.
9. G. M. Fair, E. W. Moore, and H. A. Thomas, *Sewage Works J.* **13**, 270, 756, 1209 (1941).
10. R. W. Edwards and H. L. J. Rolley, *J. Ecol.* **53**, 1 (1965).
11. M. Owens and R. W. Edwards, *Proc. Soc. Water Treat. Exam.* **12**, 126 (1963).
12. A. Schumacher, *Arch. Fi. Wiss.* **14**, 48 (1963).
13. H. Ambühl, *Schweiz. Z. Hydrol.* **21**, 133 (1959).
14. E. B. Phelps, *Stream Sanitation,* Wiley, New York, 1944.
15. H. W. Streeter, and W. H. Frost, *Public Health Bull.* **143**, 184 (1924).
16. F. W. Kittrell and S. A. Furfari, *J. Water Pollution Control Federation* **35**, 1361 (1963).
17. H. Rische, G. Keiger and J. Stempel, *Gesundh. Ing.* **82**, 210 (1961).
18. H. Herrig, *Gas- Wasserfach* **110**, 1385 (1969).
19. D. L. King and R. C. Ball, *J. Water Pollution Control Federation* **36**, 650 (1964).
20. J. Illies and L. Botosaneau, *Mittl. Int. Ver. Limnol.* **12** (1963).
21. K. Wuhrmann, J. Ruchti and E. Eichenberger, Adv. Water Poll. Research. Proc. 3rd Int. Conf. Munich 1966, **1**, 229. WPCF Washington (1967).
22. Tennessee Valley Authority, *The Prediction of Stream Reaeration Rates* Chattanooga, Tennessee, 1962.
23. A. F. Bartsch, in *Oxygen Relationships in Streams*, Doc. W58-2, U.S. P.H.S., 1958.
24. H. Schmassmann, *Arch. Hydrobiol. Suppl.* **22**, 504 (1955).
25. K. Wuhrmann, in *Europäisches Abwassersymposium*, Edit. Ges. Förd. Abwassertechnik *Ber. Abw. Techn. Ver.* **23**, 15 (1969).
26. E. B. Pike, A. L. A. Garneson and D. J. Gould, *Rev. Int. Oceanog. Méd.* **18–19**, 97 (1970).
27. F. W. Kittrell and O. W. Kochtitzky, *Sewage Works J.* **19**, 1032 (1947).
28. E. Waser, W. Husmann, and G. Blöchliger, *Ber. Schweiz. Bot. Ges.* **43**, 253 (1934).
29. P. Kraus and G. Weber, *Z. Bakt. Abt.* I, **171**, 509 (1958).
30. H. W. Jannasch, *Ber. Limnol. Stat. Freudenthal* **7**, 21 (1956).
31. H. W. Streeter and E. B. Phelps, *Publ. Health Bull.* **146** (1925).

7 Biodegradation of Hydrocarbons in the Sea

G. D. Floodgate, Marine Science Laboratories, University College of North Wales, Anglesey, Wales, U.K.

The rapid increase during the last hundred years or so of the use of oil and its many derivatives all over the world has led inadvertently to a considerable amount of this valuable material finding its way into the sea. Blumer (1) estimates that the oil lost into the sea during transportation is around 10^6 metric tons per annum and that the total figure may be higher by one or two orders of magnitude. However, the total amount of oil that is present in the sea at any one time is very difficult to estimate, partly due to the fact that considerable amounts of hydrocarbons find their way into the sea from natural sources.

7-1 Origin of Oil and Hydrocarbons in the Sea

A. Fossil Fuels. There are several routes by which oil and refined oil products may be added to the marine environment.

1. CATASTROPHIC ACCIDENTS. First, there are shipping disasters of which the wrecking of the *Torry Canyon* (2) is probably the most notorious, but

153

by no means the only example. Since the early part of the century oil tankers have been lost, releasing their cargoes into the sea. During the World Wars these losses rose steeply, as might be expected, but even in peace time ships continue to meet occasionally with accidents of various kinds. Any spilled oil rapidly spreads over a large area, forming a slick which is moved by winds and waves. Since marine accidents are most likely to happen in crowded, narrow sea lanes close to the coast, any such calamity will cause some damage to the local plant and animal life and possibly pollute the shore itself.

A second kind of accident which can result in large quantities of crude oil being released into the sea is an uncontrolled upwelling of oil which can occur when a new oil well is being sunk on the Continental Shelf. The Santa Barbara incident is a well-known example of an oil spill of this type (3). Again the proximity of the coast greatly increases the potential dangers to amenities and to the organisms living in the area.

2. CHRONIC POLLUTION. Both of the above kinds of disaster result in a sudden catastrophic influx of oil into a particular area. There is also a continuous flow of oil into the sea from ships, from ports and oil terminals, and from freshwater runoff by way of rivers and sewage systems. The amount of oil that the sea receives by this route is difficult to estimate, but it is almost certainly considerable.

The load-on-top method of tanker operation has greatly reduced the amount of oil deposited in the sea (4, 5). However, loading and unloading of the cargo still constitute a hazard. Approximately 0.0001% of the total cargo handled at the large Milford Haven terminal is lost, most of the accidents being attributable to human error (6). Emulsified industrial oils from machine tools, wastes from vegetable oil processing and soap manufacturers, together with washings from road tanker cleaning are among the important oily effluents that can find their way into the seas. The resulting steady chronic pollution probably constitutes as large, but perhaps a more insidious danger, to marine ecology as the more dramatic disasters.

3. NATURAL OIL SEEPAGES. A further route by which oil is introduced into the sea is by way of natural seepage in places where oil deposits approach the surface (7, 8). These seeps may cause a localized heavy contamination but their effect on the total burden of oil pollution is small.

B. Natural Sources. Nevertheless, fossil fuels are not the only source of hydrocarbons in seawater. Those formed by the biota constitute a minor but perpetual part of the normal natural organic matter in the marine environment. Again it is difficult to assess with accuracy the quantities of hydrocarbons of various types released annually into the sea from plant life, but ZoBell (7) has calculated that it could be several million tons per year.

The work of Clark and Blumer (9) showed that in one species of Chlorophyaceae (*Chaetomorpha linum*) and the five species of Rhodophyaceae they examined, the major normal paraffin was heptadecane, while among the Phaeophyceae, both Fucales (five species) and Laminarales (two species), pentadecane was predominant. In all cases there was no marked predominance of odd or even numbers of carbon atoms in the chain for paraffins between C_{20} and C_{32} (Carbon Preference Index about 1.0).

Various hydrocarbons are found in marine animals from zooplankton up to whales (for references, see 10). The hydrocarbons may have passed up the food chain unchanged from plant life, or they may have originated in the animals themselves. Bacteria have also been reported to contain these compounds (11, 12). The presence of paraffinic hydrocarbons in *Desulphovibrio desulfuricans* (13, 14) is very interesting in view of the great importance of this species as a geochemical agent in marine sediments. The predominant chain length was C_{25} and C_{34} with no preference for odd or even lengths. In the recently laid down sediments examined by Clark and Blumer (9) the major normal paraffins had chain lengths of C_{17} and C_{25}, each accounting for 14% of the total paraffin. There was a marked predominance of odd numbered chains in the C_{20}–C_{32} range (C.P.I. = 4.0). This is in contrast to fossil fuels which showed no preponderance of either kind (15).

A few of these natural hydrocarbons have been the subject of detailed investigations. One of these, pristane (2,6,10,14-tetramethylpentadecane), whose origin is probably the phytol group of chlorophyll, tends to accumulate in zooplankton and passes unchanged up the food chain until it appears in the final predators (10). However, it is also biodegradable (16) so that it will be steadily oxidized when released into seawater. This material is also found in sediments, though in this case the source is geochemical breakdown products of chlorophyll. The polycyclic carcinogenous hydrocarbon, 3,4-benzpyrene, has been the subject of a detailed study by French workers. Plankton, molluscs, and fish can accumulate the compound, sometimes in fairly high concentration (17–20). It is also found in sediments. Industrial or domestic effluent is usually stated to be the most important source of this hydrocarbon, although it can be found at a considerable distance from all possible pollution sources. An alternative biosynthetic origin seems likely (21).

7-2. Breakdown of Oils and Hydrocarbons

A. Nonbiological Factors. Crude oil and its derivatives undergo various spontaneous changes when spilled into the sea, so that the substrate for microbial oxidation is not constant. Evaporation accounts for much of the change by removing the volatile compounds. The rate of evaporation is variable, depending upon the prevailing hydrographical and meteorological conditions, and also upon the type of oil. According to Dean (22), two-thirds of Nigerian oil will have evaporated in a few days at sea whereas only two-fifths of Venezuelan crude will have gone. In general, fractions with a boiling point less than about 370°C disappear by evaporation in a matter of days. In addition the oil will undergo a complex process of autoxidation (23, 24). The presence of various substances within the mixture can either inhibit or accelerate autoxidation; some of the sulfur-containing compounds and phenolic materials act as inhibitors, while various metal-containing substances within the oil mixture, as well as metallic complexes found in seawater itself, act as accelerators. The position is further complicated by the decomposition brought about by light, especially of wavelength below 4000 Å (22).

Moreover, the physical form in which the oil is present in the marine environment can be expected to influence the extent and rate of decomposition profoundly. The degree to which hydrocarbons dissolve in seawater may be a major controlling factor in their biodegradation. Nyn (25) indeed goes so far as to state that "le vrai problème n'est donc pas l'accessibilité de l'hydrocarbure, mais le remplacement des molécules consummées, c'est-à-dire, la vitesse de dissolution dans l'eau." The solubility of hydrocarbons in water is, of course, low and decreases with increased molecular weight. Johnson (26) has calculated the solubility of various moderate length n-alkanes at 25°C. A saturated solution of tetradecane, for example, has a molar concentration of only 9.8×10^{-10} which is approximately 2.0×10^{-6} ppm. Furthermore, Aliakrinskaya (27) showed empirically that seawater depresses the solubility of petroleum compared with freshwater so that the amount of hydrocarbon in solution per unit volume for bacterial attack is very small. The ability of the organism to mobilize the dilute substrate may therefore limit the oxidation rate, which could account for the observation (28) that the mean generation time of bacterial cultures was related to the solubility of the hydrocarbon used as growth substrate. Nevertheless, the volume of water available in the sea is very large, so that as Dean (22) has pointed out, the whole cargo of the *Torry Canyon* would, if evenly distributed at a concentration of 1 ppm, be distributed over a volume of only 20 square miles by 500 feet deep. Unfortunately, the mechanism by which the higher alkanes enter the cell is unknown, but an alternative possibility to entering in solution is for the cell to attach itself to a globule of oil and for the alkane molecule to pass into the cell by way of the phospholipids of the cell wall. Hence the degree of dispersion of the oil into fine droplets can be expected to affect the speed of oxidation. Dispersion may be obtained by the use of detergents which can themselves be bacteriocidal (29) or mechanical by using, for example, a ship's wake (30).

A very thin film on the water surface presents a very large area for the bacteria to act upon, and there is free access of oxygen and nutrient salts dissolved in the water. In contrast, an oil/water emulsion would be unable to replenish the nutrients used during bacterial growth by exchange with the surrounding water, and toxic waste products would not be washed away, thus slowing down the rate of biodegradation. Oxygen concentration in the emulsion is unlikely to be the limiting factor since oxygen dissolves readily in oil (25).

B. Microbiological Factors.   The earliest accounts of the biodegradation of hydrocarbon material go back to the close of the last century when Myoshi (31) reported that *Botrytis cinera* would grow on paraffin wax. By 1946 ZoBell (32) was able to report that over 100 microbial species were capable of growth on hydrocarbons and the number has continued to increase. Lists of various bacteria, yeasts and molds capable of degrading hydrocarbons of many types are given by Fuhs (33) and Ponsford (34). Markovetz, Cazin, and Allen (35) list a large number of fungi, approximately 90% of which were shown to grow on alkanes or alkenes, often producing considerable growth.

A survey of organisms from type culture collections was made by McKenna and Kallio (16) who found that between 6 and 20% of each group of bacteria, yeasts, and molds that they tested were able to grow on *n*-alkanes. The ability to degrade hydrocarbons is therefore a common property of microorganisms.

Most of the work mentioned above has been carried out with pure cultures using pure single hydrocarbons as sources of carbon and energy. Perry and Scheld (36), using a more ecological approach, found that between 1 and 3% of the heterotrophic bacteria isolated from different soil samples which grew on nutrient agar would also degrade hydrocarbons. When non-hydrocarbons were used as isolation substrates, those substances whose decomposition was affected by means of an oxygenase produced more hydrocarbon splitting strains than those which did not; a phenomenon that is possibly related to the cardinal role of this group of enzymes in hydrocarbon biochemistry. Jones and Edington (37), who surveyed a moorland soil and a neighboring band of shale, found that between 0.5 and 20% of the strains isolated on soil extract agar from samples taken at several depths were capable of growth on a series of hydrocarbons.

As might be expected, soils with a high oil content contain a bacterial population which has a high proportion of hydrocarbon decomposers. This finding offered a possible means of exploration for underground oil and gas deposits (8).

The range of temperatures at which hydrocarbon-splitting bacteria will grow is wide and varies from a little above 0 to 60°C (32). Some thermophilic hydrocarbon-degrading bacteria have also been found in lake mud (38).

There is an interesting report that several strains of *Azotobacter, Pseudomonas,* and *Mycobacterium* are able to fix elemental nitrogen while using methane, *n*-butane, *n*-tetradecane, toluene, and naphthenic acid as carbon and energy sources (39). Harper (40) found an increase in soil fertility in the region of a natural gas leak and this he believed was due to nitrogen fixation by a methane utilizing *Clostridium*.

The marine microflora, like its terrestrial counterpart, also includes many active hydrocarbon-decaying organisms (41–43). A large number of samples of mud were collected by ZoBell and Prokop (44) from the Bantaria Bay area bordering on the Gulf of Mexico. Using the minimum dilution method, these authors were able to show that almost all the samples contained bacteria which degraded U.S.P. white paraffin oil. In about 10% of the samples 10^6 oil-oxidizing bacteria were present per gram soil. However, large amounts of oil find their way into this area, probably from the neighboring oil fields, so that the number of oil oxidizing bacteria per gram sediment is probably very much higher than is normal for coastal sediments.

It is interesting to note that, in contrast to soil, molds and actinomycetes do not appear to be important oil decomposers in the marine environment although both taxonomic groups are found in the oceans (45, 46) and it therefore appears that bacteria are the main agents of hydrocarbon breakdown in the sea. It should also be noted that some hydrocarbons, particularly the short-chain alkanes, are toxic to certain bacteria (26).

C. Biochemical Features. The biochemical routes by which microorganisms can metabolize hydrocarbons has been the subject of many investigations. Details of these biochemical pathways are outside the scope of this chapter. For detailed information the reader should consult review articles by Trecanni (47), Foster (48), van der Linden and Thijsse (49), Humphrey (50), and Traxler and Flannery (51). Nevertheless, some features of the metabolic processes described in the above papers are of ecological interest, though it must be borne in mind that the biochemical work has involved the study of the effect of pure strains of bacteria upon pure hydrocarbons, whereas in the marine environment a complex mixture of many hydrocarbons and non-hydrocarbon compounds is attacked by mixed cultures of microorganisms.

1. AEROBIC PATHWAY.   The first step in n-alkane oxidation appears to involve a membrane-bound oxygenase, that is, an enzyme which activates molecular oxygen. This is not surprising since the major metabolic systems of bacteria involve oxygen-containing compounds. It follows, therefore, that an adequate supply of oxygen must be available for biodegradation to take place.

2. ANAEROBIC PATHWAY.   In contrast to the aerobic mechanism, the evidence for a possible anaerobic pathway remains meager. It is usually assumed that in the absence of molecular oxygen the hydrocarbon loses hydrogen to form a double bond between adjacent carbon atoms. Oxygen is introduced into the molecule by hydration across the double bond while the nitrate or sulfate ion acts as the terminal hydrogen acceptor. However, Hansen and Kallio (52) concluded that nitrate would not serve as terminal oxidant for hydrocarbon breakdown by *Pseudomonas stutzeri*. Moreover, Updegraff and Wenn (53) report that hexadecane was not attacked anaerobically by *Desulphovibrio* strains with sulfate as the final electron acceptor.

In contrast, Traxler and Bernard (54) describe the uitilization of n-octane and n-hexadecane by *Pseudomonas aeruginosa* under anaerobic growth conditions with nitrate as terminal hydrogen acceptor. Indeed, growth even under aerobic conditions was improved by reducing the aeration efficiency.

The ubiquitous and anaerobic *Desulphovibrio* was able to attack some of the long-chain aliphatic hydrocarbons according to Novelli and ZoBell (55). Neither the normal alkanes with less than 10 carbon molecules in the chain nor the aromatics supported the growth of these organisms. A more detailed investigation of the problem by Davis and Yarbrough (56) revealed that a washed cell suspension of *Desulphovibrio desulphuricans* decolored methylene blue in the presence of alkanes more quickly than the control which was presumably oxidizing endogenous substrates only. This was followed in several hours by blackening in the culture tubes, presumably due to the reduction of sulfates. When grown on McPherson and Millers medium (ammonium ion, lactate, and mineral salts) in the presence of ^{14}C-labeled methane, ethane, or n-octadecane, some evidence of very slow oxidation was obtained. In connection with these experiments it is important to realize that the K_m of many oxygenases is extremely low (57) so that unless special precautions are taken to maintain the oxygen tension of the culture at a sufficiently low value, then oxidation may take place, even though oxygen is not detectable by normal chemical means.

Observations on anaerobic decomposition of crude oil and petroleum products in the natural environment are also very scanty. Rosenfeld (58) inoculated various American crude oils and several refined products with *Desulphovibrio* spp. which grew anaerobically. ZoBell and Prokop (44) discovered that many of the mud samples collected around Barataria Bay contained bacteria which anaerobically oxidized U.S.P. mineral oil. The formation of sulfide indicated that sulfate was the final hydrogen acceptor. It would appear therefore that in some way *Desulphovibrio* spp. may be involved in the anaerobic degradation of hydrocarbons in the marine environment in spite of the doubts that exist as to whether pure strains of this genus reduce sulfates with pure hydrocarbons as hydrogen donors (59).

The possibility of anaerobic decomposition of oil in natural deposits was considered by Andreev and his colleagues (60), and they concluded that although the possibility exits, the process is unlikely to significantly affect natural oil deposits.

The smell of hydrogen sulfide was detected in some patches of oil buried in sand after the *Torry Canyon* incident (2) presumably due to sulfate reduction, although the sulfide may have arisen from other compounds in the crude oil. Therefore while an anaerobic degradative pathway cannot be ruled out, the most rapid and important route of petroleum degradation in the sea is strongly aerobic.

3. SUBSTRATE SPECIFICITY.   Various authors (7, 16, 26, 61, 62) have emphasized the specificity shown by different bacteria towards hydrocarbons. The specificity depends upon the enzymic repetoire of the organism involved and the molecular configuration of the hydrocarbon. There is therefore the probability that a number of different strains of bacteria would be simultaneously involved in the oxidation of oil at one place. These organisms would be in competition for the available nitrogen, phosphate, and other nutrients so that the course of biodegradation could depend upon the relative efficiencies of different strains to mobilize available non-hydrocarbon nutrients, especially if these are in short supply.

Moreover, the non-toxic fractions such as the medium-chain length alkanes tend to be the more easily degraded parts of the oil mixture leaving behind the toxic substances such as the phenols and pyridines. Such an increase in toxicity could have a deleterious effect upon a polluted beach.

4. ACCUMULATED PRODUCTS.   The major end products of bacterial oxidation of oil and its derivatives are carbon dioxide, water, and bacterial cells. The presence of the latter are themselves an ecological factor of some magnitude. Bacteria are known to affect the number and distribution of certain animals and protozoa in beach sediments (63, 64) and one

result of an oil spill will be a temporary change in both the meiofauna as well as the microflora of a beach. In addition a number of incompletely degraded end products from hydrocarbons have been listed in the literature. One striking feature of several paraffin oxidizing cocci isolated from soil was their tendency to accumulate cetylpalmitate (65–67) while carboxylic acids as well as esters accumulated during the biodegradation of petroleum (bp 127–217°C). These substances may themselves act as substrates for further bacterial attack. Increased capsular material containing carbohydrates has also been noted in an obligate hydrocarbon-utilizing bacterium (68). Dunlap and Perry (69) found that the major fatty acids of the bacteria they studied were of the same chain length as the alkane upon which the cultures had been grown, the carbon chain having been assimilated apparently after oxidation of the terminal carbon only.

5. COOXIDATION. Another biochemical system which may be important in the natural environment is cooxidation. This is well established in artificial systems and has been exploited commercially (70). Thus washed cells of *Mycobacterium rhodocroes* grown on *n*-decane were capable of oxidizing a number of methyl-substituted aromatic compounds. *Cellumonas galba* was found to form *trans*-cinnammic acid from alkyl benzenes when grown on various *n*-alkanes (71) and *Nocardia corallina* accumulated dimethyl-*cis,cis*-muconic acid under cooxidation conditions employing *n*-hexadecane for growth and *p*-xylene as cooxidizable substrate (72). Clearly if such a mechanism operates in the marine environment, it will affect the removal of substances which would not normally be utilized. However, the system may be sensitive to chemical (71) and physical conditions (73).

D. Growth Requirements

1. CHEMICAL REQUIREMENTS. The essential part played by oxygen in the degradative process has already been described. Other chemical compounds are also necessary for bacterial development and oil oxidation. Not least of these is water itself. Baier (74) noted that dry oils killed bacteria more readily than wet ones. Similarly Hill (75) found that dry kerosene killed *Nocardia* spp. and *Pseudomonas* spp. in a few hours and *Cladosporum resinae* in a few days, while Cooney, Edmonds, and Brenner (76) noted that their isolates failed to grow without a water phase.

A nitrogen and phosphate source is also required for good bacterial growth. Gunkel (43) found enhanced growth when his media were supplemented with phosphates and ammonium chloride. Le Petit and Barthelemy (77) confirmed this result and noted a very rapid utilization of nitrogen between the second and fifth day of their experiments, a period which corresponds to the maximum loss of oil.

These workers also showed that peptone inhibits oil degradation to a considerable extent, but they found that "substance des depots," that is, dried deposits from sewage works, stimulated biooxidation of alkanes. A similar inhibition by peptone was found by Gunkel (43). However, only the initial rate of oil oxidation was affected. When peptone or peptone and yeast extract were added to the medium, the rate of oil breakdown lagged over the first few weeks. After 8 weeks the total percentage destroyed was much the same as in the unsupplemented medium.

2. TEMPERATURE. The rate of bacterial oxidation of oil might be expected to be temperature dependent, and this is in fact found to be the case. Ludzack and Kinkhead (78) found that after 6 weeks' incubation, the amount of oil decomposed rose from zero at 5°C to between 50–80% at 25°C. Ludzack, Ingram, and Ettinger (79) discovered a marked reduction of oil in river water in summer compared to winter. This they attributed to accelerated bacteriological activity. Gunkel (43) observed a similar temperature effect in seawater experiments. A series of flasks were incubated at various temperatures between 2 and 30°C for 8 weeks. The amount of oil destroyed varied with the temperature; 30% being oxidized in the coldest flask and 50% in the warmest.

E. The Extent of Biodegradation. It is perhaps surprising that in spite of the interest generated in oil pollution in recent years, very little detailed information is available on the extent of biodegradation of oil in the marine environment and what little there is tends to be confused. One difficulty is that there are several methods of measuring the course of the degradation process so that it is difficult to compare one paper with another.

1. OBSERVATION. The easiest way of following the removal of oil and its products is simply to observe the visual changes, or even the loss of odor (27). Tanson, cited by ZoBell (80), reported that layers of Baku crude oils were perforated in a few days and by 10–18 days the oily layer was broken up and disappeared soon after. Zobell and Prokop (44) noticed during laboratory experiments that oil as a surface layer tended to become "stringy" and disintegrated into small droplets on the underside. The droplets then sank. The films themselves were changed in physical appearance and color. Sometimes they sank, while on other occasions the oil surface was penetrated in a month or so. Ten milliliters of a series of crudes were all found to be no longer visible after 20–30 weeks at

25°C while in two further cases only 1 ml of oil had disappeared in that time. The same authors have also carried out experiments in which open-ended cylinders were partly immersed in mud and sandy saltwater marsh. One hundred milliliters of crude oil was added which disappeared over 9 weeks. During this period the oil-oxidizing bacteria increased by a million-fold.

The microscope has not been used a great deal to study oil breakdown. However, Gunkel (43) has produced some photomicrographs which show bacteria lying on the surface of an oil droplet.

2. BACTERIAL COUNTS.   A traditional way of measuring bacterial activity is by following the changes in bacterial numbers. Buchnell and Haas (81) grew several strains of *Pseudomonas* on a mineral salts solution with ammonium or nitrate ion as nitrogen source. They found that when either kerosene, light and heavy medicinal oils, or paraffin wax was the carbon and energy source, the numbers of bacteria increased by two orders of magnitude in about 10 days. A commercial mixture of low boiling point hydrocarbons (Skellysolve) and gasoline caused a reduction in numbers in some cultures with subsequent recovery. A marine *Pseudomonas* attacked mineral oil in seawater fortified with nitrate (7). The bacterial number increased by about a million in 9 days.

Gunkel (43) has made some detailed investigations into the changes in bacterial numbers in natural mixed cultures of marine bacteria under laboratory conditions using seawater-based media with ammonium ion as the nitrogen source and various oils as carbon source. The logarithmic phase growth rates were rather faster than those found for pseudomonads by Buchnell and Haas (81). It is simple to show by calculation that the Mean Generation Time (M.G.T.) was about 20 hours. The M.G.T. was approximately 10 hours for the *Pseudomonas* spp. studied by ZoBell (7). Gunkel and Trekkel (82) also developed a technique for counting bacteria in oil beach material by using a high-speed homogenizer in conjunction with a non-toxic emulsifier and a defoaming agent. The resulting emulsion was sufficiently stable to allow serial dilutions to be made for use with the Most Probable Number technique. Using this method Gunkel (29) was able to show that oil polluted beaches contained many more oil-oxidizing bacteria than nonpolluted ones.

Although bacterial counts are a useful index to bacterial activity, they suffer from a number of serious difficulties (83, 84). Further, the growth conditions governing what is essentially a batch culture are very different from those applying to the beach or open oceans, so that it is very difficult to interpret the counts in terms of the natural situation. Since the organisms are well supplied with nitrogen and phosphate, and incubated at a favorable temperature, they can be said to represent the fastest rate of growth that is likely to be encountered.

3. WEIGHT LOSS. An alternative means of estimating degradation is to measure a physical or chemical change rather than a biological one. The simplest physical parameter to follow is the weight of oil lost during oxidation. Early workers (for references, see 80) recorded losses of between 0.4 and 1.2 g oil/day/m^2 of surface exposed to the organism for crude and lubricating oils, while studies made at the Scripps Institution indicated that soil and marine bacteria destroyed many different crudes at rates varying from 0.03 to 0.5 g/day/m^2. Thus the reported rate of disappearance of crude oils varies by a factor of 40, presumably due at least in part to the different chemical mixtures. ZoBell and Prokop (44) also followed the weight lost by a series of eight crude oils in which the biological and physical conditions were standardized and autoxidation effects taken into account. The oil was suspended on ignited asbestos and added to 100 ml of seawater enriched with ammonium phosphate. The inoculum was a mixed culture of oil-oxidizing bacteria. The flasks were incubated at 25°C with aeration for 30 days. After this period a mean value of 55.6% by weight of the oil had been oxidized with a range of between 26.5 and 97.2%. So wide a variation in results is very suggestive of a wide variation in the chemistry of the substrate. It would of course be misleading to extrapolate these figures to calculate the time for 100% oxidation since the most easily degraded fractions disappear quickly, while the remainder is more refractory and may take a very long time to finally disappear.

4. OXYGEN UTILIZATION. Since, as indicated above, the introduction of oxygen into the hydrocarbon molecule is an essential part of the biodegradation process, several authors have used this phenomenon in their investigations.

Aliakrinskaya (27) measured the 5 day and 20 day Biochemical Oxygen Demand (BOD) of seawater taken from the area of Novorossysk Bay (Black Sea). The bacteria involved in the degradation were those naturally found in the seawater. Since the water of Novorossysk Bay is heavily polluted with domestic and industrial wastes, the oxygen demand was high even without added petroleum, so that a considerable increase in added petroleum had very little effect on the BOD_5, presumably because of the limits set by petroleum solubility in seawater and because the system was already loaded to its maximum capacity. The minimum amount of added petroleum that it is claimed could be detected was 0.006 ml petroleum/liter at 6.5°C. The rate of oxygen absorption at a temperature of 17–18°C was greatest on the second day and fell off rapidly on succeeding days.

ZoBell and Prokop (44) determined the 35 day BOD in seawater at

25°C of a series of pure hydrocarbons and associated compounds, inoculated with enrichment cultures of hydrocarbon-oxidizing bacteria. The ratio between the observed and theoretical BOD indicated that 25 to 79% of the pure hydrocarbon had been oxidized. With so prolonged an incubation period, there is a possibility that factors other than the ability to degrade the hydrocarbon set the limit for oxygen uptake. Such factors include autolysis of the cells and release of cell constituents, and the formation of degradation products with different oxidation rates to the hydrocarbon.

These workers also measured the 28 day BOD of a series of crude oils inoculated with 10^4 oil-oxidizing bacteria/ml, the substrate being carried on ignited asbestos and sand. The BOD_{28} of the crudes varied between 0.75 and 2.13 mg oxygen consumed/mg added oil, which by calculation indicates that between 50% and almost 100% of the available carbon and hydrogen was oxidized to carbon dioxide and water. However, the work failed to reveal any consistent relationship between the source of the crudes and their BOD. Again it is difficult to interpret this work in terms of the actual marine environment. The drawbacks associated with the BOD test are well known (85) and therefore simply extrapolating the results of these laboratory experiments to the sea could be misleading. Unhappily the Russian and American papers are not directly comparable, not least because ZoBell and Prokop (44) defined BOD in terms of milligrams of oxygen consumed per milligram substrate added while Aliakrinskaya (27) expressed BOD in terms of oxygen consumption per liter of seawater, so that BOD varies with the volume of substrate added. Moreover, since neither author indicates the time at which the BOD plateau was reached, the different time scales used may or may not be important.

An alternative oxygen method to the BOD technique is the use of manometers either in the conventional Warburg apparatus or a modified form of the Söhngen apparatus. The former arrangement was used by Jones and Edington (37) and Jones (86) in their study of moorland soils. They found that the method was limited by vapor pressure problems with certain hydrocarbons, and that reproducibility was difficult. Larger scale experiments were found to be more successful in this respect. In most cases the addition of hydrocarbons stimulated the flora contained in the soils to take up significantly more oxygen than endogenous respiration alone. The stimulation of CO_2 was also measured. Some hydrocarbons, notably benzene, toluene, and cyclohexane, depressed the endogenous respiration. ZoBell and Prokop (44) adapted the Söhngen apparatus for field work to measure oxygen uptake during crude oil oxidation in seawater. All the CO_2 evolved remained in the buffered seawater medium. By this means the oxygen uptake up to periods of 28 days was measured.

Since the inoculum consisted of 10 cells/ml, the reported figures of oxygen uptake depend as much upon the growth characteristics of the culture as on the degradation rate of the oil.

Buswell and Jurtshuk (87) used an oxygen electrode to measure the oxidation of a series of hydrocarbons by resting cell suspensions of a pure culture of *Corynebacterium*. The hydrocarbons were added as an emulsion formed by sonic oscillation. They noted that the concentration became rate limiting at high and low levels. The rate of oxidation increased from *n*-pentane up the homologous series to *n*-octane and then decreased to *n*-heptadecane. *n*-Decane showed an anomalous increase. The *n*-alkenes up to C_{10} were oxidized at a slower rate than the corresponding alkanes, while above C_{10} the reverse was true. Halogenated alkanes, iso-alkanes, cyclo-alkanes, and aromatic compounds were oxidized at much slower rates.

5. INFRARED ANALYSIS. Infrared spectrophotometry at wavelengths of between 3.3 and 3.5 μ has often been recommended as a quantitative method of measuring oil decomposition (88–91). The method is useful for low concentrations of oil in water and depends upon the absorption due to CH_3, CH_2, and CH groupings. It is therefore particularly useful in oils rich in aliphatic material but is less able to detect the lower members of the aromatic series. The method was successfully used by Gunkel (43) to follow the breakdown of oil in the laboratory under various conditions of growth.

6. CHROMATOGRAPHIC TECHNIQUES. Gas-liquid chromatography and thin layer chromatography are also of value in oil biodegradation analysis (88). Le Petit and Barthelemy (77) followed the breakdown of gas–oil by means of the gas chromatogram under conditions which resolved the alkanes and alkenes against a background of unresolved aromatics and cyclic paraffins. The oil was degraded by the organisms in freshly sampled seawater in a growth medium containing ammonium phosphate as nitrogen and phosphate source. Between 2 and 5 days were necessary to effect a very considerable reduction in the alkanes present. The alkanes in the oil ranged from *n*-undecane to hexacosane and of these only *n*-heptadecane and *n*-octadecane remained after 5 days, and even these latter compounds were much reduced in concentration (Figure 7.1). They noted that the time taken for 65–75% of the oil to disappear was in general about 5 days, but it was subject to variation depending upon the place and time at which the water sample was taken.

More recently Blumer and his coworkers (92) have examined the consequences of a fuel oil spill along the Massachusetts coast during a strong gale. The turbulence was sufficient to mix the oil with the sediments to

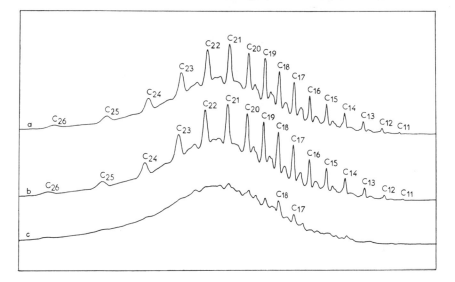

Figure 7.1 Evolution during incubation of a gas-oil in a culture medium: (a) Initial undegraded gas–oil; (b) gas–oil after 2 days of incubation; and (c) gas–oil after 5 days of incubation (77).

at least 10 m in depth. The distribution and changes in the oil composition were followed by gas chromatography. It was found that the oil was slowly released from the sediments, but showed very little change in composition up to a period of 4 months. There was some decrease in the lower molecular weight hydrocarbons and an increase in the proportion of branched alkanes. Such changes as were observed could be attributed to physical and chemical rather than to biological factors. Unfortunately the environmental conditions are not mentioned, and it is possible that the sediment had a tendency toward anaerobiosis with a consequent slowing down of the oxygenation rate. Perhaps more surprising was the presence of alkanes from dodecane to heptacosane, in a Mediterranean "tar ball," presumably derived from weathered crude oil. These alkanes include many that are known to be biodegradable and it seems probable that oxygen or other essentials for oil oxidation had been excluded at an early stage (Figure 7.2).

These workers also examined polluting oil present in scallops (*Aequipecten irradians*) and oysters (*Crassostrea virginia*) some 2 months after the accident (Figure 7.2). The hydrocarbons in the oysters showed the same general distribution as the spilt oil and the oil recovered from sediments, especially in the region above *n*-heptadecane. Below that there is a further decrease in the ratio of normal to branched alkanes.

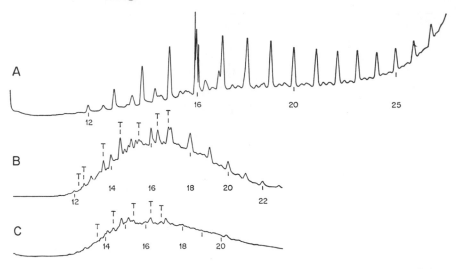

Figure 7.2 Gas chromatograms: (a) Tar ball from Mediterranean, *n*-hexadecane added as standard; (b) oysters (*Crassostrea virginia*) 2 months after oil spill; and (c) scallops (*Aequipecten irradians*) 2 months after oil spill (92).

The distribution of hydrocarbons in the scallops had undergone a change in the same direction as in the oyster but to a greater extent. The normal alkanes had largely disappeared after 2 months and the isoprenoid alkanes had become dominant. The aromatic and cyclic hydrocarbons appeared to be only slightly changed.

Scallops from an uncontaminated area contained small amounts of hydrocarbon but with a very different distribution of the constituents, *n*-heptadecane and *n*-heneicosane being the only aliphatic alkanes present in anything but very small amounts. These alkanes and the olefin *n*-heneicosahexene are typical of marine algae (9).

7-3. Conclusions

It will be clear from the above discussion that many bacteria found in the sea are capable of ozidizing a considerable part, if not all, of the oil or oil products which are spilled into the marine environment. It is also well established that oxygen, nitrogen, and phosphate are necessary to support the growth of oil-oxidizing bacteria, and that warm conditions produce a faster rate of oil breakdown than cold.

What is not so clear is the extent to which the potential breakdown is realized under the conditions prevailing in the sea at any one time. Nor is it known how fast oil can be added to the sea without overwhelming its biodegradative capacity. Recent reports of large amounts of oil in marine areas far from the major commercial sea lanes are disturbing. Future research may well have to concentrate on finding out what happens to the oil in the ocean itself rather than in the laboratory, and deriving ways whereby the latent degradative ability is realized to its fullest extent.

REFERENCES

1. M. Blumer, *Oceanus* **15**(2), 2 (1969).
2. J. E. Smith, *Torry Canyon; Pollution and Marine Life*, Cambridge Univ. Press, Cambridge, 1969.
3. R. W. Holmes, in *Oil on the Sea* (D. P. Hoult, Ed.), Plenum, London, 1969.
4. J. H. Kirby, in *Third International Conference on Oil Pollution of the Sea, Rome* (P. Barclay-Smith, Ed.), Wykeham Press, Winchester, 1968.
5. K. G. Brummage, in *Third International Conference on Oil Pollution of the Sea, Rome* (P. Barclay-Smith, Ed.), Wykeham Press, Winchester, 1968.
6. G. Dudley, in *The Biological Effects of Oil Pollution on Littoral Communities* (J. D. Carthy and D. R. Arthur, Eds.), Field Studies Council, London, 1968.
7. C. E. ZoBell, *Advan. Water Pollution Res.* **3**, 85 (1964).
8. J. B. Davis, *Petroleum Microbiology*, Elsevier, Amsterdam, 1967.
9. R. C. Clark and M. Blumer, *Limnol. Oceanog.* **12**, 79 (1967).
10. B. Blumer, M. M. Mullins, and D. W. Thomas, *Helgoländer Wiss. Meeres.* **10**, 187 (1964).
11. J. Oró, T. G. Tornabene, D. W. Nooner, and E. Gelpi, *J. Bacteriol.* **93**, 1811 (1967).
12. J. Han and M. Calvin, *Proc. Nat. Acad. Sci., U.S.* **64**, 436 (1969).
13. G. J. Jankowski and C. G. ZoBell, *J. Bacteriol.* **47**, 447 (1944).
14. J. B. Davis, *Chem. Geol.* (*Neth.*) **3**, 155 (1968).
15. E. E. Bray and E. D. Evans, *Geochim. Cosmochim. Acta* **22**, 2 (1961).
16. E. J. McKenna and R. E. Kallio, in *Principles and Applications of Aquatic Microbiology* (H. Heukelekian and N. C. Dondero, Eds.), Wiley, New York, 1964.
17. J. Perdriau, *Cahiers Oceanog.* **16**, 193 (1964).
18. J. Bourcart and L. Mallet, *C. R. Acad. Sci., Paris, Ser. D.* **260**, 3729 (1965).
19. L. Mallet and M-L. Priou, *C. R. Acad. Sci., Paris, Ser. D.* **264**, 969 (1967).
20. C. de Lima-Zanghi, *Cahiers Oceanog.* **20**, 203 (1968).
21. P. Niausatt, J-P. Erhardt, and J. Ottenwalder, *C. R. Acad. Sci. Paris, Ser. D.* **267**, 1772 (1968).
22. R. A. Dean, in *The Biological Effects of oil Pollution on Littoral Communities*, (J. D. Carthy and D. R. Arthur, Eds.), Field Studies Council, London, 1968.
23. C. E. Frank, *Chem. Rev.* **46**, 155 (1950).
24. A. C. Nixon, in *Autoxidation and Antioxidants*, Vol. 2. (W. O. Lundberg, Ed.), Wiley, New York, 1962.
25. E. J. Nyn, *Rev. Questions Sci.* **138**, 189 (1967).
26. M. J. Johnson, *Chem. Ind.* (London) **1964**, 1532.

27. I. O. Aliakrinskaya, *Oceanology* **6**, 71 (1966).

28. R. S. Wodzinski and M. J. Johnson, *Appl. Microbiol.* **16**, 1886 (1968).

29. W. Gunkel, in *The Biological Effects of oil Pollution on Littoral Communities* (J. D. Carthy and D. R. Arthur, Eds.), Field Studies Council, London, 1968.

30. W. M. Kluss, *Tanker and Bulk Carrier* **15**, 236 (1968).

31. M. Myoshi, *Jahrb. Wiss. Botan.* **28**, 269 (1895).

32. C. E. ZoBell, *Bacteriol. Rev.* **10**, 1 (1946).

33. G. W. Fuhs, *Arch. Mikrobiol.* **39**, 374 (1961).

34. A. P. Ponsford, *Mon. Bull. Brit. Coal Utd. Res. Assoc.* **30**, 41 (1966).

35. A. J. Markovetz, J. Cazin, and J. E. Allen, *Appl. Microbiol.* **16**, 487 (1968).

36. J. J. Perry and H. W. Scheld, *Can. J. Microbiol.* **14**, 403 (1968).

37. J. G. Jones, and M. A. Edington, *J. Gen. Microbiol.* **52**, 381 (1968).

38. M. J. Klug, and A. J. Markovetz, *Nature* **215**, 1082 (1967).

39. V. F. Coty, *Biotechnol. Bioeng.* **9**, 25 (1967).

40. H. I. Harper, *Soil Sci.* **48**, 461 (1939).

41. C. E. ZoBell, C. W. Grant, and H. F. Haas, *Bull. Amer. Assoc. Petrol. Geol.* **27**, 1175 (1943).

42. J. Brisou, *La microbiologie du milieu marin,* Flammarion, Paris, 1955.

43. W. Gunkel, *Helgoländer Wiss. Meeresuntersuch.* **15**, 210 (1967).

44. C. E. ZoBell, and J. P. Prokop, *Z. Allgem. Mikrobiol.* **6**, 143 (1966).

45. H. Weyland, *Nature* **223**, 858 (1969).

46. T. W. Johnson and F. K. Sparrow, *Fungi in Oceans and Estuaries*, Cramer, Weinheim, 1961.

47. V. Trecanni, *Prog. Ind. Microbiol.* **4**, 3 (1962).

48. J. W. Foster, in *Oxygenases* (O. Hayaishi, Ed.), Academic Press. London, 1962.

49. A. C. van der Linden and G. J. E. Thijsse, *Advan. Enzymol.* **27**, 469 (1965).

50. A. E. Humphrey, *Biotechnol. Bioeng.* **9**, 3 (1967).

51. R. W. Traxler and W. L. Flannery, in *Biodeterioration of Materials; Microbiological and Allied Aspects* (A. H. Walters and J. J. Elphick, Eds.), Elsevier, Amsterdam, 1968.

52. R. W. Hansen and R. E. Kallio, *Science, N. Y.* **125**, 1198 (1957).

53. D. M. Updegraff and C. B. Wenn, *Appl. Microbiol.* **2**, 309 (1954).

54. R. W. Traxler and J. M. Bernard, *Int. Biodeterior. Bull.* **5**, 21 (1969).

55. G. D. Novelli and C. E. ZoBell, *J. Bacteriol.* **47**, 447 (1944).

56. J. B. Davis and H. F. Yarbrough, *Chem. Geol.* (*Neth.*) **1**, 137 (1966).

57. J. A. Davies and D. E. Hughes, in *The Biological Effects of Oil Pollution on Littoral Communities* (J. D. Carthy and D. R. Arthur, Ed.), Field Studies Council, London, 1968.

58. W. D. Rosenfeld, *J. Bacteriol.* **54**, 664 (1947).

59. J. Postgate, *Ann. Rev. Microbiol.* **13**, 505 (1959).

60. P. F. Andreev, A. I. Bogomolov, A. F. Dobryanskii, and A. A. Kartsev, *Transformation of Petroleum in Nature*, Pergamon, London, 1968.

61. K. Shimara and H. Yamishita, *J. Ferment. Technol. Osaka* **45**, 1172 (1967).

62. J. J. Perry, *Antonie van Leeuwenhoek J. Microbiol. Serol.* **34**, 27 (1968).

63. P. S. Meadows, *J. Exp. Biol.* **41**, 499 (1964).

64. J. S. Gray and R. M. Johnson, *J. Exp. Mar. Biol. Ecol.* **4**, 119 (1970).

65. J. E. Stewart, R. E. Kallio, D. P. Stevenson, A. C. Jones, and D. O. Schissler, *J. Bacteriol.* **78**, 441 (1959).

66. J. E. Stewart and R. E. Kallio, *J. Bacteriol.* **78**, 726 (1959).

67. W. R. Finnerty, E. Hawtrey, and R. E. Kallio, *Z. Allgem. Mikrobiol.* **2**, 169 (1962).

68. O. Wyss and E. J. Moreland, *Appl. Microbiol.* **16**, 185 (1968).

69. K. R. Dunlap and J. J. Perry, *J. Bacteriol.* **94**, 1919 (1967).

70. J. H. Atkinson and F. H. Newth, in *Microbiology* (P. Hepple, Ed.), Institute of Petroleum, London, 1968.

71. J. D. Douros and J. W. Frankenfeld, *Appl. Microbiol.* **16**, 320 (1968).

72. V. W. Jamison, R. L. Raymond, and J. O. Hudson, *Appl. Microbiol.* **17**, 853 (1969).

73. R. L. Raymond, V. W. Jamison, and J. O. Hudson, *Appl. Microbiol.* **17**, 512 (1969).

74. C. R. Baier, *Kiel. Meeresforsch.* **2**, 149 (1937).

75. E. C. Hill, in *Microbiology* (P. Hepple, Ed.), Institute of Petroleum, London, 1968.

76. J. J. Cooney, P. Edmonds, and Q. M. Brenner, *Appl. Microbiol.* **16**, 569 (1968).

77. J. le Petit and M. H. Barthelemy, *Ann. Inst. Pasteur* **114**, 149 (1968).

78. F. J. Ludzack and D. Kinkhead, *Ind. Eng. Chem.* **48**, 263 (1956).

79. F. J. Ludzack, W. M. Ingram, and M. B. Ettinger, *Sewage Ind. Wastes* **29**, 1177 (1957).

80. C. E. ZoBell, *Advan. Enzymol.* **10**, 443 (1950).

81. L. D. Buchnell and H. F. Haas, *J. Bacteriol.* **41**, 653 (1941).

82. W. Gunkel and H. H. Trekkel, *Helgoländer Wiss. Meeres.* **16**, 336 (1967).

83. G. G. Meynell and E. Meynell, *Theory and Practise in Experimental Bacteriology*, Cambridge Univ. Press, Cambridge, 1965.

84. T. D. Brock, *Principles of Microbial Ecology*, Prentice-Hall, Englewood Cliffs, N.J., 1966.

85. H. A. C. Montgomery, *Water Res.* **1**, 631 (1967).

86. J. G. Jones, *Arch. Mikrobiol.* **67**, 397 (1969).

87. J. A. Buswell and P. Jurtshuk, *Arch. Mikrobiol.* **64**, 215 (1969).

88. C. Rübelt, *Helgoländer Wiss. Meeresuntersuch.* **16**, 306 (1967).

89. R. G. Simard, J. Hasegawa, W. Bandaruk, and C. E. Headington, *Anal. Chem.* **23**, 1384 (1951).

90. F. J. Ludzack and C. E. Whitefield, *Anat. Chem.* **28**, 157 (1956).

91. J. B. Rather, *Anal. Chem.* **30**, 36 (1958).

92. M. Blumer, G. Souza, and J. Sass, *Marine Biol.* **5**, 195 (1970).

8 Approaches to the Synthesis of Soft Pesticides

D. D. Kaufman and J. R. Plimmer, Agricultural Research Service, U.S. Department of Agriculture, Beltsville, Maryland

8-1. Introduction

Chemical control of pests is not a twentieth century development, but has been practiced in some form for at least as long as man has been recording his history. The first recorded mention of a chemical used for crop protection refers to one which is still in use today (1). Homer, the Greek epic poet, mentioned the "pest averting sulfur with its properties of divine and purifying fumigation" in the year 1000 B.C. Arsenic, which is used in various forms today, was actually recommended by Pliny (1, 2). Thus, it is evident that the people of ancient Greece and Rome not only were aware of crop pests, but also tried to prevent crop losses.

Much of our knowledge on the use of inorganic chemicals has been derived through centuries of usage. A variety of inorganic compounds have been used as insecticides and fungicides. We know that the majority of these compounds are very persistent. Fortunately, the problems associated with residues of heavy metals which may contaminate crops are well known and readily avoided.

Extensive development of organic chemicals for use in pest control did not begin until early this century. Considerable impetus was given to their development by the demands for chemical controls needed in World War I and II, and man's desire to eliminate worldwide disease problems associated with insect pests. The organic insecticides, especially the chlorinated hydrocarbons, were extremely effective and had very low toxicity toward mammals. Their use became increasingly widespread, and their prolonged effectiveness resulting from their slow breakdown in the environment was recognized and heralded as beneficial. It was not recognized until after a long period of use had elapsed, however, that such results might be other than beneficial. An increasing concentration of chlorinated hydrocarbons in soils and sedimentary mud was noted. The movement of these residues into food chains gave cause for alarm when it became known that their rate of concentration by biota surpassed their rate of degradation. The susceptibility of organisms to chlorinated hydrocarbons differs; consequently, their effects become manifest in species at various levels of the food chain.

Data are available indicating that many of the chlorinated hydrocarbon insecticides are found in measurable quantities in man, animals, plants, soil, water, and air (3). Persistent pesticides are capable of cycling in the environment: as chemically stable molecules they are adsorbed to dust and move in air currents; as water-soluble organic compounds of low vapor pressure they frequently codistill with water. Their lipid solubility favors concentration in biota and their subsequent role in the food chain may depend on their toxicity to particular species.

Persistence and toxicity towards nontarget organisms are features which currently characterize undesirable pesticides. Persistence, with its accompanying prolonged effectiveness, was a desirable feature of pesticides as recently as 10 or 15 years ago. Toxicity towards nontarget organisms has always been a concern of responsible users of biological toxicants, hence the general care and recommended procedures for handling such chemicals. The current concept of the desirable organic pesticide, however, is that it (a) control or eradicate the pest, (b) is nontoxic to nontarget organisms, (c) is degradable to ecologically acceptable products within a reasonable period of time, (d) has the ability to remain relatively stationary at the application site, and (e) has low potential for accumulation in any portion of the environment.

The problems associated with persistent pesticides arise from three major causes: (a) their widespread use, (b) their ability to be transferred from one phase of the environment to another, and (c) their resistance to biological or chemical degradation (4).

The first problem cannot be solved simply. Pesticides are necessary to modern farming and to maintain production by reverting to hand labor is almost inconceivable at the present time. Other methods of pest control, such as the program of sterilization used to control the screw worm or the use of predators to control insects and pests, although necessitating the accompanying use of pesticides, can reduce the quantities of pesticides required. The development of more effective pesticides and more efficient methods of pesticide placement close to or on the target organism can also achieve this effect.

The problems of transfer from one phase of the environment to another and of ready breakdown by environmental forces are related questions which may only be answered by consideration of the chemical structure of the pesticide.

In the past the criteria for selection of effective pesticides has been based on economic, toxicological, and selectivity factors. Most agricultural products are produced with a thin margin for profit. The economics of pesticide production, distribution, and usage must be such as to assure a suitable return for the monies and efforts invested. The primary toxicological requirement for a pesticide was that it selectively control or eradicate the target organisms. Our present concern for the ecology and pollution of our environment necessitates our reconsideration of the desirability of these characteristics.

In this chapter it is proposed to examine the relationship between the ease of degradation of several pesticides and their chemical structures. Both the rate of breakdown and the products are significant. It is important that the breakdown products do not themselves become an environmental problem and that ultimately they will be irreversibly bound to inert structures or degraded to simple molecules normally utilized in common metabolic cycles. For example, DDT is easily transformed to DDE or DDD, but both DDE and DDD have similar biological and chemical properties to DDT and present the same problems of magnification through the food chain and resistance to breakdown.

8-2. Structure–Activity–Degradability Interrelations

The ultimate question is whether or not pesticide molecules can be designed that will remain stationary at the site of application, degrade rapidly once released into the environment, and still selectively control or eradicate target organisms. The design of pesticides having such characteristics necessitates that we understand the mode of action and the structure–activity relationships underlying toxicity and degradability. Although we have identified large classes of pesticidal chemicals such as the organophosphates, cyclodienes, carbamates, phenoxyalkanoates, and phenylureas, unfortunately the mode of action and structure—activity relationships relative to effectiveness or degradability have not been adequately investigated. Few investigations have ever compared the structural features necessary for toxicity to target organism with those permitting degradation in the environment. Some information exists, however, for some of the major classes of pesticides.

A.   Aliphatic Acids.   Aliphatic acids considered for herbicidal use include halogenated acetic, propionic, butyric, and isobutyric acids. Of these only TCA (trichloroacetic acid) and dalapon (2,2-dichloropropionic acid) are currently recognized by the Weed Science Society of America (5). The mechanism of action, believed to be at least partially responsible for phytotoxic action of aliphatic acids, involves the inhibition of the enzyme which forms pantothenic acid from β-alanine and pantoic acid (6). Hilton *et al.* (7–10) observed an interesting structural effect on this inhibition mechanism. They observed that the α-chlorosubstituted herbicides were competitive with pantoic acid (10), whereas β-alanine protected the pantothenate synthesizing enzyme against inhibition by unchlorinated aliphatic acids, monochloroacetate, and β-chloropropionate. Chlorine substitution reduced the inhibitory effects of the aliphatic acids antagonized by β-alanine. In contrast, pantoate protected the enzyme against inhibition by α-chloropropionate and di- or trichloro-α-substituted acids of the acetic and propionic series. In this case, toxicity to the enzyme increased with additional chlorine substitutions on aliphatic acids antagonized by pantoate.

The structure–activity and structure–degradability relationships of certain of the halogenated aliphatic acids are quite similar. That is, the most herbicidally-active structures are also the most readily degraded structures. Within the propionic acid and butyric acid series, β-chlorination may weaken the effects of the α-chlorination. Similarly, biodegradation of α-halogenated propionic and butyric acids in soil and by isolated cultures occurred more rapidly than degradation of propionic and butyric acids halogenated in other positions (11–13). Halogenation in other positions either alone or in combination with α-halogenation decreased the biodegradability of these compounds.

Increasing or reducing chain length reduced herbicidal activity, even with α-chlorination. Chloroformic acid, chloroacetic acid, and dichloroacetic acid were all ineffective as selective grass control herbicides (14).

The 2,2-dichlorohexanoic acid was inactive. Changes in chain length also affected the biodegradability; three carbon compounds generally being more readily degraded than their two- or four-carbon homologs (11, 13). A microbial enzyme (15) hydrolyzing 2,2-dichloropropionate was less active on 2-chloropropionate, dichloroacetate, and 2,2-dichlorobutyrate.

The effect of halogen type on structure–activity and structure–degradability is not completely clear. According to Leasure (14) the substitution of other halogens for chlorine generally reduces herbicidal activity throughout the aliphatic acid series. Hirsch and Alexander (13) examined the degradation of several fluoro-, chloro-, bromo-, and iodo-substituted aliphatic acids by two microbial isolates. Although both chloro- and bromo-substituted compounds were degraded, the bromo-substituted compounds were degraded somewhat more rapidly than chloro-substituted compounds by both organisms. A *Nocardia* sp. actually degraded iodo-substituted compounds more rapidly than some of the corresponding bromo- or chloro-substituted compounds. Both organisms, however, degraded the β-chloro-substituted propionate more readily than the corresponding bromo- or iodopropionate, suggesting a dependence on both halogen type and position. Unique features of the carbon–halogen bond (energy, length) or the halogen itself (molecular size) may influence the activity of the microbial dehalogenases in degrading these compounds. Individual differences occur among various isolates metabolizing homologous structures. These differences, however, are probably the exception rather than the rule.

We should also be concerned with the effect of the inducing substrate on the nature of the enzyme induced. The steric effect on the rate of dehalogenation in one induced enzyme system may not be the same in another induced system. This was illustrated effectively by Goldman *et al.* (16). Gemmell and Jensen (17) observed that organisms isolated by the enrichment process on TCA were quite effective in degrading TCA but only slightly effective on monochloroacetate, dichloroacetate, 2-chloropropionate, and 2,2-dichloropropionate.

B. Halogenated Phenoxyalkanoates.   The phenoxyalkanoate herbicides are commonly referred to as the "hormone herbicides." Their hormone-like activity at low concentrations has long been known. At higher concentrations they have been used extensively as herbicides. Surprisingly little is known or understood about their mode of action. Although several theories have been proposed (6), none have been conclusively proven or are widely accepted. The following discussion is based largely on information regarding structural characteristics related to hormonal activity.

Considerable effort has been expended to establish structure–activity relationships within the phenoxyalkanoate herbicides (18). The early classical work of Koepfli *et al.* (19) revealed that for biological activity a growth substance should have an unsaturated ring system with a side chain having a carboxyl group connected to the ring through at least one carbon atom. Veldstra (20, 21) modified these requirements to (*a*) a ring system with a high potential for adsorption to a protoplasmic membrane, and (*b*) a carboxyl group in a very definite spatial position relative to the ring system. Muir *et al.* (22) and Hansch and Muir (23) suggested an additional structural requirement—that the compound should have a free *ortho* position. Leaper and Bishop (24) examined the phytotoxicity of all possible mono-, di-, and trichloro-substituted phenoxyacetates. They concluded that still another requirement be added—that for high herbicidal potency there should be another free position in the benzene ring opposite the free *ortho* position postulated by Muir *et al.* (22) and Hansch and Muir (23).

In other structure–activity relationships, Hitchcock and Zimmerman (25) observed that the root-inducing activity of phenoxy compounds varied from low activity to very high activity depending upon the kind, number, and position of substituents in the ring and to the relative length of the side chain. Activities of α-propionic and α-butyric homologs of the phenoxy compounds were 30–100 times greater than the corresponding monochloro-, dichloro-, and dibromophenoxyacetic acids. The descending order of activity for mono-substituted phenoxyacetic acids was: *para* > *ortho* > *meta;* Cl > I at the *ortho* position; Cl > NH_2 > NO_2 at the *meta* position; and Cl > Br > NH_2 at the *para* position.

Several investigations have been made (26–28) to determine the feasibility of replacing the oxygen of the ether linkage with an —S— or —NH— linkage. Although there is evidence to show that the aryloxy acids are generally more active than their arylthio analogs (26), the *N*-2,4-dichlorophenylglycine is highly active in some tests (27, 28).

Two major pathways are apparent for the microbial degradation of phenoxyalkanoic acids: (*a*) degradation via the corresponding phenol, and (*b*) degradation via a hydroxylated phenoxyalkanoate (Figure 8.1). Degradation via the first pathway may occur through oxidation of the side chain with the subsequent liberation of the phenol, or with immediate cleavage of the ether linkage. The structural effects on this metabolic sequence have been fairly well established. Very little is known concerning the effects of phenoxyalkanoate structure on their metabolism via a hydroxylated phenoxyalkanoate intermediate. Several microorganisms are known to metabolize monochloro-substituted phenoxyalkanoates, and pathways have been projected (29). Ring cleavage and further degradation of dichloro-substituted phenoxyalkanoates subsequent to hydroxylation has been observed.

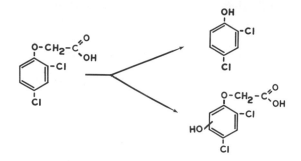

Figure 8.1   Major degradation pathways of phenoxyalkanoate herbicides.

Taylor and Wain (30) examined the side-chain degradation of certain ω-phenoxyalkanoates by *Nocardia coeliaca* and other organisms isolated from soil. The metabolite pattern demonstrated by *N. coeliaca* when growing on each of 10 different ω-phenoxyalkanoates was consistent with β-oxidation. Evidence for an α-oxidation pathway was also obtained with homologs containing 10 and 11 carbons in the side chain. ω-Oxidation has also been proposed (31). The introduction of substituents onto either the phenoxy ring or the side chain considerably reduces side-chain oxidation in both plants and microorganisms (29).

The introduction of a methyl substituent on the γ-carbon of phenoxybutyric acids (32) considerably retarded the β-oxidation of these compounds, although increasing the chain length had a negating effect on methyl substitution. This observation is somewhat analogous to what has been experienced in the degradation of hard and soft detergents. Straight-chain detergents are subject to rapid biodegradation via β-oxidation and therefore considered soft, whereas branched-chain or substituted-chain detergents were resistant to oxidation and therefore hard (33).

Certain types of ring substitution also influence plant and microbial oxidation of phenoxyalkanoate (30, 31) side chains. The presence of either an *o*-chloro-, or *o*-methyl-substituent on the phenoxy ring inhibits β-oxidation (30, 31, 34). This inhibition is reduced, however, by the addition of *p*-chloro-substituent to the phenoxy ring. Audus (35, 36) reported the following effects of ring substitution on the biodegradation of phenoxyacetate homologs:

1. Chlorine substitution in the *para* position increases the lability of the molecule.
2. Chlorine substitution in the *ortho* position inhibits degradation.

3. Chlorine substitution in the *meta* position also inhibits degradation but the effect is not as pronounced as with *ortho* substitution.

4. The activating effect of *para* substitution largely overcomes the inhibitory effect of other substitutions in the di- and trichlorophenoxy-acetates.

5. The methyl substituent has a greater inhibitory effect than the chloro group.

Alexander and his collaborators (37–40) have also investigated the effect of molecular structure on the persistence and microbial decomposition of phenoxyalkanoates, chlorophenols, and related compounds. They stressed the importance of *meta* substitution in conferring resistance of these compounds to microbial degradation. The type of aliphatic side chain and the nature of its linkage to the ring were also important factors affecting degradation. Generally, ω-substituted compounds were degraded more readily than α-substituted compounds.

The structural effect on the subsequent metabolism of the corresponding phenols during phenoxyalkanoate metabolism is similar to that observed for the parent molecule (37, 39). *Meta*-substitution lends resistance to degradation of the mono-, di-, and trichlorophenols. *Ortho*-substituted phenols are moderately resistant to degradation, whereas *para*-substituted phenols are readily degraded. The unsubstituted phenol was the major exception; the free phenol being degraded quite rapidly (37), whereas phenoxyacetate was somewhat resistant to degradation. (36, 37).

Several correlations between structure–activity and structure–degradability of phenoxylakanoates can be adduced; whereas *meta*-substitution (*para* to a free *ortho* position) confers resistance to biodegradation, it also eliminates herbicidal activity. 2,4,5-T is moderately resistant to biodegradation but is highly active, having the two necessary free positions for activity. In contrast, 2,3,4-trichlorophenoxyacetate is both inactive and highly resistant to biodegradation. Halogenation in the *para* position increases both herbicidal activity and biodegradability of phenoxyacetates. *Para*-substitution also has a sparing effect on certain di- and tri-substituted phenoxyacetates. Increasing the length of the side chain affects both herbicidal activity and biodegradability. Although other correlations may be possible, they will not be elaborated here. Suffice it to say that structural characteristics necessary for activity are similar to those favoring degradability.

C. Carbamate Insecticides and Herbicides. Carbamates have found wide application as insecticides, herbicides, fungicides, and nematicides. Information on the relation of structure to insecticidal activity of the methylcarbamates has been summarized by Metcalf (41). The highest activity is found in N-methyl- and N,N-dimethylcarbamates, whereas N-ethyl- and N,N-diethylcarbamates are much less active, and N-benzyl, N-phenyl, and other substitutions are nearly nontoxic. Considerable latitude is allowable in the alcoholic portion of the molecule, which may consist of phenol, pyrazolol, pyrimidol, or other heterocyclic alcohols. However, for highest activity, the alcohol should contain a relatively compact group such as methy, ethyl, isopropyl, *tert*-butyl, or dimethylamino at a site 2 or preferably 3 atoms (*meta* in phenol esters) from the ester link. Substitutions of sulfur for either of the carbamate oxygen atoms reduces insecticidal activity considerably while substitution of both to form a dithiocarbamate abolishes activity. Unlike phenyl N-methylcarbamates, methyl N-phenylcarbamates are not appreciably active as anticholinesterase toxicants (42). For optimum contact activity the molecule should be nonpolar and lipid soluble, avoiding quaternary ammonium or amine salt structures. This is well illustrated by *m*-dimethylaminophenyl N-methylcarbamate which was 50 times as toxic to *Heliothrips haemorrhoidalis* as the corresponding methiodide but only 0.002 times as effective as an inhibitor of fly cholinesterase (42). These structural requirements are reflections of the necessity for the toxic carbamate ester to possess a configuration which closely resembles that of acetylcholine so that the carbamate will be strongly attracted to the anionic and esteratic sites of the cholinesterase molecule and yet will exhibit maximum stability to enzymatic hydrolysis, thus blocking the active enzyme site to the access of acetylcholine.

The methylcarbamate insecticides are relative newcomers to the pesticide field; hence, very little is known concerning their degradation in our environment. Preliminary investigations indicate that the principal mode of microbial degradation initially involves hydrolysis of the carbamate linkage (11, 43). Metcalf and Fukuto (44) have observed that relative rates of hydrolysis of methylcarbamates, to a certain extent, parallel the ability of these compounds to inhibit cholinesterase. Subsequent degradation and persistence patterns of the phenols produced during hydrolysis of N-methylcarbamates are analogous to those produced during the degradation of the phenoxyalkanoates.

The carbamate herbicides fall into several categories, methylcarbamates, phenylcarbamates (or carbanilates), and thio- and dithiocarbamates. Pre-emergence crabgrass activity is specific to the type of substituents in the 2-, 4-, and 6-positions of the phenyl groups in the phenyl methylcarbamates, but some correlations of chemical structure with biological activity have been made (45). A methylcarbamate is apparently required. Neither 2,6-di-*tert*-butyl-*p*-tolyl ethyl- nor phenyl carbamates were active.

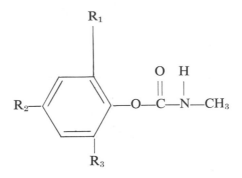

When R_1 and R_3 are both *tert*-butyl, appreciable activity is obtained when $R_2 = CH_3$, C_2H_5, Br, Cl, or CH_3O. As the alkoxy substituent in the R_2 position increases in molecular weight, activity decreases. However, if R_1 and R_3 are both methyl, isopropyl, or *tert*-amyl, there is little or no activity. When R_1 is *tert*-butyl or *tert*-amyl and R_2 is CH_3—, Br—, Cl—, or CH_3O—, then R_3 may vary from C_3H_7— to C_5H_{11}— and activity is still retained. Likewise, the hydrocarbon chain may be straight or branched. R_3 may also be an unsaturated group such as methylallyl or methylpropenyl.

Activity of methylcarbamate insecticides has been correlated with molecular weight (46). This relationship does not hold true for methylcarbamate crabgrass herbicides, because 2,4-di-*tert*-butyl-*o*-tolyl methylcarbamate, an isomer of terbutol (2,6,-di-*tert*-butyl-*p*-tolyl methylcarbamate) is inactive.

Structure–activity and structure–biodegradability information are available for the phenylcarbamate herbicides. Phenylurethane was known as a germination inhibitor as early as 1929 (47). Templeman and Sexton (48), Shaw and Swanson (49), Moreland and Hill (50), and Good (51) have made comprehensive investigations of structure–activity relationships on a number of arylcarbamic esters.

The phenylcarbamate herbicides propham (isopropyl carbanilate), chlorpropham (isopropyl *m*-chlorocarbanilate), barban (4-chloro-2-butynyl-*m*-chlorocarbanilate), and swep (methyl 3,4-dichlorocarbanilate) are generally active as mitotic poisons and are capable of producing striking cytological aberrations in some plant species. A high degree of correlation exists between structural configuration and herbicidal activity in the *N*-phenyl and substituted *N*-phenylcarbamates (49, 50). The greatest activity of the derivatives examined was associated with isopropyl and secondary butyl esters of both the *N*-phenyl- and substituted *N*-phenylcarbamates. Although the greatest activity appears to be in these two esters, other esters such as 2-chloroethyl, *N*-butyl, 2-ethylhexyl, propyl, lauryl, and ethyl possess activity also. Within the isopropyl series, substitution in the *meta* position of the aromatic ring produces the greatest activity. Substituents placed in the *meta* position do not contribute equally to activity and the following order was noted: chloro > methyl > methoxy > trifluoromethyl > nitro > acetyl > hydroxy. Active compounds also resulted from substitution in the *para* position but very little activity is obtained by substitution in the *ortho* position (50, 51). When substitutions are made in the *ortho* or *para* positions, the activity contributed by the substituted groups examined remains in the same relative order as for substitution in the *meta* position.

In addition to acting as mitotic inhibitors, disubstituted *N*-phenylcarbamates have the property of inhibiting chlorophyll production (49). Substitution in the 3- and 6-positions (2 and 5) produced the greatest activity followed by substitution in the 3- and 4-, 2- and 4-, and 3- and 5-positions in order of decreasing activity in the disubstituted isopropyl *N*-phenylcarbamates. Substitution in the 3- and 6-positions produces less total activity than substitution in the 3-position alone, but the property of inhibiting chlorophyll production is enhanced. Substitution of combinations of unlike groups such as methyl or methoxy with chlorine in the 3- and 6-positions produces activity equal to, and chlorophyll inhibition greater than, that obtained by the substitution of two chloro groups in the same positions.

Similar results with respect to ring substitution have been obtained with barban (52). Placement of the nuclear chlorine in the *ortho* or *para* position eliminates barban's selective action against oats. Reduction of the triple bond to a double or a single bond has a similar result. Replacement of the alkyl chlorine with other substituents has a similar effect.

The introduction of a chlorine atom into propham to form chlorpropham prolongs its residual effectiveness in soil. Several investigators have shown that chloropham is more readily degraded than propham in isolated culture systems. Perhaps the apparent difference in soil persistence (*i.e.,* resistance to degradation) could be explained by differences in adsorptivity.

Information describing the structure and activity relationships of thio- and dithiocarbamate herbicides is lacking. Out of several hundred thiocarbamates synthesized and tested by Stauffer Chemical Company, only EPTC (*S*-ethyl dipropylthiocarbamate), pebulate (*S*-propyl butylethylthiocarbamate), vernolate (*S*-propyl dipropylthiocarbamate), and R-2060 (*S*-ethyl butylethylthiocarbamate) were selected for further testing (53). It would seem apparent that activity within this group is confined to combinations of ethyl, propyl, and butyl groups as substituents on the nitrogen or sulfur linkages. Volatilization is a major factor in their dissipation from soil. Once incorporated, biodegradation is believed to be important. Their mechanism of degradation has not been demonstrated; however, some evidence suggests that their soil persistence increases with increasing complexity of the alkyl groups (54).

Kaufman (55) investigated the biodegradation of phenylcarbamates with mixed populations and pure cultures of soil microorganisms, whereas Kearney (56) examined the effect of phenylcarbamate structure on enzymic hydrolysis. Generally, their results are in close agreement. Both alkyl and ring substitution affect decomposition of phenylcarbamates. The rate of decomposition decreases with increased number of carbon atoms and complexity of the side chain. Position, type, and number of ring substituents also affected decomposition. Studies of the effect of electron-withdrawing groups on the ring have mechanistic implications among various systems. The electron-withdrawing groups on the ring lower the electron density in the vicinity of the carbonyl carbon and thereby facilitate hydrolysis. Kearney (56) found reaction rates to decrease with the following *meta* substituents on the ring: $NO_2 > CH_3CO > Cl > CH_3O > H$. These results suggest that both similarities and differences may exist in structure–activity–degradability-relationships of this group.

D.   Organophosphates.   The insecticidal activity of the organophosphate insecticides was first realized by G. Schrader in 1936 with the discovery of the activity of tetraethyl pyrophosphate (TEPP). In 1965, there were approximately 40 commercial organophosphorus compounds with structures based generally on the formula

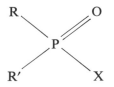

where R and R′ are short chain alkyl, alkoxy, and alkylthio or amido groups (57) and X is a labile leaving group.

The effectiveness of an organophosphate insecticide depends on its ability to inhibit cholinesterase (58). The effectiveness of an insecticide is not necessarily directly correlated with its *in vitro* activity as an inhibitor but subsequent metabolism may convert an apparently weak inhibitor to an extremely active compound. For example, parathion (*O,O*-diethyl *O*-*p*-nitrophenylphosphorothionate) and other organophosphates containing a P—S link are converted to the much more active P—O bond *in vivo*. Conversely, a highly effective inhibitor may not penetrate effectively to the enzymic site of action (*e.g.,* passage of insecticides is impaired by highly polar groups) or may be decomposed before reaching the site of action (*e.g.,* a systemic insecticide may be hydrolyzed by plant juices).

Inhibition at the enzymic surface occurs through interaction of a cationic substrate which binds to the enzyme through a combination of Van der Waals and coulombic forces. At an adjacent locus on the enzyme surface, hydrolytic cleavage of the ester occurs with the formation of a new covalent bond from the enzyme to a phosphorus atom.

Two alternative criteria for effectiveness of an organophosphate insecticide have been established. The first group of compounds contain a suitably reactive bond at the phosphorus atom or a bond capable of being transformed to a reactive bond. The second group exhibits its activity through its resemblance to acetylchloline, either by possessing ester and cationic groups separated by the requisite distance or by having a similar spatial arrangement of atoms and bonds. These criteria have been applied separately to the design of organophosphorus insecticides. Combination of these principles has also produced potent inhibitors.

The stability of the insecticide and its activity towards specific insects depends on the nature of the substituents. Either broad spectrum or highly specific insecticides are possible and the duration of activity may also be prolonged by a suitable choice of substituents. Some phosphates have high mammalian toxicity whereas the toxicity of others is so low that they can be used in feeds as systemic insecticides. Systemic insecticides for animals or plants are sufficiently stable to resist enzymic breakdown or hydrolysis by biological fluids. The advantage of plant systemics is that translocation in lethal amounts minimizes the inequalities of spray coverage. Other advantages are the protection afforded to new growth after application and the reduction in injury to beneficial insects.

Phosphate insecticides display a wide range of stability towards environmental degradation but are generally regarded as not persistent. However, it should be borne in the mind that the departure of a leaving group X by hydrolytic cleavage leaves, in addition to a phosphoric acid residue, an organic compound which may not be readily subject to further breakdown.

The rates of hydrolysis of organophosphates are normally dependent on pH. Parathion is hydrolyzed slowly in water, as 50% is decomposed after 120 days, whereas at mild alkaline pH in lime-water 50% is decomposed in 8 hours. Dichlorvos (DDVP) (2,2-dichlorovinyl dimethylphosphate) is 50% hydrolyzed at pH 7 in 8 hours (57).

Malathion [*O,O*-dimethyl *S*-(1,2-dicarboxyethyl) phosphorodithioate] is hydrolyzed relatively rapidly at pH 7 or 5 (57). The soil fungus *Trichoderma viride* decomposed malathion by hydrolysis of the ester groups to carboxylic acid functions (Figure 8.2). Hydrolysis of the *O*-methyl group attached to phosphorus also appeared probable (59). The attack on the

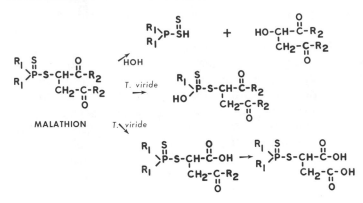

Figure 8.2 Microbial degradation of malathion. $R_1 = CH_3O^-$ and $R_2 = {}^-OC_2H_5$.

carboxylate group represents an alternative degradation pathway, transformation of the residue X, which occurs before hydrolysis of the P—X linkage. The rate of hydrolysis is slow.

Three other degradation pathways of organophosphates are of sufficient generality for inclusion: isomerization (*e.g.,* R—O—P(OR)$_2$ → R—S—P (OR)$_2$; conversion of P—S to P—O; and dealkylation reactions.

The conversion of P—S to P—O is exemplified by the conversion of parathion to paraoxon. Oxidation results in an increase in both water solubility and in anticholinesterase activity.

Under moist conditions the organophosphate insecticides do not generally persist for any considerable length of time in soil. There are exceptions, however, and reports of parathion in subterranean water and guthion [*O,O*-dimethyl *S*-4-oxo-1,2,3-benzotriazin-3-(4*H*)ylmethyl phosphorodithioate] accumulation in fish point to the need for continual monitoring of all pesticides (60, 61).

E. The Hansch Approach.   The recent investigations of Hansch and his associates (62–69) have provided considerable new insight into our understanding of chemical structure–activity relationships. Explanations of results in earlier investigations have generally been limited to consideration of steric and electronic effects. Using substituent constants and regression analysis, they (62–69) have demonstrated that the substituent effects can be factored into three groups: electronic, steric, and hydrophobic. They have developed an equation using experimentally based variables (*Es*, σ, and π) for correlating the effect of a given substituent on the biological activity of a parent compound. *Es* is the steric substituent constant, σ is the Hammett substituent constant and π is an analogous constant representing the difference in the logarithms of the partition coefficients of the substituted and unsubstituted compounds ($\pi = \log P_X - \log P_H$) in octanol and water. The value of their equations has been tested on numerous types of biologically active compounds including antibiotics, pesticides, growth regulators, and carcinogens. Substituent characteristics of importance to biological activity of several pesticide classes are shown in Table 8.1.

Table 8.1 Substituent Characteristics of Importance to Biological Activity of Selected Pesticides. Correlations Based on the Hansch Approach.

Chemical Class	Activity	Principal Effective Substituent Characteristic	Ref.
Methylcarbamates	Cholinesterase inhibition	Hydrophobic	69
Diethylphenylphosphate	Cholinesterase inhibition	Electronic	65, 68, 69
Alkylphosphonic acid esters	Cholinesterase inhibition	Electronic	69
2,4,5-Trichlorophenyl-*N*-alkylphosphoramides	Cholinesterase inhibition	Electronic–steric	69
Phenylureas	Hill reaction	Hydrophobic	67
Phenylcarbamates	Hill reaction	Hydrophobic	67
Acylanilides	Hill reaction	Hydrophobic	67
Phenoxyacetic acids	Phytohormone	Electronic	63, 64

In general, their correlations have proven exceptionally good. They have successfully demonstrated the feasibility of predicting biological activity of chemicals within a given class by basing their approach on extrathermodynamic considerations. They contend that their approach can (*a*) provide clues to the mechanism of action, (*b*) correlate biological activity with chemical constitution, (*c*) delineate the stereospecific nature of hydrophobic bonding between enzyme and substrate, (*d*) afford the first means of separating with some assurance the steric effects and lipophilic binding effects, and (*e*) determine the effect of substituents on reaction rates during enzymatic hydrolysis.

Apparently the application of their approach has been limited to already established or existing classes of chemicals showing bioactivity. The authors are aware of no instance in which this approach has been used in the development of new, previously untried, pesticides. The primary requirement of any pesticide is that it controls or eradicates the target pest.

Investigations leading to the selection of our present day pesticides have been largely empirical. Classes of compounds having the necessary toxicity have emerged only through massive and costly screening programs. Structure–activity studies within these classes of compounds have led to further refinement in choice of pesticides for selective control of species within a population. Such processes have been laborious, time consuming, and quite costly. The real value of the Hansch approach, therefore, would appear to be in the same selective development of new pesticides. The selection of specific unique processes within the target organism and the design of pesticides for that process should be a feasible objective with such an approach. Incorporated into such an approach should be an awareness of "structure–activity" relationships relative to the biodegradability of the pesticide. As seen in preceding discussions, many of the structure–activity relationships important for toxicity are similarly important to biodegradability. Similar trends occur within other classes of pesticides, but notable exceptions exist. Although sufficient structure–activity–biodegradability correlations are limited for the chlorinated hydrocarbons, these excellent insecticides are largely recalcitrant to degradation. Methoxychlor is one of the few chlorinated hydrocarbon insecticides that is readily degraded in our environment.

F. Hard and Soft Linkages.   Some generalizations can be made concerning the effect of various substituents and linkages. Halogens are a common entity to many pesticides. Chloride, bromide, fluoride, or iodide is an integral part of 71 of the 116 common herbicides currently recognized by the Weed Science Society of America. The early, more persistent insecticides were nearly all chlorinated, that is, the chlorinated hydrocarbons. Halogenated compounds are also common among the nematicides, fungicides, miticides, and molluscicides.

In general, the number, position, and type of halogen substituents affect the rate of microbial decomposition of pesticides. The relative importance of these three characteristics is somewhat dependent upon the class of compounds. MacRae and Alexander (40) concluded that number, rather than position, of chlorines on the aromatic ring determined the susceptibility of benzoates to microbial decomposition. Audus (35), however, demonstrated that with the phenoxyalkanoates the position of chlorines could be more important than the number.

The effects of other substituents on pesticide degradation are also similar from one chemical class to another. Substitution of alkyl groups for halogens tends to increase the resistance of pesticide molecules to environmental degradation. The replacement of the *ortho* chlorine of 2,4-D with a methyl group (MCPA) increases the resistance of this molecule to microbial degradation (35). Substitution of methylthio or methoxy groups for chlorine in the *s*-triazines also increases persistence (70) and resistance to microbial degradation (11). In contrast, substitutions of alkoxy or hydroxy groups in the DDT molecule, for example, methoxychlor [1,1,1-trichloro-2,2-bis(*p*-methoxypheny 1)-ethane], dicofol [1,1-bis(*p*-chlorophenyl)-2,2,2-trichloroethanol], chloropropylate (isopropyl 4,4′-dichlorobenzilate), and chlorobenzilate (ethyl 4,4′-dichlorobenzilate), are more susceptible to degradation (71, 72). Increasing the size of alkyl substituents generally leads to a greater persistence and recalcitrance (55). Although replacement of chlorine groups with iodine or bromine groups generally increases resistance to degradation (55), some problems have been encountered with bromine toxicity to crops following repeated usage of brominated pesticides (73).

Some substituents may themselves be subject to alterations prior to their removal in the degradative process. Under certain conditions the —NO$_2$ groups of the toluidines (74) and PCNB (75) may undergo reduction prior to further degradation of the parent molecule. Nitrile groups are frequently converted through an amide to a carboxyl group (76). The methylthio group of methylmercapto-*s*-triazines progresses to the sulfoxide and the sulfone (79) (Figure 8.3). The —CF$_3$ group has found usage in

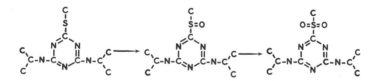

Figure 8.3   Progressive oxidation of prometryne.

several of the newer pesticides, for example, fluometuron[1,1-dimethyl-3-(*a,a,a,*-trifluoro-*m*-tolyl)urea], trifluralin[*a,a,a,*-trifluoro-2,6-dinitro-*N*-*N*-dipropyl-*p*-toluidine], and benefin[*N*-butyl-N-ethyl-*a,a,a,*-trifluoro-2,6-dinitro-*p*-toluidine]. The specific nature or contribution of this group to pesticide activity is not completely understood. In the case of trifluralin, it is eventually converted to a carboxyl group.

Hansch *et al.* (63) have suggested the use of a —SF$_5$ substituent. According to their observations, its high σ and π values should make it very useful for increasing biological activity of molecules having low lipophilic character. The group is apparently quite resistant to metabolic change and has no special toxic qualities.

Kearney and Plimmer (78) suggested that in light of present information several linkages were known to be readily susceptible to degradation and listed the following linkages:

1. $R-NH-CO_2 R^1$
2. $R-NH-CO R^1$
3.
$$\begin{array}{c} -O \\ \diagdown \\ P = S - R \\ \diagup \\ -O \end{array}$$

4.
$$\begin{array}{c} -O \\ \diagdown \\ P = O - R \\ \diagup \\ -O \end{array}$$

5. $R-CHCl-COO-$
6. $R-CCl_2-COO-$

Carbamates, anilides, phosphates, and aliphatic acids generally are degraded in a reasonable time following application. The rate at which linkages are hydrolyzed will depend on the nature of R and R^1. For example, replacement of the alkoxy group of a phenylcarbamate with an alkylamine group results in a phenylurea, which is generally more persistent than the phenylcarbamate.

Other linkages might also be included under specific conditions. Although in the methoxy-s-triazines, chloroxuron [3-(p-[p-chlorophenoxy] phenyl)-1,1-dimethylurea], dicamba [3,6-dichloro-o-anisic acid], and tricamba [3,5,6-trichloro-o-anisic acid], the ether linkage appears relatively stable in soils; in the phenoxyalkanoates it is generally susceptible to attack. It would seem logical with these guidelines to explore additional structural variations of these types in future development of soft pesticides.

In addition to considering the chemical structure–activity–degradability relationships within individual chemical classes of pesticides, there is some advantage to a general consideration of the processes involved in degrading the various chemical classes (Table 8.2). Biochemical processes known to be involved in the degradation of pesticides in soils include: dealkylation, dehalogenation, amide or ester hydrolysis, oxidation, reduction, ring cleavage, and condensate or conjugate formation (49). Several observations can be made. Pesticides whose initial degradative reaction is ester hydrolysis are relatively short-lived in soil. Pesticides which initially undergo dealkylation generally tend to be somewhat more persistent. Pesticides which are initially dehalogenated are variable in their persistence, apparently depending on the complexity of the basic molecule. Halogenated aliphatic acids are readily degraded and thus not very persistent, whereas chlorinated hydrocarbons are generally quite persistent. Halogenated benzoic acids appear intermediate in their persistence. Such ob-

Table 8.2 Relative Persistence and Initial Degradative Reactions of 11
Major Pesticide Classes

Chemical Class	Use	Persistence	Initial Degradative Process
Chlorinated hydrocarbons	Insecticides	2–5 years	Dehydrohalogenation or epoxidation
Ureas	Herbicides	4–10 months	Dealkylation
Triazines	Herbicides	3–18 months	Dealkylation or dehalogenation
Benzoic acids	Herbicides	3–12 months	Dehalogenation or decarboxylation
Amide	Herbicides	2–10 months	Dealkylation
Phenoxy	Herbicides	1–5 months	Dealkylation, ring hydroxylation, or β-oxidation
Toluidine	Herbicides	6 months	Dealkylation (aerobic) or reduction (anaerobic)
Nitrile	Herbicides	4 months	Reduction
Carbamate	Herbicides, fungicides, insecticides	2–8 weeks	Ester hydrolysis
Aliphatic acids	Herbicides	3–10 weeks	Dehalogenation
Organophosphates	Insecticides	7–84 days	Ester hydrolysis

servations, although useful, may not be the ultimate solution to attaining
degradability within a given class of compounds. Other physicochemical
characteristics of the molecule and its environment may preclude a given
chemical's availability for degradation.

Synthetic reactions (condensate or conjugate formation) should also be
a concern in the development of pesticides. Numerous pesticides are
known to form conjugates with various sugars and amino acids. Metabolites of other pesticides may recondense and form new, possibly hazardous
residues. The ultimate fate and toxicity of such biosynthetic products of
pesticides should also be determined.

G. Soil Sorption. In the synthesis of "soft pesticides" attention
should also be given to the effect of chemical structure on sorption
and movement of pesticides in soil. The Lambert (80) approach to the
soil-sorption phenomenon is similar to the Hansch approach to structure–
activity relationships. By utilizing an equilibrium which considers soil-
organic matter as the sorbing medium, Lambert developed a functional
relationship between soil sorption and chemical structure. The relationship
is based upon extrathermodynamic linear free energy approximations and

uses the parachor as an approximate measure of the molar volume of the chemical under consideration. According to Lambert (80) the significance of this correlation is that the distribution coefficients which describe sorption equilibria for certain classes of compounds are predictable functions of molecular structure. For such compounds one can predict where in the soil the chemical will reside under the influence of specific environmental conditions. The importance of this type of prediction depends upon the fact that the sorption of a pesticide in the soil mediates its biological activity and that the loci of a chemical in the soil will determine what factors are capable of operating upon it to effect its disappearance.

8-3.   Hazard Potential

Pesticides vary in their toxicity to man, but the greatest danger lies in accidental exposure to large doses of a toxic chemical. Some of the organophosphates are extremely toxic and require experienced operators for their safe application. There is no direct relationship between persistence in the environment and toxicity to man, but the combination of persistence and even a low degree of toxicity presents dangers through unforeseen possibilities of biological concentration and subsequent entry into man's food chain.

The development of herbicides might be directed towards chemicals capable of interfering with a target system uniquely present in higher plants, such as the photosynthetic process. The selection of a target system peculiar to the organism would reduce the chances that other species might be affected. However, even if a pesticide is highly specific in its action, in the absence of full knowledge of its fate in the environment and its acute and chronic effects on wildlife and other organisms, a conservative policy of use provides some safeguard against unpredictable effects.

Practically all pesticides have toxic properties which kill or upset some living organism. Three factors of risk may be recognized: (a) the ability of certain pesticides to appear far from the site of their application; (b) the property of pesticides to persist and accumulate in soils, plants, animal tissues, and other locations; and (c) our ignorance of all risks and consequences to man and his environment which may result from their use. We may attempt to minimize these hazards by further development along the lines suggested in this chapter; but whether pesticides are "soft" rather than "hard," their use must be accompanied by the same studies of toxicology and fate in the environment.

Toxicological problems in pesticide use have arisen largely from accidental exposure through spills in handling or application. Many have involved children. The problem of education in safe handling, storage, and working practices must still continue whatever the safety margins are. The use of safer pesticides will diminish, not eliminate, the risks.

Hayes (81) has listed three recognized effects of pesticides in humans: mortality, morbidity, and storage. He states that in certain instances storage may not be harmful. However, until we are absolutely certain that there are no harmful consequences from low levels of pesticides of relatively low acute toxicity stored in man, exposure should be minimized. Clearly, one consequence is the possible accumulation in tissue until potentially dangerous levels are reached.

The death rate due to accidental poisonings has shown no significant change since the introduction of organic pesticides. Hazards occur when food or clothing has been contaminated by concentrated formulations of pesticides. Transportation and application of pesticides appear to offer situations of greatest danger. To reduce the occurrence of poisoning, it would appear preferable to control situations in which pesticides are used rather than to develop pesticides of lower mammalian toxicity than those currently available.

This is not to say that the factors which affect pesticidal toxicity to man should not be studied. O'Brien (82) has commented on the safety of the chlorinated insecticides compared with the organophosphates and suggests that the inability of the former to pass the dermal barrier may be signicant in this respect: Research along these lines would be of value. Some chlorinated insecticides are quite toxic; others are neither toxic nor persistent (*i.e.,* their presence is not revealed in environmental monitoring studies). A general condemnation of the chlorinated insecticides should be tempered by the knowledge that some members of this group are effective insecticides of low toxicity to man and low environmental persistence.

The long-term effects of pesticides have not been overlooked. For the development of new pesticides, in addition to establishing low acute toxcity, it is essential that the effect of the compound on mammalian cell division should be thoroughly investigated. Carcinogenesis, mutagenesis, and teratogenesis may be undesirable side effects of synthetic organic compounds and their metabolites. A long testing period is necessary to demonstrate these potential hazards.

A number of chemicals have been identified as agents responsible for the production of human cancer. Many others have been shown to produce tumors in experimental animals. There is no structural pattern common to chemical carcinogens, and in the absence of a valid relationship between structure and carcinogenesis there is a need to subject every compound which may potentially contaminate man's food supply to testing in experimental animals. As the process of carcinogenesis is slow, identification of positive effects may require a considerable portion of the life span of the test animal. Long-term studies of this type may be an important factor in the development of a pesticide.

A number of groups of organic and inorganic chemical carcinogens have been identified. Industrial law in Great Britain has recognized their significance. Mineral oils used in cotton spinning are carcinogenic. The maufacture, presence, and industrial use of 2-naphthylaminobenzidine, 4-aminobiphenyl, 4-nitrobiphenyl, and their salts were prohibited by the Carcinogenic Substances Regulations Act of 1967. The employment of personnel in the manufacture and use of 1-naphthylamine, o-tolidine, dianisidine, dichlorobenzidine, and their salts were also regulated. These amines had been recognized as carcinogens for some time and their use in industry led to extremely high incidence of bladder cancer in exposed workers (83).

The earliest carcinogens recognized were complex aromatic hydrocarbons. It seems unlikely that these will be of potential use in pesticide synthesis or that they will occur as metabolites, but other groups of carcinogens bear a closer relationship to pesticidal structures. Urethane (ethyl carbamate) is carcinogenic, as are many N-nitroso compounds. Alkylating agents such as diazomethane, β-propiolactone, and several compounds which can generate cross linkages between macromolecules by addition to a reactive ethylene oxide or ethyleneamine ring are also carcinogens (84). Several azo dyes, such as 4-dimethylaminoazobenzene (formerly used as a food coloring), are carcinogenic. The structural requirements for carcinogenicity of some related azo compounds have been discussed by Williams (85).

The aromatic amines, alkylating agents, and N-nitroso compounds comprise large groups. Many other instances continually arise of synthetic and naturally occurring chemical carcinogens which bear no relationship to previously recognized carcinogens. Cycasin and aflatoxin are but two examples of naturally occurring carcinogens. Some inorganic compounds have also been implicated as carcinogens: elemental cadmium, cobalt, nickel (as powders), mercury, some compounds of lead, chromium, arsenic, beryllium, and other inorganic substances are suspected of being carcinogenic (84).

The relation between mutagens and carcinogens remains unclear. Certain biological alkylating agents and ethylcarbamate have both properties, but the correlation is not general and many mutagens do not appear to be carcinogens. However, the implications of mutagenicity and its undefined relationship with carcinogenesis suggest that the effects of pesticides on biological processes should be constantly subject to critical review and evaluation.

Many compounds are currently being evaluated for teratogenicity in test animals. The extrapolation to man of results obtained with test animals is not always reliable (86). It is essential that the problem of teratogenicity receive more critical attention.

8-4. Application and Formulation

Since a pesticide is defined as an economic poison, a safe but effective dose, with no excess, should destroy the target organism. The rate of degradation of a pesticide by environmental forces should preferably be rapid, but such a desirable characteristic is of little value if the pesticide is used inefficiently.

The method of application must be designed to make a pesticide available at the site of action in adequate concentration and for a sufficiently long period of time to function effectively. The process of application requires a background in a variety of technologies such as chemistry, physics, engineering, physiology, and biochemistry. Methods of application are many. For example, herbicides may be applied as sprays and dusts or they may be incorporated in soil as granules. Insecticides require a wider range of techniques based on consideration of the life cycle and habitat of the target insect.

To avoid the liberation of large quantities of pesticide into a particular environment, controlled release techniques are used to maintain low effective doses of nonpersistent pesticides in a given situation. The pesticide is incorporated or encapsulated in an inert medium from which it is slowly released to provide a toxic concentration. In this way nonpersistent pesticides may be protected from breakdown but at no time is there an extremely high concentration available to disturb the ecological balance or overwhelm natural mechanisms of environmental degradation.

Pesticides are rarely used alone or applied in their pure chemical form. They are generally applied as "formulations." Formulations may include either combinations of two or more pesticides or individual pesticides in combination with one or more adjuvant materials. The formulation of a pesticide is as important as the selection of the active ingredient. The isooctyl and butylether ester formulations of 2,4-D are more susceptible to washoff than the amine salts (87). Phenoxyesters that contain five carbons or less in the alcohol moiety are considered "highly volatile," whereas those that contain more than five are "low volatile" esters (88). The isooctyl, the propylene glycol butyl ether, and the butoxyethanol are low volatile esters. These are about 10 to 20 times less volatile than the lower alkyl esters of 2,4-D, which include the isopropyl and N-butyl ester (89). Thus it is important to use a suitable formulation for the conditions of the job to be done.

Both beneficial and deleterious interactions may occur when two or more pesticides are used in combination. Deleterious interactions may result from increased toxicity toward nontarget organisms, increased residual life in soil, or pesticide inactivation. These interactions are critical to safe usage of pesticides and are generally alleviated with adequate experimentation. Nash (90) and others have made extensive studies of deleterious interactions. Beneficial interactions may result from increased or prolonged toxicity toward target organisms, or enhanced degradation. Considerable attention should be given to capitalizing on these types of interactions.

8-5. Ecological Effects

The ecological effect of pesticides is a controversial subject of current concern. There is literally nothing man can do that will not in some way measurably change his environment. His very existence is the result of successive ecological changes in the environment which favored his development, and having existed he will unavoidably affect future ecological changes. His very footstep may temporarily alter the ecology of soil micro-organisms by virtue of the soil compaction and subsequent reduced aeration it may have caused. The conversion of virgin soils to cropped land itself has caused far greater ecological changes than those caused by most pesticides. Throughout his development man has affected his environment by perfecting agricultural and industrial practices that enable his continued existence in increasing comfort. Man, however, is not the only consumer of agricultural and industrial goods. Quite naturally, thousands of other living organisms (pests) also respond with increasing numbers to the increased food supply provided by mass production in concentrated areas. Thus if man is to avoid succumbing to the Malthusian principle, he must learn to control not only himself but his competitors as well.

By deliberate design pesticides are used to alter the ecology, that is, elimination or restriction of undesirable species in favor of species considered desirable for man's continued existence. The ubiquitous nature of so many biological and biochemical processes make it unlikely that even highly specific pesticides will not affect some other nontarget organisms. It is therefore imperative that man determine what ecological changes pesticides may produce, which changes are permanent or temporary, and decide which are acceptable or unacceptable. The ultimate problem is not perpetuation of a specific community for abstract reasons, but rather maximizing productivity while minimizing environmental pollution.

The important role of microorganisms in cyclic processes (nitrogen cycle, carbon cycle, sulfur cycle, etc.) that are essential to continued functioning of nature's ecosystem demands careful study of the effects of pesticides to avoid permanent disruption. The intricate relations of these delicately balanced systems are extensively described in microbiology (91) and biochemistry texts (92). A short but excellent discussion of the subject is given by Audus (93) in his treatment of the effects of herbicides on soil microorganisms. Other reviews have been prepared by Fletcher (94–96), Newman and Downing (97), Bollen (98), Audus (93). The reader is also referred to the sections in reviews on the behavior of chlorinated aliphatic acids (99) and s-trazines (70). From these discussions it is easy to recognize the vast range of possible effects on microorganisms when the number of beneficial and pathogenic organisms are considered. Careful examination of these reviews reveals that with a few exceptions concentrations of pesticides that result from normal rates of application for pest control have no known long lasting effect on microorganisms. Transitory effects are frequently observed. Populations of some organisms are frequently stimulated, whereas others are inhibited. Domsch (100) suggested that the ecological significance of any temporary change in the composition of microbial populations can only be judged when the role of groups of microorganisms within the ecosystem is properly identified and quantified. Interruptions of the energy flow and inhibition of essential functions within the ecosystem are important criteria for the secondary effects of pesticides. Most microorganisms are enzymatically well equipped. The elimination or reduction of pesticide sensitive members generally triggers a sequence of replacements within the population, resulting in the preservation of the soil's metabolic integrity. This, therefore, should be one of the major criteria in the selection and use of pesticides whose ultimate fate rests in the soil environment. Pesticides which permanently alter or destroy the soil metabolic integrity should be avoided or severely restricted in their usage.

When considering the effect of pesticides on microorganisms it is important to consider just where in the environment the effect may be occurring. Chemical and physical properties of both the environment and the pesticide may interact to limit any ecological effect of a given pesticide. Pesticides strongly absorbed to soil particles may remain concentrated in a limited portion of the soil profile, thus representing a concern only in a small confined area. DDT, for example, is not readily leached and remains at or near the soil surface when applied to soil. Repeated applications of DDT to orchard soils has resulted in DDT concentrations of several hundred parts per million in soil surface layers ($\frac{1}{2}$–1 in.). Although these DDT concentrations are far in excess of those known to inhibit nitrification in soil, the confinement of DDT to such limited areas by soil absorption sharply reduces the overall effect of DDT throughout the soil environment of greatest microbiological significance (0–6 in., or plow-depth).

Certain of the important microbiological processes can be temporarily alleviated by such practices as fertilization and disease control, whereas others necessitate extreme caution, e.g. in legume crops where nodulation is a necessary prelude to symbiotic nitrogen fixation. Similarly, pesticides inhibiting mycorrhizal fungi may prevent establishment of the mycorrhizal symbiosis.

Benefits in addition to control or eradication of the target organism are frequently encountered with the use of some pesticides. For example, increased plant disease control has been observed with the use of some herbicides (101, 102).

Such information, combined with knowledge on application time, concentration, and method, could lead to a significant reduction in the number, type and application rate of pesticides needed for controlling or eradicating multiple pest problems.

Rotation of pesticides used in the control of pests in monocrop situations has several ecological advantages. It avoids perpetuating indefinitely any deleterious effects of the pesticide on nontarget organisms. It would reduce the rate at which undesirable residues, if any, may accumulate. It also could facilitate control of additional pests not sensitive to, or which have developed resistance to, other commonly used chemicals, thus preventing their buildup and becoming major pest problems. Disadvantages could come in the creation of multiple-pesticide residue problems, or increased damage to the ecosystem already sensitive to the stress created by the first pesticide. Such advantages and disadvantages must be continually monitored and evaluated to maintain a pollution-free environment.

8-6. Summary

The manufacture and use of pesticides has changed considerably within the last 30 years. Whereas broad spectrum biological activity and prolonged environmental persistence were early criteria for desirable pesticides, the criteria for today are quite different. Today a pesticide must be ecologically desirable or acceptable, degradable within the effective growing season, and have a narrow spectrum of biological activity. The early insecticides such as DDT, lindane (1,2,3,4,5,6-hexachlorocyclohexane), aldrin (1,2,3,4,10,10-hexachloro–1,4,4*a*,5,8,8*a*-hexahydro-1,4-*endo-exo*-5,8-dimethanonaphthalene), dieldrin (1,2,3,4,10,10-hexachloro-6,7-epoxy-1,4,4*a*,5,6,7,8,8*a*–octahydro-1,4-*endo-exo*-5,8-dimethanonaphthalene), heptachlor (1,4,5,6,7,8,8-heptachloro-3*a*,4,7,7*a*-tetrahydro-4,7-methanoindene), and toxaphene all had low mammalian toxicity, but were generally effective against both beneficial and destructive insects. They are also resistant to degradation and subject to biomagnification. Although parathion and methylparathion are readily degraded, they are toxic to all forms of animal life. Carbaryl and malathion combine both degradability and low toxicity to mammals, but their spectrum of activity against invertebrates is not sufficiently narrow. Many of the herbicides can be fitted into similar categories. Although some of the fungicides and nematicides are toxic to higher animals, they do not present serious problems with respect to resistance to degradation or long-term residual effects.

The search for soft pesticides has barely begun. Unfortunately, progress in the development of pesticides has not always been accompanied by total information. Insecticide research has been strong in establishing mode of action and the modification of anticholine esterase inhibitors. Very little information is available describing degradability or the overall ecological effects of most insecticides. Herbicide research has revealed some information regarding site and mode of action, but has been stronger in ecological, persistence, and degradation information. Correlations of structural requirements necessary for activity with those enhancing degradability generally have not been made.

The rapidly developing field of pesticide biochemistry, ecology and toxicology have suggested numerous avenues of research and development capable of leading to the discovery of compounds having the desired characteristics. Industry annually evaluates 100,000 new organic compounds as potential pesticides (4). With the current knowledge in degradability and toxicology we should be readily capable of selecting pesticides with suitably narrow specific activity. It has been estimated that there are at least 25,000,000 potentially effective organophosphate insecticides (4). Application of the Hansch approach could provide the rationale for selection of the most suitable candidates. Ultimate suitability can only be determined through actual use accompanied by continued monitoring of the environment to evaluate accumulative effects.

Although many problems remain, there are a number of lessons to be learned from past experience. Several aspects should be considered in future development of and use of pesticides:

1. The fate of a pesticide in the environment should not be a matter for conjecture. Its mechanism and rate of breakdown in a variety of situations should be known and the major metabolites and transformation products should receive the same study as the parent compound.

2. Increased effectiveness of a pesticide will result in a decrease in the amount used. "Effectiveness" includes a highly selective response by the target organism, efficient application near the site of action, and a maximum response to a minimum dose.

3. An effective monitoring system linked with a thorough study of the toxicology of the pesticide and its possible transformation products provide safeguards for man and wildlife. In this study, the environmental fate of contaminants that may arise during a manufacturing process should also be given consideration.

REFERENCES

1. E. G. Sharvelle, *The Nature and Use of Modern Fungicides*, Burgess, Minneapolis, Minn., 1961.
2. W. Perkow, *Die Insectizide*, Huther, Heidelberg, Germany, 1956.
3. C. A. Edwards, *Critical Reviews in Environmental Control* (R. C. Adams, Jr., Ed.), Vol. 1, Chemical Rubber Co., Cleveland, Ohio, 1970.
4. Report of President's Science Advisory Committee, *Restoring the Quality of Our Environment*, The White House, Washington, D.C., 1965.
5. Weed Science Society of America, *Herbicide Handbook of the Weed Society of America*, Humphrey, Geneva, New York, 1967.
6. J. L. Hilton, L. L. Jansen, and H. M. Hull, *Ann. Rev. Plant Physiol.* **14**, 353 (1963).
7. J. L. Hilton, L. L. Jansen, and W. A. Gentner, *Plant Physiol.* **33**, 43 (1958).
8. J. L. Hilton, *Science* **128**, 1509 (1958).
9. J. L. Hilton, J. S. Ard, L. L. Jansen, and W. A. Gentner, *Weeds* **7**, 381 (1959).
10. J. L. P. van Oorschot and J. L. Hilton, *Arch. Biochem. Biophys.* **100**, 289 (1963).
11. D. D. Kaufman, Unpublished Data (1971).
12. C. L. Foy, in *Degradation of Herbicides* (P. C. Kearney and D. D. Kaufman, Eds.), Dekker, New York, 1969.
13. P. Hirsch and M. Alexander, *Can. J. Microbiol.* **6**, 241 (1960).
14. J. K. Leasure, *J. Agr. Food Chem.* **12**, 40 (1964).
15. P. C. Kearney, D. D. Kaufman, and M. L. Beall, Jr., *Biochem. Biophys. Res. Commun.* **14**, 29 (1964).
16. P. Goldman, G. W. A. Milne, and D. B. Keister, *J. Biol. Chem.* **243**, 428 (1968).
17. C. G. Gemmell and H. L. Jensen, *Arch. Mikrobiol.* **48**, 386 (1964).
18. R. L. Wain, *Advan. Pest Control Res.* **2**, 263 (1958).
19. J. B. Koepfli, K. V. Thimann, and F. W. Went, *J. Biol. Chem.* **122**, 763 (1937).

20. H. Veldstra, *Enzymologia* **11**, 97 (1944).
21. H. Veldstra, *Biochim. Biophys. Acta* **1**, 364 (1947).
22. R. M. Muir, C. H. Hansch, and A. H. Gallup, *Plant Physiol.* **24**, 359 (1949).
23. C. H. Hansch and R. M. Muir, *Plant Physiol.* **25**, 389 (1950).
24. J. M. F. Leaper and J. R. Bishop, *Botan. Gaz.* **112**, 250 (1951).
25. A. E. Hitchcock and P. W. Zimmerman, *Contrib. Boyce Thompson Inst.* **12**, 497 (1942).
26. C. H. Fawcett, R. L. Wain, and F. Wightman, *Ann. Appl. Biol.* **43**, 342 (1955).
27. C. H. Fawcett, D. J. Osborne, R. L. Wain, and R. L. Walker, *Ann. Appl. Biol.* **40**, 232 (1953).
28. H. Veldstra and H. L. Booij, *Biochim. Biophys. Acta.* **3**, 278 (1949).
29. M. A. Loos, In: *Degradation of Herbicides.* (P. C. Kearney and D. D. Kaufman, Eds.), Dekker, New York, 1969.
30. H. F. Taylor and R. L. Wain, *Proc. Roy. Soc., Ser. B* **268**, 172 (1962).
31. C. H. Fawcett, J. M. A. Ingram, and R. L. Wain, *Proc. Roy. Soc., Ser. B* **142**, 60 (1954).
32. D. M. Webley, R. B. Dugg, and V. C. Farmer, *Nature* **183**, 748 (1959).
33. American Chemical Society Committee on Chemistry and Public Affairs, *Cleaning Our Environment. The Chemical Basis for Action*, A Report by Subcommittee on Environmental Improvement, American Chemical Society, Washington, D.C., 1969.
34. D. M. Webley, R. B. Dugg, and V. C. Farmer, *J. Gen. Microbiol.* **18**, 733 (1958).
35. L. J. Audus, in *Herbicides and the Soil* (E. K. Woodford and G. R. Sagar, Eds.), Blackwell, Oxford, 1960.
36. L. J. Audus, in *The Physiology and Biochemistry of Herbicides* (L. J. Audus, Ed.), Academic Press, London, 1964.
37. M. Alexander and M. I. H. Aleem, *J. Agr. Food Chem.* **9**, 44 (1961).
38. K. Burger, I. C. MacRae, and M. Alexander, *Proc. Soil Sci. Soc. Amer.* **26**, 243 (1962).
39. M. Alexander and B. K. Lustigman, *J. Agr. Food Chem.* **14**, 410 (1966).
40. I. C. MacRae and M. Alexander, *J. Agr. Food Chem.* **13**, 72 (1965).
41. R. L. Metcalf, *Organic Insecticides, Their Chemistry and Mode of Action*, Wiley-Interscience, New York, 1955.
42. M. Kolbezen, R. L. Metcalf, and T. Fukuto, *J. Agr. Food Chem.* **2**, 864 (1954).
43. J.-M. Bollag and S.-Y. Liu, *Bacteriol. Proc.* **70**, 9 (1970).
44. R. L. Metcalf and T. Fukuto, *J. Agr. Food Chem.* **15**, 1022 (1967).
45. A. H. Haubein and J. R. Hansen, *J. Agr. Food Chem.* **13**, 555 (1965).
46. J. R. Kilsheimer and H. H. Moorefield, *Abstracts*, 139th National Meeting of the American Chemical Society, St. Louis, Mo., 1961.
47. G. Friesen, *Planta* **8**, 666 (1929).
48. W. G. Templeman and W. A. Sexton, *Nature* **156**, 630 (1945).
49. W. C. Shaw and C. R. Swanson, *Weeds* **2**, 43 (1953).
50. D. E. Moreland and K. L. Hill, *J. Agr. Food Chem.* **7**, 832 (1959).
51. N. E. Good, *Plant Physiol.* **36**, 788 (1961).
52. T. R. Hopkin, *A Historical Development of Carbyne*, Spencer Chem. Co., Minneapolis, 1959.

53. A. S. Crafts, *The Chemistry and Mode of Action of Herbicides*, Wiley-Interscience, New York, 1961.
54. D. D. Kaufman, *J. Agr. Food Chem.* **15**, 582 (1967).
55. D. D. Kaufman, *ASA Special Publication No.* **8** Soil Science Society of America, 1966.
56. P. C. Kearney, *J. Agr. Food Chem.* **15**, 568 (1967).
57. R. L. Metcalf, in *Kirk-Othmer Encyclopedia of Chemical Technology*, 2nd ed., Vol II (A. Standen, Ed.), Wiley-Interscience, New York, 1966.
58. T. R. Fukuto, *Advan. Pest. Control Res.* **1**, 147 (1957).
59. F. Matsumura and G. M. Boush, *Science* **153**, 1273 (1966).
60. L. E. Mitchell, in *Organic Pesticides in the Environment* (R. F. Gould, Ed.), Advances in Chemistry Series No. 60, American Chemical Society, Washington, D.C., 1966.
61. U.S. Department of Interior, F. and W. S., Bureau of Sport Fisheries and Wildlife Cir. 178 (1963).
62. C. Hansch, *Accounts Chem. Res.* **2**, 232 (1969).
63. C. Hansch, R. M. Muir, T. Fujita, P. P. Maloney, F. Geiger, and M. Streich, *J. Amer. Chem. Soc.* **85**, 2817 (1963).
64. T. Fujita J. Iwasa, *J. Amer. Chem. Soc.* **86**, 1616 (1964).
65. C. Hansch and T. Fujita, *J. Amer. Chem. Soc.* **86**, 1616 (1964).
66. C. Hansch, K. Kiehs, and G. L. Lawrence, *J. Amer. Chem. Soc.* **87**, 5770 (1965).
67. C. Hansch and E. W. Deutsch, *Biochim. Biophys. Acta* **112**, 381 (1966).
68. C. Hansch, E. W. Deutsch, and R. N. Smith, *J. Amer. Chem. Soc.* **87**, 2738 (1965).
69. C. Hansch and E. W. Deutsch, *Biochim. Biophys. Acta* **126**, 117 (1966).
70. C. I. Harris, D. D. Kaufman, T. J. Sheets, R. G. Nash, and P. C. Kearney, *Advan. Pest Control Res.* **8**, 1 (1968).
71. S. Miyazaki, G. M. Boush, and F. Matsumura, *Appl. Microbiol.* **18**, 972 (1969).
72. C. M. Menzies, *Metabolism of Pesticides*, Bureau of Sport Fisheries and Wildlife Special Scientific Report—Wildlife No. 127, Washington, D.C., 1969.
73. J. P. Martin and P. F. Pratt, *J. Agr. Food Chem.* **6**, 345 (1958).
74. G. W. Probst and J. B. Tepe, in *Degradation of Herbicides* (P. C. Kearney and D. D. Kaufman, Eds.), Dekker, New York, 1969.
75. C. I. Chacko, J. L. Lockwood, and M. L. Zabik, *Science* **154**, 893 (1966).
76. M. A. Zaki, H. F. Taylor, and R. L. Wain, *Ann. Appl. Biol.* **59**, 481 (1967).
77. J. R. Plimmer and P. C. Kearney, *Abstracts*, 158th National Meeting of the American Chemical Society, New York, A30 (1969).
78. P. C. Kearney and J. R. Plimmer, in *Pesticides in the Soil: Ecology, Degradation and Movement* (International Symposium on Pesticides in Soil), Michigan State Univ., East Lansing, Mich., 1970.
79. D. D. Kaufman, in *Pesticides in the Soil: Ecology, Degradation and Movement* (International Symposium on Pesticides in Soil), Michigan State Univ., East Lansing, Mich., 1970.
80. S. M. Lambert, *J. Agr. Food Chem.* **15**, 572 (1967).
81. W. J. Hayes, in *Scientific Aspects of Pest Control*, National Academy of Sciences Publication 1402, Washington, D.C., 1966.
82. R. D. O'Brien, *Advan. Pest Control Res.* **4**, 75 (1961).
83. C. E. Searle, *Chem. Brit.* **6**, 5 (1970).

84. J. A. Miller and E. C. Miller, *Carcinogenesis* **15**, 217 (1966).
85. R. T. Williams, *Detoxication Mechanisms,* 2nd ed., Wiley, New York, 1959.
86. H. Kalter, *Supplement to Teratology Workshop Manual,* University of California, Berkeley, Calif., 1965.
87. A. P. Barnett, E. W. Hauser, A. W. White, and J. H. Holladay, *Weeds* **15**, 133 (1967).
88. A. D. Baskin and E. A. Walker, *Weeds* **2**, 280 (1953).
89. G. W. Flint, J. J. Alexander, and H. P. Funderburk, *Weed Sci.* **16**, 541 (1963).
90. R. G. Nash, *Agron. J.* **59**, 227 (1967).
91. M. Alexander, *Soil Microbiology,* Wiley, New York, 1961.
92. A. D. McLaren and G. H. Peterson, *Soil Biochemistry,* Dekker, New York, 1967.
93. L. J. Audus, in *The Physiology and Biochemistry of Herbicides* (L. J. Audus, Ed.), Academic Press, London, 1964, p. 163.
94. W. W. Fletcher, in *Herbicides and The Soil* (F. K. Woodford and G. R. Sagar, Eds.), Blackwell, Oxford, 1960, p. 20.
95. W. W. Fletcher, *Proc. Brit. Weed Control Conf.* **8**, 896 (1966).
96. W. W. Fletcher, *Landbouwk. Tijdschr.* **78**, 274 (1966).
97. A. S. Newman and C. E. Downing, *J. Agr. Food Chem.* **6**, 352 (1958).
98. W. B. Bollen, *Ann. Rev. Microbiol.* **15**, 69 (1961).
99. P. C. Kearney, C. I. Harris, D. D. Kaufman, and T. J. Sheets, *Advan. Pest Control Res.* **6**, 1 (1965).
100. K. H. Domsch, in: *Pesticides in the Soil: Ecology, Degradation and Movement* (International Symposium on Pesticides in Soil), Michigan State Univ., East Lansing, Mich., 1970.
101. R. Heitefuss and H. Bodendorfer, *Z. Pflanzenkrankh. Pflanzenschutz* **75**, 641 (1968).
102. L. T. Richardson, *Can. J. Plant Sci.* **37**, 196 (1957).

Part III
Intestinal Pathogens as Pollutants

9 Water-Borne Pathogens

Edwin E. Geldreich, Division of Water Hygiene, Office of Water Programs, Environmental Protection Agency, Cincinnati, Ohio

9-1. Public Health Significance of Warm-Blooded Animal Pollution
A. Salmonellosis
B. Shigellosis
C. Leptospirosis
D. Enteropathogenic *E. coli*
E. Tularemia
F. Cholera
G. Tuberculosis
H. Human Enteric Viruses
I. Parasitic Protozoa
J. Parasitic Worms
9-2. Transport of Pathogenic Microorganisms to the Stream
A. Treated Sewage Effluents
B. Urban Stormwater
C. Rural Stormwater
D. Wildlife Refuges

9-1. Public Health Significance of Warm-Blooded Animal Pollution

The access of fecal pollution to water may add a variety of intestinal pathogens at any time, and at one time or another, pathogenic organisms will be present in water degraded by a variety of pollutional discharges from warm-blooded animals. The most common pathogens include strains of *Salmonella, Shigella, Leptospira*, enteropathogenic *Escherichia coli, Pasteurella, Vibrio, Mycobacterium*, human enteric viruses, cysts of *Endamoeba histolytica*, and hookworm larvae. *Salmonella* strains have frequently been detected in sewage, streams, irrigation waters, wells, and tidal waters. However, the isolation of the other pathogenic organisms from the water environment has been less frequent, probably because laboratory methods of isolation and identification remain too cumbersome for routine use.

207

The feces of man and the sewage wastes he creates are a major source of the pathogens that are carried in water. Monitoring sewage for pathogens has been demonstrated to be an excellent epidemiological tool for determining what diseases may be prevalent in the community at the moment (1). Human pathogens are also found to frequent the intestinal tract of the other warm-blooded animals, including animal pets (2), livestock (3), poultry (4), and the wild animal community (5), these organisms being acquired through contaminated food and water sources (6). Such animals may themselves become infected by these pathogens or serve merely as natural carriers. Among the cold-blooded animals, freshwater fish may harbor human pathogens after exposure to contaminated water or food sources and carry these organisms to clean stream recreational areas (7). This frequent occurrence of pathogens in domestic animals and wild life illustrates and adds support to the concern about fecal pollution from all warm-blooded animals, not just from man.

A. Salmonellosis. In humans, salmonellosis most commonly occurs as an acute gastroenteritis with diarrhea and abdominal cramps. Fever, nausea, and vomiting are frequently found as additional symptoms (8). The incidence of recognized human salmonellosis, although seasonal, is fairly low, and a minimum population is necessary before isolation of *Salmonella* strains is possible. The average number of individuals excreting *Salmonella* at a given time is not known for certain. In the absence of a known outbreak, however, this baseline value has been reported to vary from less than 1% in the United States (9) and Great Britain (10) to 3% in Australia (11) and 3.9% in Ceylon (12). If these figures are indicative of the general population, negative *Salmonella* results would be expected from domestic sewage from a small population, although significant levels of fecal coliforms and possible other pathogens might be present at the moment. Since other pathogens could be present in that polluted water during the absence of *Salmonella*, there is no complete assurance that the incidence of other waterborne diseases will correlate with *Salmonella* occurrence alone. Therefore, a test to detect *Salmonella* should not be interpreted to signify concurrence or absence of other pathogens.

The total number of *Salmonella* serotypes known to be pathogenic to man exceeds several hundred and their frequency of isolation from man varies from year to year and from country to country. As an illustration, the 10 most frequent *Salmonella* serotypes isolated in 1965 from humans in the United States were: *S. typhimurium, S. heidelberg, S. newport, S. infantis, S. enteritidis, S. saint-paul, S. typhi, S. derby, S. oranienburg,* and *S. thompson* (13). In Denmark during the period 1960–1968, the

10 most common serotypes isolated from humans were: *S. typhimurium*, *S. paratyphi B*, *S. enteritidis*, *S. newport*, *S. typhi*, *S. infantis*, *S. indiana*, *S. montevideo*, *S. blockley*, and *S. muenchen* (14). Human salmonellosis may not necessarily be the most serious waterborne disease but it is the disease whose causative agent is easiest to isolate from water, food, and feces and to relate to the pollution indicator bacteria.

Salmonellae are frequently found in clinically healthy farm animals. Studies on large groups of cattle indicate the percentage of such latent infections is about 13% in the United States and about 14% in the Netherlands (15). Between 3.7 and 15% of clinically healthy sheep have also been reported to be carriers. With respect to pigs, the percentage of symptomless carriers has been found to range from 15 to 20% in the Netherlands, 7% in France, 12% in England, 13.4% in Norway, and 22% in Belgium (16).

The *Salmonella* strains most frequently isolated from both diseased and healthy farm animals include the following 13 serotypes: *S. typhimurium*, *S. derby*, *S. dublin*, *S. oranienburg*, *S. java*, *S. choleraesuis*, *S. anatum*, *S. newington*, *S. infantis*, *S. stanley*, *S. abony*, *S. chester*, and *S. meleagridis* (3, 17, 18). A number of epidemics have been observed in the human population which were caused by 6 of these 13 *Salmonella* serotypes frequently found in farm animals. Typhoid is specific for man, that is, it does not occur in other animals. The etiologic agent for this disease is *S. typhi*.

Salmonellae are frequently detected in the polluted water environment including sewage (19), stabilization ponds (20), irrigation water (21), streams (22), stormwater (23), and tidal water (14, 24). Cattle feedlots, livestock salesyards, and abattoirs (25) can be major sources of salmonellae that enter streams and lakes used for recreational purposes. The survival of salmonellae in the aquatic environment is influenced by the same factors that control the persistence and die-away of bacterial pollution indicators. Nutrient-rich wastes, low stream temperatures, and a source of salmonellae can produce an impact that will depress the stream self-purification process. As an example, salmonellae were isolated in the Red River of the North, 22 miles downstream of sewage discharges from Fargo, North Dakota, and Moorhead, Minnesota, during September and prior to the sugar beet processing season (26). By November, salmonellae were found 62 miles downstream of these two cities. With increased sugar beet processing, wastes reaching the stream in January under cover of ice brought high levels of bacterial nutrients, and salmonellae were then isolated 73 miles downstream or 4 days flow time from the nearest sources of warm-blooded animal pollution. In farm pond water stored at 21 to 29°C, several salmonellae survived for periods of 14 to 16 days (27).

Where farmyard waste water was used to supplement irrigation water supplies, holding tank storage for a 20-day retention period in summer or 60 days retention in winter did produce a "*Salmonella*-free" supply (28). Similar survival patterns were observed for 52 samples of urban stormwater collected over a 2-year period. In these studies (23), *S. typhimurium* cells were added to individual runoff samples stored either at 10 or 20°C to approximate the water temperature at time of collection. Results showed a 99% kill of *S. typhimurium* in those stormwater samples held 10 days at 20°C, while 5% of this strain still persisted beyond 14 days in runoff waters stored at 10°C.

B.   Shigellosis.   The most commonly identified cause of acute diarrheal disease in the United States is exposure to *Shigella* (29) through person-to-person contact, poor quality drinking water, or contaminated food. Symptoms of shigellosis may vary from a mild transitory diarrhea to severe prostrating attacks accompanied by high temperatures, vomiting, and profuse bloody stools. This disease is endemic in certain custodial institutions, Indian reservations, and in lower socioeconomic urban communities. During a recent 5-year study (1964–1968) in the United States, it was reported that two-thirds of 45,263 isolates were from children under 10 years of age (30). The incidence was also high in the age group 20–40 years, particularly among women of child-bearing ages (30). Estimates of the average number of individuals excreting *Shigella* in the population has been reported to be 0.46% in the United States (30), 0.33% in England and Wales (31), and 2.4% in typical rural villages in Ceylon (12). Summer and early autumn are peak periods for the spread of this disease, but in areas of year-round cold climate, the incidence peak occurs in February and March. Following an epidemic, most of the individual cases may be expected to excrete shigellae for at least 1 week and in one study a positive culture of *S. flexneri* 2a was obtained from a healthy carrier after 15 weeks (32).

Shigellosis surveillance indicates there are at least 32 *Shigella* serotypes of which *S. sonnei* and the subgroups of *S. flexneri* account for over 90% of all isolates from the human population. *S. dysenteriae* and *S. boydii* comprised less than 1% of the total *Shigella* isolates in the United States and were frequently traced to exposure of travelers in other countries (30).

Shigellae are rarely found in animals other than man. The National Communicable Disease Center has reported (30) that only 0.5% of isolations over a 5-year period were from other animal sources. Most of those animals were nonhuman primates which occasionally transmit shigellosis to man. A study of sera from white perch netted during late summer in rivers flowing into the Chesapeake Bay was occasionally found to contain specific antibodies to *S. flexneri*. Those fish that gave positive results were caught in the Potomac, Magothy, Back, and Middle Rivers, and occurred with a 2.3% positive frequency among the 263 fish specimens examined from 11 streams of varying water quality (7). In cold-blooded animals, such as freshwater fish, the occurrence of shigellae or other pathogens and fecal coliforms is related directly to the water quality and available food in their environment.

Most shigellosis epidemics are food-borne or spread by person-to-person contact. However, there have been a significant number of epidemics which resulted from poor quality drinking water. These outbreaks frequently result from accidental breaks in water treatment systems (33), flood-borne excreta contamination of well supply, through inadvertent cross connection of contaminated water pipes with potable water supply lines (34), in public water supplies receiving no treatment (35), and through sewer line seepage into water supply lines (32).

Although shigellae have been found in various polluted waters (36, 37), methodology that has been applied to examination of natural waters is still considered to be low in sensitivity and only qualitative in nature. Current techniques are complicated and time consuming and require careful screening of many suspect colonies on various media. As a result of problems in methodology, *Shigella* is rarely studied in the freshwater and estuarine environments. When better laboratory procedures are available for *Shigella* detection and enumeration from polluted water, the real magnitude of the occurrence and distribution of this pathogen will be better understood.

The survival of *Shigella* strains, once they leave the human intestinal tract and enter the water environment, is limited by many ecological factors. As observed with *Leptospira* and enteroviruses in water, persistence of *Shigella* is significantly better when the total bacterial population is low (36, 37). When wells were intentionally dosed with *Shigella* organisms, these organisms could be recovered 22 days later (38). Other studies in which *Shigella* was added to river water resulted in a maximum of 4 days persistence. With aeration of these river water samples, survival time was drastically reduced to as short as 30 minutes (39). *Klebsiella* strains and a majority of coliform organisms tested in *in vitro* experiments at 37°C together with *S. flexneri* II demonstrated interruption of *Shigella* exponential growth within 12 to 15 hours. The production of formic and acetic acids by coliform organisms in mixed cultures apparently exerted an effect on the *Shigella* strain that ranged from bacteriostatic to bactericidal (40).

Water pH is another important factor that affects *Shigella* survival. Experiments on possible survival and multiplication of *S. flexneri* in the intestinal tract of carp and bluegill indicate a regrowth of this organism when incubated at 20°C in a 1% fecal suspension which was free of coliforms and had a pH of 7.6 to 8.3. Similar experiments with 1% fecal suspensions from either carp or bluegill that had an initial pH of 7.2 resulted in a rapid die-away of the pathogen within 2 to 3 days (41).

Persistence of *Shigella* is also influenced by water temperature, that is, it survives longer at low temperatures. Laboratory experiments involving survival studies on three types of *Shigella* in untreated farm pond water incubated at 20°C, indicated *Shigella* persistence was about 12 days (27). In domestic sewage, persistence was observed to be longer at 15°C, under winter discharge conditions, than during the summer when the sewage effluent temperature was 25°C (36). In Siberia the long duration of low temperatures to −45°C extended *Shigella* persistence to 145 days in feces, 135 days in soil, and 47 days in a frozen river (42). A natural self-purification process does occur during the brief Arctic summer period of intense sunlight. However, the rate of natural self-purification is slow and repeated fecal contamination in the permafrost around a settlement may maintain a reservoir of *Shigella* and other pathogens that could recycle through man again. Periodic testing of cultures isolated from the Arctic environment showed that virulence was not lost after prolonged freezing. These pathogens did undergo some modification of biochemical characteristics during their exposure to subzero temperatures. Colony morphology for *Shigella* was also observed to change after several days in farm pond water incubated at 20°C (27). This instability of some biochemical characteristics that can occur in *Shigella* strains introduced into the water environment illustrates the difficulty of detection by conventional laboratory procedures.

Estuarine and irrigation return waters contain varying concentrations of salts which can extend *Shigella* persistence in these aquatic environments. Although raw sewage is normally a very harsh environment for *Shigella*, additions of NaCl or KCl in minimum 0.035 M concentrations to sewage at pH 7.5 will reverse the otherwise rapid die-away of *S. flexneri* II (43). Since calcium chloride had an inhibitory effect, the varying chemical characteristics of a given water probably accounts for some of the erratic recoveries of shigellae in polluted water.

In the estuarine environment, shigellae were found to be quite resistant to the osmotic effects of high concentrations of sodium chloride; however, survival was dependent on temperature. At 13°C, *Shigella* persisted for more than 25 days but only 4 days at 37°C in varying concentrations of sodium chloride. Other important factors influencing *Shigella* survival in seawater are the composition of the microbial flora and strain response variations to natural seawater. One *Shigella* strain was observed to persist for 70 days whereas another strain survived only 15 days (44).

C. Leptospirosis. Leptospirosis is caused by a group of coiled-shaped actively motile bacteria (*Leptospira*) that generally gains access to the blood stream through skin abrasions or mucus membranes to produce acute infections involving the kidneys, liver, and central nervous system. Evidence of human leptospirosis has been reported in Europe, the Near East, North America, Central America, and the Far East (45–51). The annual incidence rate for the total population is considered to be less than 1% with 0.37% being the incident rate reported in Israel (51). However, when occupations or recreational activities commonly associated with animals, polluted water, and fecally contaminated soil are considered apart from the total human population, such as meat and poultry processing, livestock handling, dairy farming, irrigation farming, sewage and garbage disposal, hunting, trapping and swimming in farm ponds, and rivers, the incidence rate may be as high as 3% (46, 47, 51).

Many serotypes of *Leptospira* have been isolated from human cases. At present 18 serogroups and more than 100 serotypes are recognized from various parts of the world (8). A search of the literature indicates the most common human isolates are *L. pomona*, *L. autumnalis*, *L. australis*, *L. grippotyphosa*, *L. hardjo*, *L. canicola*, *L. ballum*, *L. bataviae*, and *L. icterohemorrhagiae*.

Leptospirosis is transmissible to man from various domestic animals including cattle, swine, horses, and dogs. Animal pests (rats, house mice) frequently are infected as are wild animals including deer, fox, skunk, raccoon, opossum, vole, and frog (8, 46, 51–53). Epidemiological investigation of animals around Israeli villages where human leptospirosis was diagnosed revealed carrier rates in cattle to be 2.3%, rats 5.9%, mice 33.0%, and dogs 26.6% (51). The carrier rate for normal pigs from a Brisbane abattoir was found to be 2.5% (52).

Although transmission of pathogenic leptospires from infected reservoir hosts may occur throughout the year, outbreaks in the United States occur almost exclusively during the recreational season (53). A small outbreak of leptospirosis was reported from Philadelphia in 1941 in which seven persons contracted the disease by bathing in polluted streams (54). Pathogenic leptospires were found in the urine of two persons and five cattle who had access to the same Iowa stream which had been used for swimming (55). An outbreak of leptospirosis in the Cedar Rapids, Iowa, area was traced to infected swine which polluted a creek 20 miles upstream of a popular swimming hole (56). A total of 50 cases of leptospirosis was reported for persons who became ill after swimming in a slow-moving stream alongside a field where cattle and swine were pastured (57). This disease was also the cause of various epidemics involving various groups of people who swam in farm ponds and streams in California (58).

Irrigation and drainage canals may also be involved in transmission of leptospirosis from cattle to man. This vehicle of transmission has been a more common problem in Europe, especially in Holland, where people frequently use the numerous canals for swimming (48–59).

Leptospira enters recreational streams and lakes through direct urination of infected cattle, swine, and wild animals that gain access to the water or from drainage of adjacent livestock pasture land (60). *Leptospira* strains were isolated in 19 of 51 water samples obtained from Iowa rivers (61). *L. pomona* has been isolated from Washington state surface waters frequented by infected cattle (62). Another pathogenic species, *L. grippotyphosa*, was recovered from a stream in Pennsylvania and also found to be common to cattle and vole populations in the area (63).

Many of the interrelated influences of the water environment on *Leptospira* are only partially understood because of their extremely slow generation time of 48 hours at 23°C (64), the more complicated procedures for isolation from polluted water, and their identification through serology. A review of the literature indicates that *Leptospira* is frequently found in slow-moving streams, creeks, canals, and small lakes. The optimum water pH for *L. icterohaemorrhagiae* appears to be between 7.0 and 7.2. Shortest survival occurs in acid waters of pH 6.2 or below and in alkaline waters of pH 8.5 (65). In another study four *Leptospira* serotypes were found to survive longer in slightly alkaline waters but there were significant differences between serotypes and their response to water pH (66).

Stream temperature is also a factor that affects the survival of *Leptospira*. At low temperatures common to winter stream data, multiplication of these organisms is retarded but the persistence is increased over that expected for summer stream temperatures. Laboratory storage experiments with water from the Charles River (Massachusetts) indicate survival times of 8–9 days at 5–6°C, 5–6 days at 25–27°C, and 3–4 days at 31–32°C (65). However, there must be other interrelated factors since it has been reported that *L. icterohaemorrhagiae* survived the summer stream temperatures in the East Indies for at least 22 days without loss of virulence (50).

Salinity of the contaminated stream or lake can also influence the survival of leptospires. These organisms have been reported to survive more than 10 days in lake water with low salinity. In lake and river waters with a salinity ranging from 70 to 6,350 ppm chloride, leptospirae survived for less than 1 week. Using seawater with a salinity of 13,000 to 17,000 ppm chloride, survival time was reduced to less than 1 day (48).

One of the interrelated biological factors that control leptospiral persistence in the water environment is the density and composition of the microbiological population. Leptospires survived for a very short time when mixed with fecal material and stored either at 5 or 37°C (63). The high microbial population in raw sewage shortens survival to between 12 to 14 hours (65). Similar studies with sterile tap water at pH 7.0 and a water temperature of 25–27°C indicate survivals of 30 to 33 days but upon addition of a heterogenous bacterial population to the tap water, survival time for *L. icterohaemorrhagiae* was reduced by about 50% (66). Other experiments on survival of *L. icterohaemorrhagiae* in laboratory-aged lake water contaminated with an air-borne bacterial population of over 1 million organisms per ml indicate survival time at 25–32°C could be as long as 55 days (67). Laboratory studies on leptospiral persistence in culture-contaminated soil indicate a 43-day survival. When the soil was contaminated with urine from infected animals, leptospires could be recovered from the soil only for 15 days (68). Stormwater runoff was simulated in these experiments by additions of rain water to the dosed soils with the result that leptospires were recovered in the runoff water at intervals ranging up to 24 days.

These research findings suggest the microbial flora is an important factor that influences leptospiral survival time in the aquatic environment. It is conceivable that stormwater runoff from cattle feedlots and pasturelands together with stream dilution will drastically alter the bacterial and chemical factors that normally limit survival of *Leptospira* in fecal wastes held on the adjacent lands. Once released to the aquatic environment, this pathogen may be carried downstream many miles to the vicinity of some recreational area.

D. Enteropathogenic E. coli. Various serotypes of *E. coli* frequently cause a gastroenteritis characterized by a profuse watery diarrhea with little mucus and no blood, nausea, prostration, and dehydration with a general absence of fever (8). The serious diarrhea among children under 5 years, particularly of the newborn, is frequently a result of the etiologic agent, enteropathogenic *E. coli* (8, 69). Adults may also succumb to a diarrhea caused by these organisms. In addition, most urinary infections of adults have been reported to be caused by pathogenic *E. coli* (70, 71). The carrier rate in the population apparently is somewhat variable. From a detailed year-round study of 172 normal infants in Houston, Texas, enteropathogenic *E. coli* was isolated from 1.2% (72), from 15.5% of 385 mothers of newborn infants and 1.8% of 114 staff members of the hospital in Cincinnati (73), and from 6.4% of 219 food handlers in Louisiana (9). Among other nations, enteropathogenic *E. coli* was reported present in 2.42% of 24,864 normal children in a 5 year period in England and

Wales (31), 3.3% among 214 3-day old healthy newborns in Trieste, Italy (69), and from 2.0% of 1500 Thai adults and children in Bangkok (74).

Although there are more than 140 specific serological "O" groups of *E. coli,* clinical disease is caused by a relatively few serotypes (75). In the United States and Great Britain, 11 serogroups of *E. coli* are the most frequent etiologic agents of infant diarrhea (8, 31). These serotypes are

026:B6	0112:B11	0126:B16
055:B5	0119:B14	0127:B8
086:B7	0124:B17	0128:B12
0111:B4	0125:B15	

As expected, there is some variation in the frequency of occurrence for these enteropathogenic *E. coli* serotypes that are isolated from year to year and the incidence of infection has been observed to relate to population density, being higher in towns than in country areas (76). Most urinary tract infections reported from five large hospitals in different geographical areas of the United States were caused by *E. coli* serogroups 04, 06, and 075 (70). In another study over half the urinary tract infections of a group of young women were due to these same three *E. coli* serogroups (71). Food handlers found to be carriers of enteropathogenic *E. coli* (6.4%) in Louisiana had eight of the serotypes most frequently found in infant diarrhea (9).

Serotypes of *E. coli* pathogenic to man have been found in some farm animals. Examination of 150 hogs that included locally reared animals in the Phoenix, Arizona area and those in separate shipments from Iowa (77) revealed a 9% occurrence of enteropathogenic *E. coli.* These strains included serotypes:

026:B6	086:B7	0124:B17	0128:B12
055:B5	0112:B13	0125:B15	

E. coli isolates reported from 15 cows belonged to 81 different serogroups with 12 serogroups predominating (78). The potential pathogenic hazard to man from these strains was not established. However, development of an OX antiserum to type the *E. coli* strains that were O group antisera negative revealed that strains identified in this serological procedure as OX 28, OX 101, and OX 117 were pathogenic to cows and their calves. Animal slurry from a pig farm was found to contain two hemolytic strains of *E. coli* (serogroup 0141). These strains and those of *S. typhimurium, S. dublin,* and *Brucella abortus* added to a slurry from a dairy farm were observed to persist with slow die–away over a period of 11 to 12 weeks (79). Applying this liquid waste over pasture land for disposal may create a health risk to dairy herds that graze these fields. Although the importance of farm animals in the transmission of *Salmonella* and *Leptospira* to humans is well documented, the epidemiological significance of enteropathogenic *E. coli* from these animals to human illness is yet to be established.

There is an increasing awareness of enteropathogenic *E. coli* serotypes as a common cause of gastroenteritis occurring among adults in addition to being a prime cause of infant diarrhea. Most of the reported instances of water-borne enteropathogenic *E. coli* infections have been related to consumption of contaminated drinking water. An outbreak of gastroenteritis among participants at a conference center near Washington, D.C., was caused by an enteropathogenic *E. coli* serotype 0111:B4. This organism was isolated from the water supply and the feces of 14 of those made ill (80). Enteropathogenic *E. coli* serotype 0126:B16 was found in sewage-contaminated drinking water that caused an outbreak of gastroenteritis among 103 persons in a logging camp on the Olympic peninsula (81). An epidemic in two residential areas of Uppsala, Sweden, occurred in 1965 with 442 cases within 14 days, reaching a peak of 261 sick on July 28–30 (82). During the epidemic, three serotypes of *E. coli* (026:B6, 0111:B4, and 0128:B12) were isolated both from feces and the water supply. All examinations for salmonellae, shigellae, and enteroviruses in patient feces were negative. The water supply common to both communities had 900 total coliforms and 300 fecal coliforms per 100 ml. Enteropathogenic *E. coli* serotype 0111:B4 was present at an estimated density of 100 organisms per 100 ml and the other two serotypes were estimated to occur at about 10 organisms per 100 ml (83). After a period of heavy rain in October 1965, a new public water supply in Gimo, Sweden, became contaminated with 30 fecal coliforms per 100 ml of which some were identified to be enteropathogenic *E. coli* serotypes 026:B6 and 0114:B. A few cases of acute gastroenteritis occurred during this time but no fecal specimens from patients were examined for *E. coli* serotypes. Upon sufficient chlorination of the water supply, results for coliforms became negative and no new cases of gastroenteritis occurred (83). An outbreak of enteritis in a children's camp in Hungary was found to be caused by enteropathogenic *E. coli* serotype 0124:K72, H:32 which was isolated from patient stools and in the reservoir of a nearby spring used for drinking water. The camp sewer, some 150 feet from the spring, was found to be leaking (84).

A water sample from the distribution lines of a public water supply in France was found to contain *E. coli* serotype 0111:B4 after a breakdown in chlorination (85). The same study also reported *E. coli* serotype 055:B5 isolated from shallow well supplies in northern France. Pathogenic *E. coli* serotypes were found in potable water supplies of varying quality in Israel. An analysis of these data collected over a 2-year investigation indicated a rise in occurrence of pathogenic *E. coli* in the autumn (86). Surveillance for *E. coli* serotypes in drinking and surface waters in Hamburg, Germany, between 1963 and 1965 revealed 0.4% isolation from well waters, 0.5% from ship potable water supplies, and 0.1% from surface waters (87).

Enteropathogenic *E. coli* are present in streams and lakes polluted with warm-blooded animal feces, the occurrence being probably less than 1% of the fecal coliform population (88). The Fyris river in Sweden was found to contain enteropathogenic *E. coli* on 12 different occasions involving 10 different "O" serogroups (89). In fact, *E. coli* serogroup occurrences were used in an investigation of a lake and stream drainage basin in northeastern Pennsylvania to trace the probable sources of microbial pollution (90). A total of 10 serotypes were found in 224 samples from 5 waste stabilization ponds in North Dakota and the number of serotypes recovered per sample apparently related to the population served for a given pond (91).

Fish do not have a permanent coliform flora in their intestinal tract. Ingested fecal coliform and fecal streptococcus tracer organisms ingested by bluegills and carp were retained in the intestinal tract for 9 to 14 days in experiments with fish exposed to these organisms in the water but not in the food (41). Thus it is conceivable that once fish temporarily acquire pathogenic organisms in their gut, they could become carriers of such bacteria to clean stream areas some distance from a polluted aquatic environment.

E. coli enters the aquatic environment from the discharge of fecal contamination introduced by some warm-blooded animal source (92). Once separated from the intestinal tract, survival of any *E. coli*, pathogenic or nonpathogenic, is influenced by a host of environmental factors including water pH (93), metal ion toxicity (94), additions of bacterial nutrients (95), water temperature (96), sunlight exposure (97), intermittent stream riffles (95), bacterial adsorption with sedimentation (98), and predation (99). Survival of *E. coli* in estuaries and the open sea is controlled by many of the same factors common to freshwater ecology (100). The more significant factors in seawater include bacterial adsorption and sedimentation, toxicity of the trace quantities of heavy metals, lower water temperature, and the competition for available nutrients between the marine flora and those bacteria derived from the intestinal tract of warm-blooded animals (100).

When *E. coli* is found in freshwater or the estuarine environment, its occurrence indicates a recent introduction of fecal contamination. Multiplication of *E. coli* is rarely observed, with growth being confined to warm water discharges of large volumes of bacterial nutrients from untreated cannery wastes, poultry processing wastes and raw domestic sewage. With dilution by better quality water downstream or through tidal action in estuarine waters, any potential multiplication of *E. coli* is suppressed (95, 101). Thus these natural self-purification forces can produce a sharp die-off for *E. coli* to only 10% viable organisms in 2 to 5 days (100).

E. Tularemia. The tularemia pathogen usually gains access to the blood stream through skin abrasions or mucus membranes to produce chills and fever, swollen lymph nodes, and a general prostrate condition. Frequently an ulcer develops at the site of initial exposure and the pathogen may be found in such lesions up to a month from onset, sometimes longer (8). In many cases a general weakness may persist for weeks before any return to a normal state of health. The infectious agent has been variously named *Francisella tularensis*, *Pasteurella tularensis*, and *Bacterium tularense*, being described as a small Gram-negative ellipsoid to elongated rod that is not considered to be transmitted directly from man to man.

Human cases most frequently occur in the hunting seasons through bites and fluid releases in contact with wood ticks (102) and in handling infected wild animals, principally rabbits (103), muskrats, and beaver (104). Apparently tick infestation is the major factor in perpetuating the natural reservoirs of this disease among wild animals, with tularemia reaching epizootic proportions and resulting in the decimation of large populations of rodents in a given season.

This disease is spread via the water route to man through drinking water contaminated by the urine, feces, and dead bodies of numerous species of rodents and wild rabbits (105). *F. tularensis* has been recovered from the intestinal tract of mountain trout living in a stream, whose water produced tularemia in guinea pigs via subcutaneous inoculations. No evidence of *F. tularensis* was found from other tissues of this fish. It was concluded that the organism was a transient resident in the fish intestine, reflecting the contamination in the water environment (104). Human infections from handling freshly caught fish have been reported. The fish were believed to serve only as a mechanical transfer of infection from polluted water, implied to be the source of *F. tularensis* (106).

Intensive studies of a small stream, Grid Creek, and one of its tributaries, Cattail Creek, near Hamilton, Montana, indicated an extensive contamination over 35 miles of the drainage basin lasting 16 months. Repeated attempts at isolation of *F. tularensis* directly from these waters by inoculation into cystine heart agar plates were negative. All positive findings were based on inoculation of guinea pigs with amounts ranging from 0.01 to 10 ml to produce an infection. Laboratory studies indicated that a total of five to seven *F. tularensis* organisms were sufficient, in subcutaneous inoculations, to produce guinea pig infections (104).

Tularemic epidemics have occurred when agricultural workers and ranchers drank water from contaminated springs and small creeks in the Soviet Union (105), from a stream in Montana (107), and from a poorly protected well on a ranch in Klamath County, Oregon (108). During World War II, tularemia epidemics in the Rostov area of southern Europe involved 8,500 cases in November 1941 and 14,000 in January 1942 through use of contaminated wells and streams by the Soviet army (109). Diversion of contaminated water into an artesian well distribution system, as a result of war destruction of the supply, caused an urban outbreak of tularemia. The epidemic ceased upon chlorination of the water supply (110).

Reports related to body contact tularemia by swimming in contaminated water containing *F. tularensis* are limited to general supposition (111). However, a case of oculoglandular tularemia occurred in Southern Alberta, Canada after direct contact with sewage (112). Workers engaged in washing and dicing of sugar beets in a plant in South Moravia were shown to have a significantly increased risk of tularemia. Using serological procedures, a total of 237 cases of tularemia were identified from employees during two sugar beet processing compaigns (years 1964–1965 and 1967–1968). *F. tularensis* was isolated from the outer slices of the sugar beets, but attempts to isolate it from the washroom atmosphere and sugar beet wash waters were unsuccessful. Workers were believed to have become infected through inhaling aerosols around the beet washing operations. An increase in the number of field mice infected with tularemia was observed in the sugar beet stock piles during the epidemics among plant workers (113).

Knowledge of the factors affecting survival of *F. tularensis* in water is limited. Much like other pathogens, *F. tularensis* persistence in water is increased with decreasing water temperature. Using naturally contaminated stream water stored at 7°C, survival was found to be at least 23 days (104). Other research has indicated survival in ice to be 32 days. Using unsterile well water inoculated with 5 million *F. tularensis* organisms per ml, survival at 9°C was about 60 days while in a split sample held at room temperature this pathogen survived for little more than 12 days (114). Extracts prepared from mud samples of a contaminated creek supported excellent growth. The extracts of mud were reported to contain moderate amounts of substances having cultural properties similar to "blood cell extracts" (115). Thus, as noted for other bacterial pathogens, the addition of complex nutrients to the aquatic environment will protect and extend the survival of these organisms and create an extended health hazard some distance downstream.

F. Cholera. The bacterial pathogen *Vibrio cholerae* can produce a serious, acute intestinal disease that is characterized by sudden diarrhea with profuse, watery stools, vomiting, suppression of urine, rapid dehydration, fall of blood pressure, subnormal temperature, and complete collapse. Death may occur within a few hours of onset unless prompt medical treatment is given to the patient (8). *V. cholerae* (*V. comma*), the El Tor biotype and Inaba and Ogawa serotypes, are all pathogenic to man.

The spread of cholera may be through person to person contact (116), consumption of food contaminated by process water or food handling (117), and in polluted drinking water (118). Healthy carriers of *V. cholerae* may vary from 1.9 to 9.0% (118), while the El Tor biotype carrier rate has been reported to be even higher, 9.5 to 25% (119). These symptomless carriers excrete vibrios intermittently with the duration of pathogen discharge being relatively short, averaging 6 to 15 days (120) with a maximum period between 30 and 40 days (118). Chronic convalescent carriers have been observed to shed vibrios intermittently for periods of 4 to 15 months (116).

Recorded evidence of cholera epidemics goes back to 1563 in a medical report from India. In the nineteenth century cholera spread from its apparent ancestral site in the Orient to other parts of the world, producing pandemics in Europe. Development of protected community water supplies and sewer systems in large cities brought a dramatic end to widespread epidemics on the European continent. At the turn of the century, cholera retreated to the Orient (121) only to return first to Egypt in 1947 (122), then to Iran and Iraq by 1964. In 1970, epidemics were reported in Astrakhan and Odessa in southern Russia and south-westward across Africa from Egypt to Guinea (123). This increasing spread of cholera in recent years may reflect a lack of international quarantine enforcement by some countries which also have primitive public water supplies and inadequate sanitary regulations, the international mobility of carriers in the world population, and the quick transport of contaminated food and water by ships and aircraft.

A variety of noncholera vibrio strains is found in both the fresh water and estuarine environment. Some strains of *V. cholerae* isolated under endemic conditions have been known to yield nonpathogenic mutants. Whether some of the noncholera vibrios will revert back to a pathogenic strain is as yet unknown (124).

Survival of vibrios in the aquatic environment relates sharply to various chemical, biological, and physical characteristics of a given stream or estuarine water. The viability of *V. cholerae* in surface waters has been observed to vary from 1 hour to 13 days (125). Autoclaving these waters increased the survival by several days in each of the studies, possibly because heat produces a breakdown of the suspended organic matter, killing the bacterial competitors, and increasing the water alkalinity to a more favorable pH range between 8.2–8.7. *V. cholera* viability is brief in water at pH 5.6 (126). In other studies using synthetic water, both chlorides and organic matter were essential for persistence. When *V. cholera* is introduced into either a synthetic water containing common saprophytes or into the high density bacterial flora of sewage or activated sludge, there is a significant suppression of the pathogen often resulting in 99% kill in 6 hours (127).

Although cholera vibrios may persist for only a short time in the grossly polluted aquatic environment, fecal contamination from victims of epidemics and the carriers may continue to reinforce their population in water. This fact is verified by isolation of *V. cholerae* from the Hoogly river and associated canals in Calcutta, India, during epidemic and nonepidemic periods (128). Chlorination (2.0 to 3.0 mg/liter for 10 min contact time) of the turbid, heavily polluted waters of the Hoogly River without any other prior treatment has not produced a potable water supply free of cholera vibrios or *Salmonella* (129). It is assumed that the pathogenic bacteria present in this poor quality raw water source are protected in particle clumps from exposure to chlorine during the disinfection period.

G. Tuberculosis. Transmission of pulmonary tuberculosis by water can occur but is relatively uncommon, partly because of the minimum 4 to 6 weeks incubation period before a primary lesion can be demonstrated (8). In many infections of this nature, the pulmonary type of disease may go unsuspected for years. Thus it is difficult to trace a given infection back to a particular exposure incidence either by swimming in polluted water or by drinking water that at some time contained a minimum dosage of virulent tubercle bacteria (130). The first documented case of human infection was reported in 1947 and involved three children who had fallen into a grossly polluted river at a location about 600 feet below a discharge of sewage from a sanatorium (131). Pulmonary tuberculosis was diagnosed in these children 4 to 5 weeks after their exposure and tubercle bacilli were found in the river. Other tuberculosis cases involving near-drowning of children in sewage contaminated water have been reported (132).

Skin infections caused by mycobacteria in swimming pools and in natural bathing waters generally develop within 2 or 3 weeks and have been more readily related to the source of infection. Six cases of skin tuberculosis following nose injuries in a swimming pool were believed to have resulted from possible contamination of pool water (133). No tubercle bacilli were found in the water but it was known that the pool was used by a person shown to have active pulmonary tuberculosis. In one study 124 cases of swimming pool granuloma were associated with use of a warm mineral pool in Glenwood Springs, Colorado (134). At the peak of the epidemic, as many as 1,000 *M. balnei* organisms per ml were recovered from the pool water. The same mycobacterium has been recovered from skin lesions of patients after swimming on the beaches in Hawaii and a swimming pool in Jacksonville, Florida (134).

The mycobacteria pathogenic to man include *Mycobacterium tuberculosis*, *M. balnei*, and *M. bovis*. There are also other atypical mycobacteria that occasionally are found to be pathogenic in specific instances. Although these acid-fast bacterial pathogens are usually detected in the sputum of diseased individuals, the same organisms can also be found in their feces (135). Tubercle bacilli have been detected in the untreated and treated sewage discharges of sanatoria (136). Quantitation of tubercle bacilli in these discharges indicates raw sewage may contain 1,000 to 150,000 organisms per 100 ml (137). Once discharged to the receiving stream, virulent tubercle bacilli are frequently detected downstream (138). Municipal sewage may also contain tubercle bacilli with occurrence calculated at 3.1 and 5.6% of the population served by the sewage systems for two different towns in Europe (139).

Investigations of mycobacterial occurrences in swimming pools indicate that various non-pathogenic types may be found (137–140). Two independent studies reported that *M. aquae* was the predominating type with this organism apparently resistant to free chlorine residuals ranging from 2.0 to 2.5 mg/l (140). The avian tubercle bacilli were isolated on four occasions but in two studies of 13 and 30 pools, neither *M. balnei* nor any human or bovine strains were found.

Examination of potable water supplies in Paris, France (140), Nashville and ten other cities in Tennessee (141) indicate that saprophytic mycobacteria may not be uncommon in potable supplies. Those found in the Paris study developed on primary culturing at 27°C with little recovery at 37°C incubation. Various unclassified mycobacteria were found in the Tennessee study. It was speculated that the source of these organisms was the Cumberland and Duck Rivers which are the source waters for municipal water treatment plants in these cities.

Sewage discharges from cattle feedlots, meat processing plants (136), and dairy wastes (139) that may be handling either diseased animals or the raw milk from an infected dairy herd could contribute virulent tubercle bacilli to some receiving waters. With efforts directed toward elimination of tuberculosis among dairy and beef herds in most countries of the world, this source of tubercle bacilli reaching the stream and subsequent bathing areas may, in time, become only of historical importance.

Tubercle bacilli can persist in the aquatic environment for weeks. When small collodion bags containing tubercle bacilli were suspended in the sewer system of a sanatorium, live bacilli were found after 36 days (142). Storage experiments involving raw domestic sewage and estuarine water held at room temperature indicate a 10% survival of the avian strain of tubercle bacilli in 73 days (143). When sputum containing tubercle bacilli was added to sewage or river water, virulent organisms were recovered after 5 months (143a). At this point it must be remembered that the pathogenicity of mycobacteria present in water must be verified by animal tests since the bacterial flora of sewage and polluted surface waters generally includes acid-fast bacilli that are not of pathogenic significance (142). Apparently pathogenic mycobacteria may lose their capability to cause disease through some mutation processes. Tubercle bacilli remained virulent in sewage storage for 124 days when tested with guinea pigs (144). However, longer storage of these organisms in sewage resulted in a decrease of virulence with no virulent strains found after 203 days. Among the factors influencing prolonged survival of the pathogenic mycobacteria in the water environment, low water temperature and available organic nutrients have been found to be very important (145).

H. Human Enteric Viruses. Any human virus that is excreted in the feces may theoretically be transmitted through water by fecal contamination (146). This group of viruses includes the infectious hepatitis agent, enteroviruses (poliovirus, coxsackie viruses, and ECHO viruses), adenoviruses, and reoviruses. This grouping of human viruses includes approximately 100 types and, with the exception of the unknown agent for infectious hepatitis, they have been found in sewage and polluted rivers (147).

Ironically, virus transmission by the water route has been best demonstrated by the one virus yet to be isolated and characterized in the laboratory, that is, the viral agent for infectious hepatitis. A careful study of published reports in the world's science literature (1895–1964) has revealed 50 incidents of infectious hepatitis that could be attributed to contaminated drinking water (148). The largest water-borne infectious hepatitis epidemic occurred in Delhi, India, during 1955–1956 and involved over 20,000 clinical cases with an estimated 10 times this number that were considered to be subclinical (149). Although the municipal supply was reported to receive "adequate" treatment, significant turbidity was present which may have enmeshed the virus particles in a protective shield from contact with chlorination (150). It is also significant that during this epidemic there was no parallel increase in enteric bacterial diseases, suggesting that perhaps treatment of water adequate for elimination of enteric bacterial pathogens is not necessarily adequate for destroying the infectious hepatitis viral agent. In the reported infectious hepatitis epidemics from 1895 to 1964 and those more recent (151), contaminated municipal supplies, wells, and spring water were indicted as the mode of transmission.

Transmission of poliomyelitis by water appears to be of minor importance. Epidemics of poliomyelitis attributed to water occurred in Sweden in 1939, 1944, 1948, and 1949, in Nebraska in 1952, and in Canada in 1953 (148). In only one of these eight poliomyelitis episodes, the 1952 outbreak involving defects in the water distribution system at Huskerville, Nebraska, was there sufficient evidence to carry conviction that it was a water-borne incident. It is therefore concluded that poliomyelitis may on occasion be water-borne but more often is transmitted through person-to-person contact.

Viral agents, yet unidentified, may also be responsible for acute gastroenteritis. For the years 1946–1960 a total of 142 epidemics of gastroenteritis and diarrheal diseases have occurred in the United States alone, involving 18,790 cases (152). Not all of these illnesses could be attributed to known bacterial pathogens or possible chemical toxicities. In a detailed, year-round study of the etiology of infantile diarrhea conducted in Houston from 1964 through 1967, viral agents were isolated from 27% of infants with diarrheal disease and from 19% of control infants (153). Although this difference was not statistically significant, Group A coxsackie viruses were recovered 3.7 times more often in diarrheal than in nondiarrheal infants. Thus the viral etiology seems likely but must await further research into the identity of these viral entities (148).

Outbreaks of pharyngoconjunctival fever may be traced to swimming pools (154). One explosive outbreak due to adenovirus 3 occurred after swimming in pool water that had not been chlorinated for 36 hours over one week-end period (155). By afternoon the chlorinator was operating and none of the school children who swam in the pool that afternoon contracted the disease. Based on both clinical illness and laboratory evidence, 26 of 36 children, age 6–15 years, were affected. Throat and eye swabs were positive for adenovirus 3 during the acute period. Stool specimens from 5 of 10 patients tested during the first 3 weeks were positive for the virus. Unfortunately, the pool water was not examined for virus until 10 to 16 days later, with the result that findings were negative in the four pool samples collected. In another report, ECHO virus 3 and ECHO virus 11 were isolated from an urban wading pool but no human enteric viruses could be found at the same period in any chlorinated swimming pools from the same area that were also examined (156).

There has been some evidence of induced human infection by exposure to other warm-blooded animals (157). Human enteric viruses have been found in the feces of approximately 10% of beagle dogs (158). If animal pets and other warm-blooded animals are eventually shown to be a significant reservoir of human enteric viruses, then our concern about stormwater pollution is further strengthened. In the urban community, fecal contamination in separate stormwater runoff is derived from the fecal material deposited on soil by dogs, cats, and rodents (22). With transport of human enteric viruses from potential reservoirs in animal pets via stormwater runoff to impoundments, streams, estuaries, and bathing beaches, another factor in the public health risk from this intermittent pollutional discharge would have to be considered.

Survival of human enteric viruses in the water environment is influenced by many of the same ecological forces previously discussed in reference to bacterial pathogens. The microbial population in water and water temperature does influence virus survival. As an example, Coxsackie virus A2 survived much longer in distilled water than in sewage, longer in river water previously autoclaved to destroy the microbial population (158a), and longer in seawater previously heated for 1 hour at 45°C (159). In other experiments using sterile distilled water, spring water, or well water to which 100,000 infective doses per liter was added either alone or with *Clostridium welchii*, *Streptococcus faecalis*, or *Proteus vulgaris* individually at a density of 1,000 cells per liter, virus survival was markedly reduced by the association with any of these bacteria (160). Storage temperature was 18–20°C.

In natural waters, Coxsackie virus could be recovered from sewage after 40 days at 20°C and 61 days at 8°C, but in Ohio River samples no virus was found after 6 days at 20°C or 16 days at 8°C (158). Polio virus, coxsackie virus and ECHO virus survived in waters from four different farm ponds at 20°C and 4°C for periods up to 9 weeks (161). In the same study viruses persisted for at least 12 weeks in these farm pond waters to which 1 *M* magnesium chloride was added, for up to 5 weeks in the presence of manganese salts, and for less than 3 weeks in the presence of iron salts.

Flowing water in streams provides a means of continuous dilution of virus, so that any accumulation of a high viral concentration at a given location is unlikely unless virus is constantly being discharged at a nearby point upstream (162). In quiet waters adsorption and sedimentation of silt can be effective in removing virus from the stream or lake. Virus inactivation on clay particles in natural waters is related to a proper electrical charge distribution leading to a clay–cation–virus bridge so that virus is removed along with removal of clay. The presence of extraneous organic matter in a stream may reverse inactivation of virus by competing with the virus for the adsorption sites on the clay (163). Desorption of enteroviruses from soil particles was found to be better when soil pH was 7.5 (164). In these experiments, survival of poliomyelitis type 1, ECHO virus 7, ECHO virus 9, and Coxsackie B-3 in soil was observed to depend on the type of soil, its moisture content, temperature, and pH. Survival of enteroviruses in loamy soils was shorter than in sandy soils. The maximum survival of 170 days occurred in sandy soil of pH 7.5 and during storage at 3 to 10° C. This type of soil environment may be found along the polluted river bank and possibly in the intertidal areas of bathing beaches. More studies on human enteric virus survival at the soil–water interface are needed before further implications can be made.

I. Parasitic Protozoa. Amebiasis is a disease of the large intestine that may vary from mild abdominal discomfort involving diarrhea alternating with constipation or a chronic dysentery with mucus and blood (8). The infectious agent is the parasitic protozoan, *Entamoeba histolytica.* Fecal contamination of drinking water supplies is a major source of transmission of amebiasis, although the oral–fecal route and consumption of uncooked contaminated vegetables must not be overlooked in epidemiological investigations.

Carrier rates for amebiasis in the general population have been estimated to be 10% in Europe, 12% in the Americas, 16% in Asia, and 17% in Africa (165). However, studies on lower economic groups show carrier rates to be much higher, correlating with poor personal hygienic practices, unprotected water supplies, and primitive waste disposal. For example, in the United States the carrier rate of students at one college was 1.5% (166), 9% in a small town (167), and 11% among Cherokee Indian children (168). Higher occurrence of *E. histolytica* may also relate to occupations. Nine percent of agricultural workers cultivating irrigated fields and 14% of municipal sewage plant operators were found to harbor *E. histolytica* in their intestinal tract (169).

Defects in plumbing systems involving cross connections between sewer lines and water supply pipes, back siphonage from toilets, drainage from defective sewer lines over an open water cooler, and leaking water lines with low pressure that were submerged in sewage have brought *E. histolytica* cysts to the unsuspecting water consumer (170). Enforcement of plumbing codes and establishment of effective cross connection control programs in metropolitan areas should prevent reoccurrence of this type of hazard.

E. histolytica is usually found at low density levels in sewage. At Haifa, Israel, the cyst counts in raw sewage were 5/liter and 1 to 2/liter in the effluent of the sewage treatment plant (171). Survival experiments indicate that cysts of *E. histolytica* may persist in good quality water such as distilled water for at least 153 days at 12 to 22°C (172). In sewage and in natural waters, cyst persistence was reduced by 30% for each 10°C rise in water temperature (173). Although cysts remain viable in sewage for a significant period of time, their initial densities are low and are drastically reduced in the receiving stream through dilution, death, and settling.

The intestinal flagellated protozoan, *Giardia lamblia*, can produce a variety of intestinal symptoms, most frequently in a protracted intermittent diarrhea. This organism, which is worldwide in distribution, has a carrier rate that ranges from 1.5 to 20% in different areas of the United States, depending on the socio-economic structure of a given community and the age group surveyed (8, 168, 174). One epidemic of giardiasis occurred in Aspen, Colorado, during the 1965–1966 ski season and affected 11.3% of 1,094 skiers (174). Studies carried out after the epidemic demonstrated contamination of well water by sewage leaking from defective pipes passing near the wells. *G. lamblia* cysts were found in the sewage from the defective pipes and in stools from 6.9% of the permanent residents served by the defective sewage system.

Amebic meningocephalitis is a fatal disease caused by a pathogenic strain of the ameba, *Naegleria gruberi*. This pathogen gains entry to the human body by way of the upper nasal cavity, with rapid migration to the brain, spinal fluid, and blood stream (175). The onset of symptoms begins about 4 to 7 days after water contact. Death ensues 4 to 5 days later. Amebic meningoencephalitis is associated with swimming and diving in the warm waters of small lakes (175), an indoor swimming pool (176), and a polluted estuary (177). Epidemics that have been reported involve 4 cases from Orlando, Florida, 6 cases in South Australia, 12 cases in Richmond, Virginia, and 17 cases in Czechoslovakia.

Naegleria species are among the numerous free living amebas, common to soil, sewage effluents, surface waters, and to a lesser extent swimming pools. They have on several occasions been found in public water supplies using rivers as source water (178). The observations that: (*a*), pathogenic *Naegleria* appears to grow better in HeLa cell cultures than on cultures of bacteria; (*b*) that there is pathogenicity of this strain to mice by nasal instillation; (*c*) that there is a lack of antiamebic effect by known chemical control agents; and (*d*) that cysts do not form in cell cultures, provide evidence that this *Naegleria* strain belongs to a different species (178).

If this pathogen is more suited to a parasitic existence in nature, the aquatic host is as yet unknown. Until control measures are developed, it seems vital to prohibit further swimming in any small lakes and swimming pools found to be sources of this fatal disease.

J. Parasitic Worms. There are several intestinal parasitic worms that can be found in sewage and are a potential hazard to public health, particularly to sewage plant operators, farm laborers employed in irrigation agriculture, and by body contact in recreational lakes polluted by either sewage or stormwater runoff from cattle feedlots. Although infections may occur from drinking water, modern water treatment methods are a very effective barrier against this route of exposure to parasitic worms. However, future plans to develop processes for reuse of treated waste effluents for potable water supplies in water-scarce areas may intensify the potential of this water route.

Infection of beef tapeworm (*Taenia saginata*) results in clinical symptoms that are variable with complaints of abdominal pain, digestive disturbance and loss of weight (8). The adult tapeworm lives in the intestinal tract and may discharge approximately 1 million eggs per day in the feces of an infected individual (179). Cattle become infected by grazing pastures recently sprayed with domestic sewage or by drinking grossly polluted water (169), either of which contain human feces infected with this parasite. Infection in cattle occurs in muscle tissue. Man may occasionally become infected by accidental intake of water polluted with parasitic ova that will mature in tissue to produce an intermediate type infection.

The incidence of this disease is highest in East Africa where 50% of the population is affected (180). Other high incident areas include Tibet, Mexico, Peru, and eastern Europe. In the United States the incidence has been estimated at 1% or less. In cattle, the percent infected may range from less than 1% to 8% (179).

The eggs of *T. saginata* have frequently been found in sewage (181) with densities estimated to be one to two eggs per 100 ml of sewage (179). *T. saginata* eggs appear to be quite resistant, and can survive for 335 days when kept under cool moist conditions (182) characteristic of liquid sewage sludges. Dried sludge was considered free of viable eggs only after storage for at least 1 year prior to use in agriculture.

Ascariasis is a disease of the small intestine which is more prevalent among young children. Symptoms are variable with passage of live worms in stools or vomitus being the first sign of infection (8). The large intestinal round worm, *Ascaris lumbricoides*, is the infectious agent with transmission of embryonated eggs through fecal contamination in water, soil, and by eating vegetables. These embryonated eggs are produced at the rate of 200,000 per day and under favorable conditions remain viable in soil for months.

A. lumbricoides has been found in 2% of the feces of sewage plant operators and in 16% of the feces from farm workers involved in cultivation of crops with sewage irrigation (169). Monitoring sewage and sewage sludge has been demonstrated in Poland to be a good index of the incidence of ascariasis in the human population (183). These studies suggested that the disease reaches a peak during the autumn–winter months.

Raw sewage from Darmstadt, Germany, was found to contain 540 eggs of *Ascaris* per 100 ml (184). The high density in this sewage was related to an epidemic of ascariasis that occurred with 90% of the population affected. Application of raw sewage and night soil to garden crops was believed to be the major reservoir for this parasitic worm. *Ascaris* was detected, at an average density of 0.11 organisms per 100 ml, in water sampled downstream of a chlorinated primary sewage effluent discharging into a river in Colorado (185).

Studies on *Ascaris* ova in stream samples indicated that survival varied with dissolved oxygen content. Larvae developed after 60 days storage of ova suspended in stream water that contained 7.0 to 9.0 mg/liter dissolved oxygen (186). In the absence of dissolved oxygen, no larvae developed, suggesting that *Ascaris* survival time was enhanced by rapid diffusion of domestic sewage into the receiving stream.

Schistosomiasis is a debilitating infection in which the adult male and female worms live in the veins of humans. Migration of this blood fluke to the liver and bladder can cause complications of a serious nature in chronic infections (8). Infection is acquired in rice cultivation, swimming, and wading through freshwater that is infested with the larval stage (cercariae) of *Schistosoma mansoni*, *S. haematobium*, and *S. japonica*. The life cycle of this parasite depends on one phase of its life being spent in an appropriate snail and the other part in man, animal pets, farm animals, or rodents. Geographical distribution of this disease includes Africa, the Middle East, the Orient, South America, and the Caribbean islands (187). None of these species of *Schistosoma* are indigenous to North America. Continuation of the parasite life cycle depends on urine and fecal wastes of infected individuals polluting streams and lakes where the eggs hatch and the larva (miracidium) finds a suitable freshwater snail host. Snail population control, development of adequate sanitation facilities, and elimination of indiscriminate defecation in rice fields and water courses is essential for significant suppression of the magnitude of schistosomiasis in endemic areas.

Schistosome dermatitis or "swimmer's itch" is a skin irritation acquired from the penetration of larvae of various schistosomes that infect birds and rodents. The larvae are released from the intermediate host, snails, which inhabit farm ponds, wildlife marshes and shallow marine coastal waters (188). Exposure to the disease is most frequently related to swimming in snail infected waters. Swimmer's itch is common to many lakes in various areas of the world and has been reported in marine coastal waters of Long Island and California. Snail population control can effectively reduce the prevalence of this form of skin dermatitis.

Hookworm disease (ancylostomiasis) is another debilitating infection associated with the bloodsucking activity of the helminths *Necator americanus* and *Ancylostoma duodenale*. Symptoms are generally vague but usually follow a pattern of malnutrition, loss of energy, and anemia. This disease is common not only to tropical and subtropical countries but is endemic in the southeastern United States (8). Hookworm disease is generally acquired by the infective larvae penetrating the bare skin, usually the foot. In India, hookworm infections occurred from using sewage in irrigation farming, transferring the infection by direct contact in the field or from eating contaminated vegetables (189).

Fish tapeworm disease (Diphyllobothriasis) is an intestinal disease transmitted to man by eating raw or inadequately cooked fish that are the intermediate host for *Diphyllobothrium latum*. This mild disease is often present without any specific symptoms and is diagnosed from laboratory identification of eggs or worm segments in the feces (8). This pathogenic worm has a life cycle that involves release of worm eggs from man into streams and lakes, where the copepod becomes infected as the first intermediate host, then to fish through consumption of the infected copepods. Control is through development of adequate treatment of domestic sewage and by proper cooking of freshwater fish to kill the parasites. Discharge of untreated domestic sewage to a reservoir impoundment in the Soviet Union was reported to cause fish tapeworm infections. The parasitic infection cycled through the nearby human population using this lake for fishing (190).

Dog tapeworm larvae can infect man with the development of cysts in many different locations in the body, most frequently the liver and lungs. Health impairment to man depends upon the number of cysts that form in tissue, and their size and location in vital organs (8). The disease is most prevalent in the Middle East, and some Mediterranean countries, the great grasslands of South America, Australia and New Zealand, where sheep raising is a major occupation for man and his dogs.

Eggs of the dog tapeworm, *Echinococcus granulosus*, are passed to man most frequently through close association with infected dogs but occasionally contamination of unprotected drinking water supplies by dogs, wolves, dingoes, and other species of *Canidae* is the source of this parasite to man. Studies on the high rate of *Echinococcus* infections in Yugoslavia have been reported to relate to areas where lakes and streams are scarce and drinking water supplies come from open pools of rainwater runoff trapped in ground depressions (191).

9-2.   Transport of Pathogenic Microorganisms to the Stream

Municipal sewage contains the major domestic input of human fecal discharges. In some cities, wastes from meat packing, dairy plant operations, and intermittent additions of stormwater runoff also form part of the domestic sewage collections. As a result of these concentrations of fecal wastes, pathogens derived from humans, farm animals, and wild animals are found in raw sewage, their occurrence and magnitude reflecting diseases prevalent at the moment.

A.   Treated Sewage Effluents.   Sewage treatment processes will reduce the pathogen content of raw sewage. The reduction is governed by length of retention time during treatment, chemical composition of the wastes and their state of degradation, antagonistic forces in the biological flora, pH, and operational temperature among other more subtle and less understood factors. The trickling filter process has been found to reduce *S. paratyphi B* densities from 84 to 99% (19), *M. tuberculosis* populations by 66% (192), enteric viruses in a range from 40 to 60% (193), tapeworm ova by 18 to 70% (194), and cysts of *E. histolytica* by 88 to 99% (195). Total coliform reductions ranged from 82 to 97% (19, 193). In activated sludge systems, coliform organisms, *Salmonella, Shigella,* and *M. tuberculosis* were generally reduced from 85 to 99% (196), poliovirus type I by 90%, and coxsackie A 9 virus by 98% (197). The extended survival of the pathogens in dry sludge could be eliminated by application of higher temperatures during storage prior to disposal. Removal of *S. typhi* during anaerobic digestion extended from 25 to 92.4% depending upon the length of retention time (198), *M. tuberculosis* reduction after anaerobic digestion was reported to be 69 to 90% (19). Although anaerobic digestion was quite effective in reducing cysts of *E. histolytica*, this method was found to be comparatively ineffective in the inactivation of parasitic ova (195). The treatment of sewage in waste stabilization ponds will generally produce total coliform reductions ranging from 50 to 99.9% (199). Virus reduction of 92% was reported in one study of an oxidation pond with a 30-day retention period (200). Home septic systems that incorporate a sand-filled seepage pit or drainage field can, under careful operation, be very effective in removing *S. typhi* from the septic tank effluent (201). Unfortunately, many of these individual systems are often poorly constructed, receive little maintenance, and are overloaded by the householder so that input of sewage receives little treatment and undergoes limited bacterial reductions.

In contrast, a marked reduction in the number of pathogenic organisms can be achieved in municipal sewage treatment plants designed for secondary treatment if they are of adequate capacity to meet a community's needs and are managed by competent operators. Unless these systems are scrupulously controlled, the process is at the mercy of inefficient plant operations and possible frequent bypass due to combined sewer overload and plant malfunction. Under proper operating conditions, the resulting treated effluents will still contain a portion of each kind of microorganism originally present in the raw sewage (202). Those pathogenic bacteria, virus, and parasites that do remain in the treated effluents constitute a potential health hazard to persons using the receiving waters for recreational purposes and will require appropriate chlorination procedures to reduce these residual populations to below demonstrable densities. Additional treatment by chemical flocculation with sedimentation will be necessary to remove parasitic ova and viruses (202).

B. Urban Stormwater.   Stormwater can be a major source of intermittent pollution to designated recreational reaches of streams and lakes. In the urban community, fecal contamination in separate stormwater systems is derived initially from the fecal material deposited on soil by cats, dogs, and rodents (22). Occasionally, urban stormwater also contains a varying input of discharge from saturated drain fields in household septic tank systems which may also be illegally connected to the storm sewer system. Thus, there is always present an opportunity for the chance occurrence of pathogens in this source of water pollution. Using experimental media for the quantitative detection of salmonellae, 4,500 *S. thompson* and 450,000 fecal coliforms per 100 ml were found in a stormwater sample from an urban business district.

C. Rural Stormwater.   In the rural community, stormwater runoff transports the fecal contamination from livestock pastures, poultry and pig feeding pens, cattle feedlots and to a lesser magnitude, the fecal contributions from wild life. Rural stormwater pollution is becoming a more serious problem in some areas as a result of changes from small individual farm operations to the concepts of concentrated animal containment within feedlots to produce the greatest weight gain in a minimum of time (203). It has been calculated that domestic animals produce over 1 billion tons of fecal wastes in a year and that as much as 50% of this waste production may be found in feedlot operations (204). A single feedlot operation involving 10,000 head of cattle will produce 260 tons of manure per day. These accumulations of manure present a serious waste disposal problem because of the quantity produced daily and because of the characteristics of the waste itself. Basically, animal wastes can be characterized as solid material containing large quantities of feed residue and undigested material. For these reasons treatment technology applicable to domestic sewage is generally not successful in animal waste treatment. Disposal by application to farm fields is generally not practical since the cost of hauling manure is greater than the cost of commercial fertilizers. Thus, land fill operations are commonly practiced to reduce the accumulation of manure piles. These accumulations of feedlot manure serve as breeding areas for fly populations, produce noxious odors, and become a source of unsavory dusts when dry. During storm periods adjacent stream water quality is dramatically degraded by additions of runoff from feedlots, which is characterized by high BOD values and high fecal content. *Salmonella*, *Leptospira*, and other pathogens can be found in the litter and runoff from cattle feedlots holding apparently normal animals (18, 60). Feed supplement, by-product materials, and wastewater from industries that reduce offal to reclaim protein for animal food have also been found to contain one or more *Salmonella* serotypes at various times (205). Degredation of stream water quality by stormwater runoff from feedlots can be controlled by designing diversion dikes to channel the wastes to stabilization ponds of sufficient storage capacity. In areas where development of waste stabilization ponds is not possible because of land restriction, other treatment methods must be developed to safeguard stream water quality and its recreational uses (206).

D.   Wildlife Refuges.   Fecal contamination from wild animals seeking drinking water or by colonies of beaver and muskrat inhabiting the water generally reaches the stream or lake through direct defecation and urination. The density and frequency of fecal contamination on soil that is related to wild animals would be more limited than expected from farm animals because of the smaller fecal output per day and the wider area of dispersal. Those fecal organisms derived from wild animals which are transported in stormwater to the streams represent recent pollution (within a 20-day interval) since there are many environmental factors that limit prolonged survival in soil (207).

Wildlife refuge areas receive significant fecal contamination during nesting and migratory seasons. These contributions may include pathogenic bacteria in addition to chemical deposits high in organic nutrients. Much of this pollution is concentrated in relatively small land-water areas designated for wildlife and is restricted to limited recreational uses. Bacterial pathogens are excreted by wild animals (5, 6) and are found on the soil and in the bottom muds of adjacent streams and lakes during periods of intense wildlife use. Pathogens survive for weeks in bottom muds because of the high concentration of nutrients. For these reasons, it is desirable to maintain a continuous buffer zone between wildlife refuge areas and those areas designated for swimming and skiing. Motorboat horsepower should also be limited in the wildlife refuge water areas to prevent water quality deterioration from resuspension of this bacterial population concentrated in the colloidal particles at the mud–water interface (208).

REFERENCES

1. B. Moore, *Monthly Bull. Ministry Health* **7**, 241 (1948).
2. M. M. Galton, J. E. Seatterday, and A. V. Hardy, *J. Infec. Dis.* **91**, 1 (1952).
3. P. M. Nottingham and A. J. Urselmann, *N. Z. J. Agr. Res.* **4**, 449 (1961).
4. K. D. Quist, *Salmonella in Poultry as Related to Human Health,* U.S. Dept. of Agriculture, Report of National Plans Conference, 24–30 November 1962.
5. C. B. Lofton, S. M. Morrison, and P. D. Leiby, *Zoonoses Res.* **1**, 277 (1962).
6. J. L. Summers, *The Sanitary Significance of Pollution of Waters by Domestic and Wild Animals: A Literature Review, U.S. Dept. of Health, Education, and Welfare, Public Health Service, Shellfish Sanitation Technical Report,* 1967.
7. W. A. Janssen, and C. D. Meyers, *Science* **159**, 547 (1968).
8. American Public Health Association, *Control of Communicable Diseases in Man,* 11th ed. (A. S. Benenson, Ed.), American Public Health Assoc., New York, 1970.
9. H. E. Hall and G. H. Hauser, *Appl. Microbiol.* **14**, 928 (1966).
10. Public Health Laboratory Service and Society of Medical Officers of Health, *Monthly Bull. Ministry Health and Public Health Lab. Service* **24**, 376 (1965).
11. I. M. Mackerras and V. M. Pask, *Lancet* **2**, 940 (1949).
12. J. Gulasekharam and T. Velaudapillai, *Z. Hyg.* **147**, 347 (1961).
13. U.S. Department of Health, Education, and Welfare, Public Health Service, National Communicable Disease Center, *Salmonella Surveillance Report, Annual Summary, 1965* (1966).
14. K. Grunnet and B. B. Nielsen, *Appl. Microbiol.* **18**, 985 (1969).
15. H. J. Rothenbacker, *J. Amer. Vet. Med. Assoc.* **147**, 1211 (1965).
16. E. Prost and H. Riemann, *Ann. Rev. Microbiol.* **21**, 495 (1967).
17. W. Pollach, *Vien Tierargtl. Mschr.* **51**, 161 (1964).
18. J. R. Miner, L. R. Fina, and C. Piatt, *Appl. Microbiol.* **15**, 627 (1967).
19. J. H. McCoy, *Proc. Soc. Water Treat. Exam.* **6**, 81 (1957).
20. R. M. Cody and R. G. Tischer, *J. Water Pollution Control Fed.* **37**, 1399 (1965).
21. S. G. Dunlop, R. M. Twedt, and W. L. Wang, *Sewage Ind. Wastes* **24**, 1015 (1952).
22. M. L. Peterson, *Proc. 10th Conf. Great Lakes Res.* **16**, 79 (1967).
23. E. E. Geldreich, L. C. Best, B. A. Kenner, and D. J. Van Donsel, *J. Water Pollution Control Fed.* **40**, 1861 (1968).

24. F. T. Brezenski and R. Russomanno, *J. Water Pollution Control Fed.* **41**, 725 (1969).
25. W. Pollach, *Vien Tierargtl. Mschr.* **51**, 161 (1964).
26. U.S. Department of Health, Education, and Welfare, Public Health Service, *Report on Pollution of Interstate Waters of the Red River of the North (Minnesota, North Dakota).* Robert A. Taft Sanitary Engineering Center, Ohio (1966).
27. D. A. Andre, H. H. Weiser, and G. W. Maloney, *J. Amer. Water Works Assoc.* **59**, 503 (1967).
28. A. Braga, *Igiene Mod.* (Italy) **57**, 635 (1964).
29. A. V. Hardy and J. Watt, *Public Health Rept.* **60**, 57 (1945).
30. L. B. Reller, E. J. Gangarosa, and P. S. Brachman, *Amer. J. Epidemiol.* **91**, 161 (1970).
31. Public Health Laboratory Service and the Society of Medical Officers of Health, *Monthly Bull. Ministry Health and Public Health Lab. Service* **24**, 376 (1965).
32. I. Nikodemusg and L. Ormay, *Arch. Inst. Pasteur Tunis* **36**, 43 (1959).
33. D. M. Green et al., *J. Hyg.* **66**, 383 (1968).
34. A. I. Rors and E. H. Gillespie, *Monthly Bull. Ministry Health* **11**, 36 (1952).
35. R. H. Drackman et al., *Amer. J. Hyg.* **72**, 321 (1960).
36. W. L. L. Wang, S. G. Dunlop, and P. S. Munson, *J. Water Pollution Control Fed.* **38**, 1775 (1966).
37. V. A. Lavrumov and L. A. Kiriushina, *Gigiena i Sanit.* **23**, 57 (1958).
38. D. Bartos, J. Bansagi, and K. Bakos, *Z. Hyg. Infektionskrankh.* **127**, 347 (1947).
39. L. B. Dolivo-Dobroval'skii and V. S. Rossovaskaia, *Gigiena i Sanit.* **21**, 52 (1956).
40. D. J. Hentges, *J. Bacteriol.* **97**, 513 (1969).
41. E. E. Geldreich and N. A. Clarke, *Appl. Microbiol.* **14**, 429 (1966).
42. G. G. Mirzoev, *Gigiena i Sanit.* **33**, 437 (1968).
43. W. L. L. Wang, S. G. Dunlop, and R. G. DeBoer, *Appl. Microbiol.* **4**, 34 (1956).
44. M. Nakamura, R. L. Stone, J. E. Krubsack, and F. P. Pauls, *Nature* **203**, 213 (1964).
45. W. F. Ashe, T. Pratt, and D. Kumpe, *Medicine* **20**, 145 (1945).
46. S. D. Greenberg, *J. Student Amer. Med. Assoc.* **4**, 19 (1955).
47. J. L. Braun, S. L. Diesch, and W. F. McCulloch, *Can. J. Microbiol.* **14**, 1011 (1968).
48. W. Schuffner, *Trans. Royal Soc. Trop. Med. Hyg.* **28**, 7 (1934).
49. B. Walch-Sorgdrager, *Leptospirosis Bull., Health Organ, League of Nations* **8**, 1943 (1939).
50. P. H. Van Theil, *Geneesk Tijdschr. v. Neder Indie.* **78**, 1859 (1938).
51. M. Torten, S. Birnbaum, M. A. Klingberg, and E. Shenberg, *Amer. J. Epidemiol.* **91**, 52 (1970).
52. P. R. Schnurrenburger, R. A. Masterson, and J. H. Russell, *Ohio State Med. J.* **64**, 462 (1968).
53. D. Howell et al., *Vet. Rec.* **84**, 122 (1969).
54. W. P. Havens, O. J. Bucher, and H. A. Reimann, *J. Amer. Med. Assoc.* **116**, 289 (1941).
55. J. Braun, *J. Amer. Vet. Med. Assoc.* **138**, 532 (1961).

56. Anonymous, "Swine Blamed for Leptospirosis," *Omaha World Herald,* Omaha, Neb., September 29, 1959.
57. M. Schaeffer, *J. Clin. Invest.* **30**, 670 (1951).
58. W. T. Hubbert, *Public Health Rept.* **82**, 429 (1967).
59. A. B. Semple, *J. Royal Inst. Public Health Hyg.* **15**, 303 (1952).
60. S. L. Diesch and W. F. McCulloch, *Public Health Rept.* **81**, 299 (1966).
61. J. L. Braun, S. L. Diesch, and W. F. McCulloch, *Can. J. Microbiol.* **14**, 1011 (1968).
62. R. W. H. Gillispie, S. A. Kengy, L. M. Ringer, and F. K. Bracken, *Amer. Vet. Res.* **18**, 76 (1957).
63. L. Clark, J. I. Kresse, R. R. Marshak, and C. J. Hollister, *J. Amer. Vet. Med. Assoc.* **141**, 710 (1962).
64. S. L. Chang, *J. Infect. Diseases* **81**, 35 (1947).
65. S. L. Chang, M. Buckingham, and M. P. Taylor, *J. Infect. Diseases* **82**, 256 (1948).
66. C. E. G. Smith and L. H. Turner, *Bull. World Health Organ.* **24**, 35 (1961).
67. W. A. Sawyer and J. H. Bauer, *Amer. J. Trop. Med.* **8**, 17 (1928).
68. D. J. Smith and H. R. M. Self, *J. Hyg.* **29**, 436 (1956).
69. G. D. Rottini and T. Zacchi, *Giorn. Microbiol.* **16**, 189 (1968).
70. M. Turck and R. G. Petersdorf, *J. Clin. Invest.* **41**, 1760 (1962).
71. M. Turck et al., *J. Infect. Diseases* **120**, 13 (1969).
72. M. D. Yow et al., *Amer. J. Epidemiol.* **92**, 33 (1970).
73. M. L. Cooper et al., *J. Diseases Children* **97**, 255 (1959).
74. S. Y. Gaines, Y. Achavasmith, M. T. Thareesawat, and C. Dunagmani, *Amer. J. Hyg.* **80**, 388 (1964).
75. L. A. Rautz, *Arch. Internal Med.* **109**, 37 (1962).
76. J. Taylor, *Bull. World Health Organ.* **23**, 763 (1960).
77. D. C. Mackel, L. F. Langley, and C. J. Prchal, *J. Bacterial.* **89**, 1434 (1965).
78. P. J. Glantz, H. Rothenbacker, and J. F. Hokauson, *Amer. J. Vet. Res.* **29**, 1561 (1968).
79. J. D. Rankin and R. J. Taylor, *Vet Record* **85**, 578 (1969).
80. S. A. Schroeder et al., *Lancet* **1**, 737 (1968).
81. S. B. Werner et al., *Amer. J. Epidemiol.* **89**, 277 (1969).
82. S. Bengtsson et al., *Lakartidningen* **63**, 4499 (1966).
83. D. Danielsson, G. Laurell, F. Nordbring, and O. Sander, *Acta Pathol. Microbiol. Scand.* **72**, 118 (1968).
84. B. Lanyi, J. Szita, B. Ringelhann, and K. Kovach, *Acta Microbiol. Acad. Sci. Hung.* **6**, 77 (1959).
85. P. Monnet et al., *Ann. Inst. Pasteur* **87**, 347 (1954).
86. Y. Yoshpe-Purer, *Israel J. Med. Sci.* **1**, 616 (1965).
87. G. Muller, *Zentr. Bakteriol. Parasitenk. Abt. I, Orig.* **203**, 464 (1967).
88. D. J. Van Donsel and E. E. Geldreich, Unpublished Data (1970).
89. D. Danielson and G. Laurell, *Acta Paediat.* **53**, 49 (1964).
90. P. J. Glantz and T. M. Jacks, *Water Resources Res.* **4**, 625 (1968).
91. A. A. Gustafson and J. B. Hundley, *Health Lab. Sci.* **6**, 18 (1969).
92. E. E. Geldreich, *Water Sewage Works* **114**, R-98 (1967).
93. T. O. Rogers and H. A. Wilson, *J. Water Pollution Control Fed.* **38**, 990 (1966).
94. G. W. Malaney, W. D. Sheets, and P. Quillin, *Sewage Ind. Wastes* **31**, 1309 (1959).

95. F. W. Kittrell and S. A. Furfari, *J. Water Pollution Control Fed.* **35**, 1361 (1963).
96. N. B. Hanes, G. A. Rohlich, and W. B. Sarles, *J. New Engl. Water Works Assoc.* **80**, 6 (1966).
97. A. L. H. Gameson and J. R. Saxon, *Water Res.* **1**, 279 (1967).
98. C. M. Weiss, *Sewage Ind. Wastes* **23**, 227 (1951).
99. M. Varon and M. Shilo, *J. Bacteriol.* **99**, 136 (1969).
100. R. Mitchell, *Water Res.* **2**, 535 (1968).
101. N. B. Hanes and R. Fragala, *J. Water Pollution Control Fed.* **39**, 97 (1967).
102. R. B. Bost, S. C. Percefull, and H. E. Leming, *J. Amer. Med. Assoc.* **137**, 352 (1948).
103. M. R. Bow and J. H. Brown, *Amer. J. Public Health* **36**, 494 (1946).
104. R. R. Parker, E. A. Steinhaus, G. M. Kohls, and W. L. Jellison, *National Institutes of Health Bulletin No. 193,* U.S. Public Health Service (1951).
105. M. I. Tsareva, *J. Microbiol. Epidemiol. Immunobiol.* **30**, 34 (1959).
106. A. Falk, *Minnesota Med.* **30**, 849 (1947).
107. W. L. Jellison, D. C. Epler, E. Kuhns, and G. M. Kohls, *Public Health Rept.* **65**, 1219 (1950).
108. Communicable Disease Center Technology Branch, *Summary of Investigations* **14** (April–Sept. 1958).
109. B. Schmidt, *Z. Hyg. Infektionskrankh.* **127**, 139 (1947).
110. M. I. Tsareva, *Zh. Mikrobiol. Epidemiol. Immunobiol.* **7/8**, 48 (1945).
111. V. S. Silchenko, *J. Microbiol.* (USSR) **28**, 788 (1957).
112. W. R. N. Lindsay and J. W. Scott, *Can. J. Public Health* **42**, 146 (1951).
113. K. Popek et al., *Zentr. Bakteriol., Parasitenk. Abt. I, Orig.* **210**, 502 (1969).
114. L. M. Khatenever, *Tularemia Infections,* Union Institute of Experimental Medicine, Moscow, 1943, Chap. 6, pp. 92–109.
115. I. W. Gibby, P. S. Nicholes, J. T. Tamura, and L. Foshay, *J. Bacteriol.* **55**, 855 (1948).
116. J. J. Dizon, "Carriers of Cholera El Tor in the Philippines," in, *Proceedings of the Cholera Research Symposium,* PHS Pub. #1328, 322 (1965).
117. A. S. Benenson, S. Zafar Ahmad, and R. O. Oseasohn, "Person to Person Transmission of Cholera," in *Proceedings of the Cholera Research Symposium,* PHS Pub. #1328, 332 (1965).
118. R. Pollitzer, *Cholera,* WHO Monograph Series No. 43, World Health Organ., Geneva (1959), p. 867.
119. C. H. Yen, *Bull. World Health Organ.* **30**, 811 (1964).
120. R. Sinha, *Lancet* **1**, 1162 (1969).
121. C. H. Yen, "Cholera in Asia," in *Proceedings of the Cholera Research Symposium,* PHS Pub. #1328, 346 (1965).
122. R. Pollitzer, "Cholera Advances in Historical Perspective," in *Proceedings of the Cholera Research Symposium,* PHS Pub. #1328, 380 (1965).
123. National Communicable Disease Center, *Morbidity and Mortality Weekly Rept.* **19**, 349 (Sept. 5, 1970).
124. H. L. Smith, Jr., *Proceedings of the Cholera Research Symposium,* PHS Pub. #1328, 4 (1965).
    Cholera Research Symposium, PHS Pub. #1328, 4 (1965).
125. T. P. Pesigan, "Studies on the Viability of El Tor Vibrios in Contaminated Foodstuffs, Fomites, and Water," in *Proceedings of the Cholera Research Symposium,* PHS Pub. #1328, 317 (1965).

126. J. F. D. Shrewsbury and G. J. Barson, *J. Pathol. Bacteriol.* **74**, 215 (1957).
127. S. C. Pillai, M. I. Gurbaxani, and K. P. Menon, *Indian Med. Gaz.* **87**, 117 (1952).
128. A. H. A. Gareeb, *J. Hyg.* **58**, 21 (1960).
129. R. Sen and B. Jacobs, *Indian J. Med. Res.* **57**, 1220 (1969).
130. A. E. Greenberg and E. Kupka, *Sewage Ind. Wastes* **29**, 524 (1957).
131. V. Gaustad, *Acta Tuberc. Scand.* **21**, 281 (1947).
132. F. J. W. Miller and J. P. Anderson, *Arch. Diseases Childhood.* **29**, 152 (1954).
133. S. Hellerstrom, *Acta Dermato-Venereal.* **31**, 194 (1951).
134. W. B. Schaefer and C. L. Davis, *Amer. Rev. Respirat. Diseases* **84**, 837 (1961).
135. A. T. Laird, G. L. Kite, and D. A. Stewart, *J. Med. Res.* **29**, 31 (1913).
136. K. E. Jensen, *Bull. World Health Organ.* **10**, 171 (1954).
137. D. Pramer, H. Heukelekian, and R. A. Ragotzkie, *Public Health Rept.* **65**, 851 (1950).
138. S. M. Kelly, M. E. Clark, and M. B. Coleman, *Amer. J. Public Health* **45**, 1438 (1955).
139. A. Koser, *Gesundh. Ing.* **75**, 392 (1954).
140. L. Coin, M. L. Menetrier, J. Labonde, and M. C. Hannoun, *Second International Conference on Water Pollution Research,* Pergamon, New York, 1964.
141. Tennessee Department of Public Health, *Epidemiologic Studies of Atypical and Fast Bacilli in Tennessee,* Public Health Service Grant No. CC00078, (1966), 133 pp.
142. E. Kroger and G. Trettin, *Zentr. Bakteriol., Parasitenk. Abt. I, Orig.* **157**, 206 (1951).
143. C. Rhines, *Amer. Review Tuberculosis,* **31**, 493 (1935).
143a. M. Mannsfeld, *Acta Soc. Biol. Latviae* **7**, 153 (1937).
144. B. Schmidt, *Z. f. Hyg.* **133**, 481 (1952).
145. W. Rudolfs, L. L. Falk, and R. A. Ragotzkie, *Sewage Works J.* **22**, 1261 (1950).
146. S. L. Chang, *World Health Organ. Bull.* **38**, 401 (1968).
147. S. Kelly and W. W. Sanderson, *J. Water Pollution Control Fed.* **32**, 1269 (1960).
148. J. W. Mosley, *Transmission of Viruses by the Water Route* (G. Berg. Ed.), Wiley-Interscience, New York, 1965.
149. R. Viswanathan, *Indian J. Med. Res., Suppl.* **45**, 1 (1957).
150. N. A. Clarke and S. L. Chang, *J. Amer. Water Works Assoc.* **51**, 1299 (1959).
151. D. T. Gavan and J. W. Nutt, *Arch. Environ. Health* **20**, 523 (1970).
152. S. R. Weibel, F. R. Dixon, R. B. Weidner, and L. J. McCabe, *J. Amer. Water Works Assoc.* **56**, 947 (1964).
153. M. D. Yow et al., *Amer. J. Epidemiol.* **92**, 33 (1970).
154. J. A. Bell et al., *J. Amer. Med. Assoc.* **157**, 1083 (1955).
155. H. M. Foy, M. K. Cooney, and J. B. Hatlen, *Arch. Environ. Health* **17**, 795 (1968).
156. S. Kelly and W. W. Sanderson, *Public Health Rpts.* **76**, 199 (1961).
157. J. A. Kasel, L. Rosen, and H. E. Evans, *Proc. Soc. Exp. Biol. Med.* **112**, 979 (1963).
158. D. L. Lundgren, W. E. Clapper, and A. Sanchez, *Proc. Soc. Exp. Biol. Med.* **128**, 463 (1968).
158a. N. A. Clarke, R. E. Stevenson, and P. W. Kabler, *J. Amer. Water Works Assoc.* **48**, 677 (1956).

159. E. Lycke, S. Magnusson, and E. Lund, *Arch. Ges. Virusforsch.* **17**, 409 (1965).
160. L. Squeri, A. Ioli, and M. L. Calisto, *Boll. Ist Sieroterap, Milan.* **47**, 595 (1968).
161. G. Joyce and H. H. Weiser, *J. Amer. Water Works Assoc.* **59**, 491 (1967).
162. D. M. McLean and J. A. Brown, *Health Lab. Sci.* **3**, 182 (1966).
163. G. F. Carlson, F. E. Woodard, D. F. Wentworth, and O. J. Sproul, *J. Water Pollution Control Fed.* **40**, R89 (1968).
164. G. A. Bagdasaryan, *J. Hyg. Epidemiol. Microbiol. Immunol.* **8**, 497 (1964).
165. D. L. Belding, *Basic Clinical Parasitology,* Appleton-Century-Crofts, New York, 1958, 469 pp.
166. M. H. Ivey and F. B. Engler, Jr., *Missouri Med.* **57**, 30 (1960).
167. G. R. Healy, I. S. Kagan, and N. N. Gleason, *Health Lab. Sci.* **7**, 109 (1970).
168. G. R. Healy et al., *Public Health Rept.* **84**, 907 (1969).
169. H. Sinnecker, *Z. Ges. Hyg.* **4**, 98 (1958).
170. C. A. LeMaistre et al., *Amer. J. Hyg.* **64**, 30 (1956).
171. H. Kott, N. Buras, and Y. Kott, *J. Protozoal. Suppl.* **13**, 33 (1966).
172. W. C. Boeck, *J. Hyg.* **1**, 527 (1921).
173. S. L. Chang, *Amer. J. Hyg.* **61**, 103 (1955).
174. G. T. Moore et al., *New Engl. J. Med.* **281**, 402 (1969).
175. R. J. Duma, H. W. Ferrell, E. C. Nelson, and M. M. Jones, *New Engl. J. Med.* **281**, 1315 (1969).
176. L. Cerva, K. Novak, and C. G. Culbertson, *Amer. J. Epidemiol.* **88**, 436 (1968).
177. M. Fowler and R. F. Carter, *Brit. Med. J.* **2**, 740 (1965).
178. S. L. Chang, 10th International Cong. of Microbiol. Mexico, 1970.
179. A. E. Greenberg and B. H. Dean, *Sewage Ind. Wastes* **30**, 262 (1958).
180. N. R. Stoll, *J. Parasitol.* **33**, 1 (1947).
181. E. J. Hamlin, *Surveyor* **105**, 919 (1946).
182. P. H. Silverman and R. B. Griffiths, *Ann. Trop. Med. Parasit.* **49**, 436 (1955).
183. I. Iwanczuk and I. Stobnicka, *Wiadomosci Parazytol.* **14**, 407 (1968).
184. W. Baumhogger, *Z. Hyg. Infekt Kr.* **129**, 488 (1949).
185. W. L. L. Wang and S. G. Dunlop, *Sewage Works J.* **26**, 1031 (1954).
186. H. Gaertner and L. Mueting, *Z. Hyg. Infektionskrankh.* **132**, 59 (1951).
187. Study Group on the Ecology of Intermediate Snail Hosts of Bilharziasis, *World Health Organ.* Tech. Rept. No. 120 (1957).
188. G. A. Swanson, H. R. Erickson, and J. S. Mackiewicz, *N.Y. Fish Game J.* **7**, 77 (1960).
189. S. C. Pillai et al., *Bull. Nat. Inst. Sec. India* No. 10 (1955).
190. A. M. Sologub, *Hyg. Sanitation, Moscow* **11**, 13 (1957).
191. H. Emili and P. S. Tomasic, "The Health Aspects of Polluted Water with Special Reference to the Epidemiology of Water-Borne Infections," Water Pollution in Europe, Fourth European Seminar for Sanitary Engineers, 22 (1954).
192. H. Heukelekian and M. Alfanese, *J. Sewage Ind. Wastes* **28**, 1094 (1956).
193. F. W. Gilcreas and S. M. Kelly, *J. New Engl. Water Works Assoc.* **68**, 255 (1954).
194. W. L. Newton, H. J. Bennett, and W. B. Figgat, *Amer. J. Hyg.* **49**, 166 (1949).
195. E. B. Cram, *Sewage Works J.* **15**, 1119 (1943).
196. W. W. Mathews, *Annual Report Sewage Treatment Works for 1956,* Gary Sanitary District, Gary, Ind., 1956.

197. N. A. Clarke, R. E. Stevenson, S. L. Chang, and P. W. Kabler, *Amer. J. Public Health* **51**, 1118 (1961).

198. R. E. McKinney, H. E. Langley, and H. P. Tomlinson, *Sewage Ind. Wastes* **30**, 1467 (1958).

199. E. E. Geldreich, H. F. Clark, and C. B. Huff, *J. Water Pollution Control Fed.* **36**, 1372 (1964).

200. B. England, R. E. Leach, B. Adame, and R. Shiosaki, in *Transmission of Viruses by the Water Route* (G. Berg, Ed.), Wiley-Interscience, New York, 1965.

201. E. I. Gonchariuk, G. V. Savchenko, and M. A. Levyant, *Gigiena i Sanit.* **31**, 13 (1966).

202. P. W. Kabler, *Sewage Ind. Wastes* **31**, 1373 (1959).

203. R. C. Loehr, *Pollutional Implications of Animal Wastes—A Forward Oriented Review,* Robert S. Kerr Water Research Center, Dept. of the Interior, Ada, Okla., July 1968.

204. C. H. Wadleigh, *Wastes in Relation to Agriculture and Forestry,* Publication No. 1065 U.S. Dept. of Agriculture, 1968.

205. J. D. Clise and E. E. Swecker, *Public Health Rept.* **80**, 899 (1965).

206. R. H. L. Howe, *Water Wastes Eng.* **6**, A-14 (1969).

207. D. J. Van Donsel, E. E. Geldreich, and N. A. Clarke, *Appl. Microbiol.* **15**, 1362 (1967).

208. D. J. Van Donsel and E. E. Geldreich, "Relationships of Salmonellae to Fecal Coliforms in Bottom Sediments," Water Research, In Press.

Part IV
Pollution and Community Ecology

10 Pollution Controlled Changes in Algal and Protozoan Communities

John Cairns, Jr., and Guy R.
Lanza, Center for Environmen-
tal Studies and Department of
Biology, Virginia Polytechnic
Institute and State University,
Blacksburg, Virginia

The environmental requirements of most species of algae and protozoa are inadequately known. Usually laboratory studies involve relatively simple, stable conditions without interspecific competition, succession, accessory organics, or other features of more complex natural habitats. Field studies of algae and protozoa usually are limited to species lists together with details of associated macro-environmental conditions, the latter for a limited area over a narrow time span.

10-1. Use of Structural Changes in Algal and Protozoan Communities to Assess Pollution

A. General Effects of Stress Caused by Industrial and Municipal Wastes and Agricultural Runoff. Stresses caused by wastes have a variety of effects on algal and protozoan communities: (*a*) reduction of the number of species present; (*b*) an increase in the *range* of numbers of individuals per species; (*c*) a reduction in the colonization rates by creating environmental conditions unfavorable to potential colonizing species both before and after they arrive at the community; (*d*) changes in selective predator or parasite pressure upon particular segments of the algal and protozoan communities resulting in a shift in balance within the community; and (*e*) a shift in dominance within the community favoring some species over others. Probably most or even all of these are simultaneously operative in most polluted aquatic ecosystems.

1. DIRECT EFFECTS.   At the moment, change in species diversity seems to be the most reliable and generally applicable means of assessing the biological effects of pollution. The response pattern of a community to all forms of pollutional stress is a reduction in the complexity of the community as evidenced by a reduction in the number of species present. This is usually accompanied in the first stages of stress by a widening of the disproportion in the number of individuals per species. That is, in a "healthy" community (*i.e.*, one unaffected by pollutional stress) the range in numbers of individuals per species is comparatively narrow. As pollutional stress is applied to the system a number of species are usually eliminated or reduced in number of individuals while other more tolerant species become even more abundant, thus widening the range of individuals per species in the system. As stress increases still further, all species are eventually affected and are either eliminated or decline severely in abundance. Thus, the community follows the same general dose-response pattern as a single population. In a population, there is usually a zone of no response (in which low concentrations of toxicants or pollutional stress have no apparent effect), a range of graded response (in which the response increases as the stress increases), and a zone in which the community is capable of no further response with increase in concentration. If one regards a species or a community as an information system about the environment, then a biological definition of pollution might be any change in environmental quality for which the species or community has inadequate information and is thus incapable of an appropriate response. This would include the introduction of a compound which had no counterpart in nature, or a range of temperature change not previously experienced in the habitat, or a temperature change which occurred at an inappropriate time in relation to other temperature changes.

Pollution could thus be defined as the appearance of some environmental quality to which the exposed community had an inadequate response. The inadequate response to the environmental quality could result from lack of previous exposure; occurrence at an inappropriate time, or concentration; or exposure to a new combination of conditions. A community or aggregation of species follows the same general dose-response pattern if one substitutes species for individuals in the foregoing sequence. That is, for most forms of pollutional stress, there is a range of concentrations (or temperatures, etc.) which has no effect upon the number of the species present. There is a zone of graded response which initially eliminates the marginal or poorly adapted species, and as concentrations increase, gradually affects all of the species present. Finally when all species have responded fully, further increases in concentration produce no additional effect. Thus, a rather simple response pattern to stress is characteristic of both populations and communities, and may be used to assess the effect of stress upon them.

The first major use of this generalized response pattern to detect pollution is shown in Figures 10.1 and 10.2 in which diatometers (a simple instrument designed to suspend glass slides slightly below the water surface) were placed in several Pennsylvania streams (1). The results clearly show both the reduction in species and the increased range of numbers of individuals per species in the community.

Patrick, in an earlier study (2), showed that pollutional stress caused reduction in numbers of aquatic species present in the Conestoga River system. Different waste discharges, ranging from sewage to effluents from industrial manufacturing plants, reduced the number of species present. The intermediate response, termed "semi-healthy" by Patrick, was characterized by a reduction in the number of species of certain "higher" forms

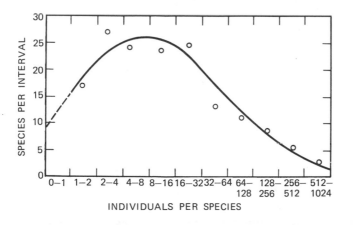

Figure 10.1 Diatom population for November 1951 from Ridley Creek, Chester County, Pennsylvania, a stream not adversely affected by pollution.

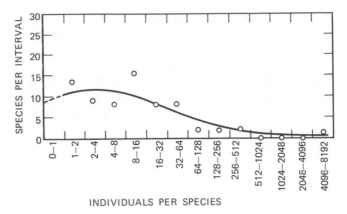

Figure 10.2 Diatom population for November 1951 from Lititz Creek, Lancaster County, Pennsylvania, a "polluted stream."

of aquatic organisms such as fish, which was accompanied by an increase in the number of species of the "lower" organisms. Presumably this increase was the result of decreased predation and increased food, in those cases where the wastes were organic. Figure 10.3 illustrates this point. Ultimately, of course, as is shown by the "very polluted" histogram, virtually all species are eliminated.

A study of the protozoa of the Conestoga River Basin (3) also indicated that under certain circumstances (*e.g.*, in low nutrient ecosystems), a small increase in organic loading of the stream might lead to an increase in numbers of individuals per species and numbers of species as well. However, further increases would produce the pattern already discussed. This merely indicates that organic compounds capable of being utilized as nutrients have an optimal concentration in terms of numbers of individuals and numbers of species maintained in each receiving system.

2. INDIRECT EFFECTS. Algae and protozoa may be affected by pollutants indirectly if the pollutants eliminate organisms which prey upon algal and protozoan communities (4). Many predators are selective and therefore may reduce populations of one species more than another. Populations with reduced predator pressure which are not susceptible to the pollution can expand. If the predators are removed or reduced in number, the competitive advantage of the species with low predator pressure is reduced and other species may become dominant. If one were not studying the entire system, an attempt might be made to interpret these changes as direct effects of pollutional stress and to correlate them directly with changes in

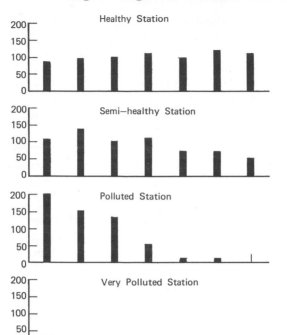

Figure 10.3 Population structure of aquatic communities under different degrees of pollution. Colum I: Species of diatoms, blue-green algae, and green algae which are known to be tolerant of pollution. 4 species = 100 per cent. (these figures represent the average obtained from sampling several hundred areas in eastern coastal plain streams). Column II: Oligochaetes, leeches, and pulmonate snails. 6 species = 100 per cent. Column III: Protozoa. 41 species = 100 per cent. Column IV: Diatoms, red algae, green algae other than those included in Column I. 81 species = 100 per cent. Column V: Prosobranch snails, triclad worms, and a few smaller groups. 11 species = 100 per cent. Column VI: Crustaceans and insects. 47 species = 100 per cent. Column VII: Fish. 20 species = 100 per cent.

the chemical and physical environment, when in fact these factors may have had little or no direct effect upon the algal and protozoan species.

Evidence for the effect of presence or absence of a predator upon invertebrates has been provided by Brooks and Dodson (5). They showed that predation by *Alosa pseudoharengus* (the alewife) upon lake zooplankton eliminated the usual large-size crustacean dominants (*e.g., Daphnia, Diaptomus*) which were replaced by small sized, basically littoral species, especially *Bosmina longirostris*. The antagonistic demands of competition and predation apparently determine the body size and therefore the species of the dominant herbivorous zooplankters present in the water. This, in turn, would cause a shift in the types and range of smaller particles, including certain algae and protozoans upon which these organisms might graze. This type of effect has also been shown by Hrbacek, *et al.* and Hrbacek (6, 7) as well as by Pennington (8), who described the rapid replacement of rotifers by *Daphnia* in tub cultures of microalgae. It has been generally recognized that changes in community structure and types of algae and protozoans may affect higher organisms. It has been less well recognized that changes in population structure in higher organisms may have a comparable effect upon the algal and protozoan populations.

B. Assessment Methodology

1. MEASUREMENT OF CHANGES.   Currently we know enough about pollu-tion-caused changes in algal and protozoan communities to develop criteria for monitoring water quality. That is, if someone pollutes a lake or river, we can use changes in microbial community structures to detect it. The evidence is usually based on identification of species (9) or a method giving similar evidence but which does not require formal identification (10a, 10b). However, we do not understand adequately the interactions operative in maintaining the structure or how these relate to the function of the entire community.

It is quite evident that changes in algal and protozoan communities may be used to assess pollution after it has occurred. However, the methods discussed have several minor drawbacks as tools for the maintenance of water quality.

1. Despite the fact that microbial communities usually respond quite rapidly to pollutional loading changes, there is a significant time lag be-tween the onset of toxic conditions and the determination of the response of the microbial community.

2. The method lacks precision in the sense that there are normal and natural fluctuations in numbers of species present, as well as in the density of individuals per species. That is, the system oscillates around a mean number of species and oscillates even further about a mean number of individuals (11). In order to separate shifts in the structural characteristics of the community caused by pollution from natural cyclic changes, one must wait until the change has significantly exceeded natural variation. This tends to reduce the precision of the method, but this reduced precision has some value in the sense that when a significant change in structure is recognized, it is quite evident that immediate action should be taken to correct the situation. One should, of course, have a reference area with which the samples from the affected area may be compared. It is quite

unlikely in a natural situation that communities at two different locations would be identical qualitatively (12) or that they would be oscillating in perfect phase with each other. It is also unlikely that they would be identical quantitatively.

2. AUTOMATION POTENTIAL. Microbial communities furnish ideal material for automating the biological assessment of pollution. Automation will probably ultimately be essential for large-scale environmental quality control. Basically, there are only two ways of describing species: by form and by function. The form of most algal and protozoan species has been used by classical taxonomists for generations and is too well understood to require explanation here. Function is used to classify bacteria as well as being increasingly used to supplement the descriptions of form of higher organisms. However, we know relatively little about the cause–effect pathways among aggregations of microbial species and at present must describe microbial communities, and the changes in them caused by pollution, in terms of the kinds of species present.

The use of species lists in describing a microbial community is essentially a description of the number of different kinds of particles present. By considering a diatom community as an aggregation of particles of different size and structure, one might generate a curve quite similar to that developed using traditional taxonomic methods without even knowing the Latin names or hierarchical positions of the species present. Unfortunately, technical difficulties have made this impractical for widespread use at the moment, but with the use of laser holography or some comparable technique, the evaluation of an array of living particles should be possible without using traditional taxonomic methods.*

The use of function as a pollution assessment technique might also be automated by using changes in an aquatic community's biochemical profile as an index of pollutional effects. This is a promising area for exploration, but unfortunately has received little attention. We are attempting some exploratory work in our own laboratory along these lines (13) but have not yet developed any operative techniques of the kind described.

10-2. Basic Categories of Pollution

It is generally well known that changes in the biological and chemical interactions of a complex aquatic ecosystem may be induced by pollutants.

---

* We have not overlooked Coulter Counters—they just do not have sufficient precision for a community of nearly 200 species, many very similar in shape and size.

Pollution can result from a variety of conditions, often appearing as the result of excesses of (a) nontoxic organic and inorganic substances, (b) thermal changes, (c) toxic substances, (d) suspended solids, and (e) radioactive materials. The careful application of research involving algal and protozoan communities can lead to productive solutions in dealing with the degradation of the aquatic environment.

It has been noted that, generally, the effect of temperature changes in the range between the extremes of biological tolerance is both less direct and evident. In nature, temperature increases result in (a) a decrease in dissolved oxygen concentration from about 14 ppm at 0°C to about 9 ppm at 20°C; and (b) increases in metabolic rates and, consequently, the amount of dissolved oxygen organisms remove per unit of time from the aquatic system. The result is an increased probability of exhausting the oxygen supply in warmer habitats containing much organic matter (14). Increases in organics, accompanied by rises in temperature in the aquatic system, could also be further complicated by the numerous other existing interactions. A few examples of the many possible direct and indirect effects upon living organisms are (a) changes in the toxic potential of ammonia resulting from pH changes in the environment or hypoxia in the test organisms (15, 16), and (b) the reaction of vital substances such as iron in the presence of increased organics and oxygen depletion (17).

It is often convenient to consider community components separately although the holistic nature of community structure and interactions must be kept in mind at all times. Algal and protozoan laboratory and field studies are useful when carefully extrapolated data from one study are utilized to supplement our knowledge of the other. Pollution induced changes in cellular physiology which manifest themselves in both dramatic and subtle changes in organisms, that is, cell division, growth, production of intercellular and extracellular toxins and other substances, can undoubtedly be expected to affect the total community structure. In our limited discussion here, we can only consider some of the major facets involved and examine selected examples from the available information.

A.   Nontoxic Organic and Inorganic Substances.   The introduction of excess organic material to an aquatic ecosystem produces a variety of effects, including increases of nutrients as organic substances, or inorganic substances resulting from decomposition and breakdown. If the amount of organic material is substantial, a major disequilibrium may result, manifested by rapid replacement of existing microbial communities by ones different in both structure and function. The established species at any one point in time and space are those whose environmental and nutrient requirements have been met, but not necessarily in that same time and space, for example, tychoplankton.

Nutrient requirements have been defined by Ketchum (18) as follows: (a) The *absolute requirement* for a nutrient is based upon the postulate that algae cannot grow, reproduce, or photosynthesize if the nutrient is lacking from the environment, and that no other nutrient can be substituted for it; (b) the normal requirement is the quantity of each nutrient contained in cells produced during active growth of a population while no nutrient is limiting (ill defined due to the variability of phytoplankton population composition); (c) the minimum requirement is the quantity of a nutrient in the cell when the nutrient is limiting the growth of the population, all other nutrients being present in excess (determines the mass of cells which can be produced for a given amount of a nutrient, provided that some other environmental factor does not limit population growth); and (d) the optimum concentration or range of concentration will permit the maximum rate of growth, reproduction, or photosynthesis of an algal population.

As pointed out previously, the interrelationship of various organisms in the community, as well as other critical factors of community interaction, must be kept in mind during any examination of research data extracted from community components in culture or from field studies. This is especially true of nutrient balances. Wright and Hobbie (19) in their study of the use of glucose and acetate by bacteria and algae in aquatic ecosystems have provided one of several good examples. In this study, specific transport systems effective at very low substrate concentrations were traced to bacteria. At the same time, a diffusion mechanism effective only at higher substrate concentrations was traced to algae. They concluded that bacteria effectively removed substrate from solution at low levels (unlike algae) and probably maintained the low substrate levels, therefore possibly preventing heterotrophic algal growth.

The arrival of organic substances, in the form of solutions and suspensions of complex organic compounds, can induce alterations in the community structure through both positive and negative influences. Of considerable significance in improving our comprehension of such influences is the recent successful cultivation of entire, though synthetic, phytoplankton algal communities in the laboratory. The effect of certain substances (glycolic acid, humic acid, octadecanol, sodium tripolyphosphate, and a proteolytic detergent enzyme) on community structure have already been evaluated (20). Such community cultures offer a potentially useful laboratory research tool for simulating complex interactions in nature. Other research on the critical role of organic substances in aquatic systems (21, 22) has also strengthened our knowledge of their influence on community structure and function.

Sugars, organic acids, and other substances in sufficient concentrations can be utilized by many algal cultures to maintain growth in the absence of light or as a sole or supplementary source of carbon in the presence of light (23). Planktonic forms such as *Chlamydomonas* and many other flagellates are also capable of heterotrophic growth in the dark. Facultative heterotrophy in certain soil algae cultured on both artificial glucose–salts media (24), and soil extract in the presence of bacteria (25) has also been demonstrated.

Certain photosynthetic unicellular algae require (*a*) vitamin B_{12}, (*b*) thiamin, and (*c*) biotin (26), and the influence of these three substances upon the community structure is evident. Vitamins appear to affect various algal groups differentially and probably influence algal succession patterns (26, 27). Field observations and *in vitro* surveys indicate that *Cytotella nana* blooms probably depend upon levels of vitamin B_{12} (28, 29).

Severe organic pollution in a natural ecosystem has drastic effects upon the algal portion of the microbial community. Hynes (30), in his comprehensive discussion of Butcher's research on glass slide communities of several rivers (Tame, Trent, Bristol, Avon) exposed to organic pollution, cites examples of fluctuations in the numbers of algae, as well as species changes in the community structure. The normal eutrophic algal flora of many species (dominated by *Cocconeis, Ulvella,* and *Chamaesiphon*) vanished. Resistant blue-green forms (*Oscillatoria* and *Phormidium*) and green forms (*Ulothrix, Stigeoclonium tenue*) could, however, be expected to survive. The diatoms, *Gomphonema parvulum* and *Nitzschia palea,* were also resistant to organic pollution, occurring in large numbers, even in the *Sphaerotilus* zone. *Navicula* and *Surirella ovata* sometimes accompanied them.

Research involving various culture techniques and field observations has demonstrated that, generally, increases of nitrogen and phosphorus to certain levels will result in increased production of certain species and a decline in others (31–35). These changes very often alter the primary producer portion of the food web, resulting in different species interactions at this and higher levels. Studies conducted on communities which have undergone changes in their ecological structure partially due to nutrient-controlled phenomena offer much useful information. The western basin of Lake Erie, for example, displayed an increase in ammonia nitrogen and total nitrogen between 1930 and 1958, as well as an increase in total phosphorus levels between 1942 and 1958 (36). The phytoplankton biomass has increased three times between 1919 and 1963. This increase has been accompanied by spring and fall maxima which were both higher and of longer duration, and with different genera of diatoms assuming dominant roles (37).

As important as nitrogen and phosphorus may be, it would be a mistake to direct all of our attention to them. Studies on populations of *Asterionella* in several English lakes by Lund (38) demonstrated that insufficient quantities of silica may restrain further growth of the population in the presence of sufficient nitrogen and phosphorus. Also, it has been suggested that increases of monovalent ions (Na and K), which are an absolute requirement for blue-green algae (a requirement pattern not shared by other fresh water types; K is required by all algae but not Na), might strongly influence the abundance of blue-greens in the algal community at certain times (26).

Among the controversies involving the importance of specific nutrients in community growth, carbon has received increasing consideration recently. Some reports indicate that carbon, rather than phosphorus, is the controlling nutrient in the production of algal blooms. The Lange-Kuentzel-Kerr hypothesis, for example, indicates the existence of an extremely efficient mutually supportive relationship between blue-green algae and bacteria. Bacteria degrade organic matter and produce CO_2 which the algae utilize in the photosynthesis of new organic matter. At the same time the algae liberate oxygen which the bacteria, in turn, utilize in the digestion of organic matter. Nitrogen, phosphorus, and other necessary substances are cycled between the algae and bacteria and the environment during the process (39).

The tremendous complexity of nutrient-controlled changes in algal and protozoan populations, and the communities which they compose, is further complicated by the great variety of existing nutritional types. Much detailed research involving cultures of algae and protozoa has resulted in a greatly expanded knowledge of their nutrient requirements and metabolism, and much of this information is useful in the analysis of community interaction and structure.

In certain instances, entire organisms (algae, protozoa, bacteria, etc.) serve as a source of organic nutrients for other members of the same community. Certain microscopic herbivores are known to utilize *Euglena* and *Chlamydomonas* as a nutrient source, thus forming the basis of a food chain in the community (14). Populations of *Stentor coeruleus* preferred *Euglena* over *Trachelomonas* and yeasts when all three nutrient sources were available (40).

An alteration in the nutrient balance of a given ecosystem undoubtedly would lead to changes in the bacterial community. It has been shown, for example, that the bacterial content of drinking water increases greatly after the addition of phosphates (41). In certain natural waters, increases in bacterial density are directly proportional to the phosphate content of the water and the increased addition of phosphates to an aquatic ecosystem could result in increases in (*a*) bacterial content, (*b*) oxygen demand, (*c*) production of growth factors for algae, and (*d*) algal growth (42). Thus organic pollutants degrade to release phosphates. This could alter bacterial and possibly fungal community structure, which would induce changes in the community of algae and protozoa which depend upon these organisms directly as food or indirectly for their metabolic products. Aquatic fungi have not been examined adequately with reference to their roles during such ecosystem changes.

Varying periods of dominance of *Vorticella* species in an activated sludge sewage disposal plant were correlated with activity patterns of their accompanying mixed microflora. This correlation was further investigated by studying clonal populations of *Vorticella* species and their accompanying mixed microflora as culture isolates in the laboratory. Under these conditions exponential phase bacterial activity stimulated *V. octava* and stationary and declining phases stimulated *V. convallaria*. These data indicated that *V. octava* was predominantly predatory on the microflora while *V. convallaria* was predominantly dependent on the by-products produced in the medium by the microflora. In addition, bacterial activity was stimulated by the predatory activity of *V. octava* (43).

Certain protozoa are auxotrophic, requiring minute quantities of specific vitamins from the environment since they are incapable of synthesizing the vitamins themselves. Some investigators feel that vitamins are frequently in short supply in natural waters so that any increase or decrease would produce population alterations within communities (14).

Changes in the nutrient balance can markedly favor one or many species of algae or protozoa. Vitamin B_{12} has been detected during a dinoflagellate bloom, though normally deficient at other times (44). This vitamin has also been found in excess during recurrent *Euglena* blooms while being in short supply at other times (45). Such evidence points to the involvement of vitamin B_{12} in these population changes. Sources of vitamin B_{12} may be found in the flora naturally occurring in the water, or land surface drainage into coastal waters, and recently the importance of rainwater as a source of vitamin B_{12} has been demonstrated (27) and correlated with *Chlamydomonas* blooms.

Studies of the green flagellates in laboratory cultures have indicated that the basic requirement for minerals is similar to that of the green plants and that these nutrients must be available in specific proportions to promote adequate growth. Nitrogen and phosphorus are, thus, capable of assuming the role of limiting factors when they exist in less than minimal amounts. In addition, at concentrations in excess of optimum, these two nutrients may repress growth of some species. Other nutrient requirements must also be filled by Zn, Mn, Ca, Mg, K, Fe, B, S, Li, Cu, and Mo, and these substances must be present in sufficient quantity and in balanced proportions.

A study of the ecology of plankton in a small artificial pond demonstrated that the amount of phosphate seemed to influence which phytoplankters appeared. When phosphate levels were above 0.02 ppm, chrysomonads were not evident while cryptomonads and diatoms increased. Euglenoids were also abundant at several times this concentration. Within each phosphate range, other factors (*i.e.,* nitrogen sources and growth factors) led to more immediate effects and the number of euglenoids appeared to be influenced by the available supply of organic nutrients. The nature and amount of food appeared to control ciliates, and many plankters were seen to exist perilously close to their minimum survival conditions so that slight environmental changes could promote significant alterations or extinction of community members (46).

B. Thermal Changes. Increased temperature as a stress factor in aquatic environments has received much attention in recent years. The rapid growth of the demand for electric power and the trend toward power grids as opposed to small self-contained electric systems are all leading in the direction of larger generating facilities. Fossil-fueled plants are now being built to produce four or five times the electricity of those built 20 years ago. Nuclear power plants, not economically feasible in small sizes, are being built to produce more electricity than conventional plants. They produce electricity less efficiently and as a result require 50% more cooling water than conventional methods (47). The increased exposure to heated waste water will have both direct and indirect effects upon aquatic ecosystems.

The influence of temperature upon microbial species may be approached from two points of view: (*a*) the effect upon species distributions, and (*b*) the effect upon the biochemical and biophysical mechanisms which allow certain species to flourish under situations which are incompatible with the survival of others (48).

Cultures of algae have demonstrated morphological and physiological alterations or differences when maintained at various temperatures. *Scenedesmus obliquus* cultures at 13.5 and 23°C produced different cell sizes, smaller cells being characteristic of the higher temperature range. The water content of the larger cell was higher so that the actual dry weights of both types were not as different as the sizes suggested. The shorter and broader small cells also had higher rates of oxygen consumption (49).

In another study, populations of *Navicula pelliculosa* and *N. seminulum* cultured at 20°C were exposed to abrupt temperature increases of 20°C for 24 hours. Overall cell content, pigmentation, and stored lipids were observed. Light microscopic examination revealed varying cellular damage, while ultraviolet microscopic surveys, utilizing 3,4-benzpyrene fluorescent stain on lipids, demonstrated a decrease in cellular fluorescence (13).

Another investigation indicated that an increase of temperature above 24°C, which provided the most abundant growth of the rhizoidal system and shoots of *Chara zeylandica,* caused the shoot growth to halt. A temperature of 32°C restricted the lateral branches to small outgrowths (50).

Temperature increases may also cause shifts in dominance. Experiments in which seeded slides (from the Sabine River) collected at temperatures of 18–20°C were exposed to a gradual increase of temperature to 40°C produced such a shift. At the beginning of the experiment, the algae on the slides were dominated by diatoms together with a few green and blue-green algae. Within the temperature range of 20–30°C, diatoms remained the dominant group component; within the range of 30–35°C, green algae dominated; and above 35°C the blue-greens, often abundant in other forms of pollution, dominated. Furthermore, when the temperature was gradually lowered, green algae and diatoms reappeared at their temperature ranges. This suggests that some organisms in each group survived when their optimum range was exceeded, though the species were incapable of successfully competing with other species better suited to a given temperature range (51).

A recent investigation of the offshore waters of 4 of Lake Michigan's 22 electric power generating facilities included studies of the effects of effluents on phytoplankton, zooplankton, and the benthos. The regional phytoplankton consisted mainly of diatoms, along with some of the typically warmer water green and blue-green algae. A slight decrease in numbers of organisms found in power plant outfall areas was noted when these were compared to reference areas. No other adverse trends were observed (52). Four years of research on the effects of condenser discharge water from the Martin's Creek Plant of the Pennsylvania Power and Light Company on river aquatic life, including microorganisms, revealed that the heated water changed species composition of microbial communities. Blue-green algae and certain members of the heat-tolerant diatom family *Fragillariacae* increased in abundance with a corresponding loss or complete supression of less tolerant forms such as green algae and many species of protozoa. There was a marked reduction in number of periphyton species when heated water was compared to the normal water of the river (53).

A preliminary report from another study on the environmental effects of condenser discharge water in southwest Lake Michigan indicated that while no changes in plankton could be attributed to plant operations, attached algae were reduced in the area of the outfall (54).

The majority of algae and protozoa are capable of living only within rather narrow temperature ranges although certain species extend their survival capacity through the process of encystment. In addition, for any one species the temperature tolerance is affected by other environmental conditions. This fact is clearly demonstrated by *Ochromonas malhamensis* which was successfully grown above 35°C, the maximum tolerated temperature under normal culture conditions, by the addition of extra vitamin B_{12}; thiamine; metals such as Fe, Mg, Mn, and Zn; and several amino acids (55). This suggests that certain organisms may be ecologically restricted, in terms of limits of temperature tolerance, because of a lack of available nutrients required at elevated temperatures. Also, it has been demonstrated that a phospholipid requirement, along with morphological abnormalities, could be induced by growth of *Tetrahymena pyriformis* at supraoptimal temperatures (56). Thus, at higher than normal temperatures, dietary requirements may be altered along with a variety of other metabolic responses and critical organism functions, that is, reproductive rate, growth and development, regeneration, encystment, etc.

A few of the other available examples of responses to temperature increases may be considered here. The responses of *Euglena* and *Chlamydomonas* to heat and light stresses provide a good example of the interlocking effects of two environmental qualities (57). At temperatures below 32.5°C the growth of *Euglena* was independent of light while *Chlamydomonas*, on the other hand, was light-dependent. At 35°C *Euglena* exhibited decreasing growth with increasing illumination. Giant, multinucleated *Euglena* cells were found at 35°C with a greater percent of such abnormal cells occurring at the higher light intensities. No such abnormal forms were noted in *Chlamydomonas*, and in the case of *Euglena* at elevated temperatures it was postulated that a dark-formed thermosensitive protein, essential for normal cellular division, is denatured. Furthermore, light could increase the effect of heat on chlorophyll and the chloroplast, possibly through conversion to intraplastidic heat through the plastid carotenoids, thus having some indirect synergistic role in the phenomenon.

Another study involved the culture of three chlorophyll-less *Euglena* substrains in light and dark at 25 and 33–35°C (58). Also employed were *Astasia longa* (a naturally colorless euglenoid) and the ciliate *Tetrahymena*. At 25°C none of the organisms were light dependent, while at high temperatures only the three *Euglena* substrains exhibited an effect of light on growth. One substrain was inhibited, as was the photosynthetic parent strain, while the other two were stimulated. Abnormal forms were produced in all three substrains at elevated temperatures, and two were multinucleated while the others had an enlarged single nucleus. Multinucleated forms were seen in a small percentage of the *Astasia*, and almost all the *Tetrahymena* cells became giants in high-temperature culture. Neither *Astasia* nor *Tetrahymena* were affected by light within the range used. While the light–temperature interactions of the nonphotosynthetic substrains of *Euglena* were unexplained, the major factor causing the abnormalities was thought to be thermal stress.

The effect of temperature on the regenerative rate of organisms is another factor of ecological significance. A study was conducted on the effects of temperature and other factors on the ability of *Blepharisma* to regenerate removed hypostomes (59). These ciliates were cultured at a constant temperature of 25°C under continuous yellow light and exposed to various temperature increases. Cells maintained at 30°C demonstrated accelerated regeneration while 35°C resulted in retarded regeneration. Brief exposures to a temperature of 40°C resulted in a retardation of regeneration. Inhibited or retarded regeneration could possibly be the result of the denaturation of enzyme protein by increased temperature. Temperature effects below 25°C were also examined, and the regeneration rate was also seen to undergo retardation (stopping at 10°C and losing pigment in a "shock" reaction with many dying at 5°C). Elevating the temperature between 13 and 30°C produced an increase in the regeneration rate.

Nutrition also affected the regenerative capacity of the cells. Starved cells had a slower regeneration rate and the diurnal rhythms of the rate of regeneration disappeared. High- and low-temperature limits for division were similar to those found for regeneration. On the basis of preliminary evidence, the investigators note that *Blepharisma* has some degree of thermal adaptation to both higher and lower temperatures than those usually tolerated.

A common and effective means of coping with unfavorable environmental conditions is cyst formation. Even the process of encystment is affected by temperature fluctuations, and *Oxytricha fallax* exhibited an accelerated rate of change into precystic forms at increased temperatures (60). For those interested in the performance of protozoan communities (*e.g.*, cropping bacteria), there is little difference between encystment and death.

The response of entire freshwater protozoan communities to heated waste waters is another significant area which has received far too little attention. The rate of species diversity restoration in freshwater protozoan communities following gross pH and temperature shocks has been determined. Freshwater protozoan communities in plastic troughs through which unfiltered Douglas Lake (Michigan) water flowed were exposed to abrupt changes, and the time required for the community to achieve a species diversity (*i.e.*, number of species) comparable to that existing before the shock was determined. Characteristics such as numbers of individuals and nutritive relationships, while important, were considered to be beyond the scope of the study.

The recovery rate was generally comparatively rapid. For example, a shock from about 20 to 50°C caused a decrease from 26 to 7 species and recovery to 18 species required 24 hours, although a recovery to the original diversity did not occur until between 72 and 144 hours. In contrast, the thermal shock from about 18 to 40°C caused a comparatively slight reduction from 31 to 22 species with recovery to approximately control diversity in 72 hours. The relatively long, but comparatively mild thermal shock from about 21°C to slightly over 30°C for about 24 hours reduced the diversity from 34 to 21 species in the initial 4-hour period and then, while this thermal shock was still in progress, there was an increase of 3 species at 12 hours.

The two most important conclusions to be drawn from this preliminary study are (*a*) the magnitude or intensity of the shock seems to be more important in reducing a protozoan species diversity than its duration, and (*b*) restoration of protozoan species diversity may require only a few hours if the shock is mild, but as much as 5 or 6 days if the shock is severe (61). The latter is probably only true of temperature, pH, and other shocks where there is no toxicant residue left in either substrate or organisms (62).

The sequence of events producing the decrease in diversity and stability of communities due to temperature increases in excess of tolerance limits is poorly understood. We have included in our discussion mostly research which demonstrates documented alterations in the structure and function of existing communities resulting from artificial temperature increases. However, some studies have indicated that heated waste water discharged into rivers may have little or no effect on algal or protozoan communities (63–65) or even on higher aquatic organisms (12, 66, 67).

C. Toxic Substances. The response to toxicants is influenced by (*a*) the nature of the toxicant; (*b*) concentration; (*c*) exposure time; (*d*) environmental characteristics of the receiving system; (*e*) age, condition, etc. of the exposed organisms; and (*f*) the presence of other toxicants. Antagonistic, additive or synergistic interactions with other substances, changes in dissolved oxygen concentration, pH, nutrient, salt balance, etc. alter the response of aquatic organisms to toxic substances. Most laboratory bioassays are carried out for short periods of time in relatively simple, stable conditions. In nature the toxic material may be altered (*e.g.*, production of methyl mercury from mercury), and may precipitate and later go into solution when environmental conditions change, or its toxicity may be modified by environmental conditions. Finally the exposure of organisms to a toxicant may be quite different in a natural situation than in a laboratory bioassay—for example, biological magnification is rare in a laboratory bioassay but common in nature. Therefore results of the type quoted here should be used as guides, not standards.

1. HEAVY METALS.   Excess heavy metals are often introduced into aquatic ecosystems through the discharge of wastes from industrial processes and acid mine drainage residues. Laboratory research involving the effects of certain metals (Zn, Cu, Cd, Li, Cr, Hg, Ni, Pb) on algae and protozoa have indicated various results at different concentrations and conditions (68–79). In addition to the various toxicity thresholds of organisms, the effects upon critical functions (*e.g.,* photosynthesis, reproduction, respiration) have also been observed and provide valuable information in understanding potential sublethal effects.

Field studies under partially controlled conditions or under natural conditions have also provided useful information about the effects of certain heavy metals on community structure. In one study (80) glass slide colonies of periphytic communities in four outdoor canals in Ohio were utilized in a study of the effects of four concentrations of zinc. The number of dominant species of the community was reduced by exposure to 1 mg/liter or more of zinc, and no species was seen which could serve as an indicator of zinc pollution. A geometric regression of the average number of dominant algal species was demonstrated, commencing with no added zinc and proceeding to that with the highest concentration of zinc. Exposure to zinc in relatively high concentrations produced comparatively low biotic diversity. Community restructuring in the form of an increase in the standing crop of fungi and slime-forming bacteria occurred at the highest zinc concentration, apparently resulting from the utilization of zinc-killed phytoplankton as their food supply.

Another study was initiated because of sporadic fish kills which occurred in the Upper Sacramento River below Keswick Reservoir in California. The results of a biological survey (81), including diatom communities, indicated that Spring Creek, one of the lower tributaries of Keswick Reservoir, and two of its branches were severely degraded by acid mine drainage. Tests on the waters and delta muds of Spring Creek confirmed the presence of high concentrations of iron, copper, and zinc along with a low pH which, undoubtedly, were the main contributors to the existing biological degradation (81).

Copper pollution was discussed by Hynes (30). Copper levels were more than 1.0 mg/liter in a river undergoing recovery from organic pollution. Serious effects upon the existing algal community structure resulted. Species above the outfall in the recovery area were sampled by using glass slides for 3 week periods and found to support *Stigeoclonium*, *Nitzschia palea*, *Gomphonema parvulum*, *Chamaesiphon*, and *Cocconeis* (in quantities of approximately 1000/mm² on the slides). Below the outfall, the number decreased abruptly to 150–200/mm² on the slides with *Chlorococcum* and the diatrom *Achnanthes affinis*, both unusual species, dominating the flora. This community structure existed for at least 3 miles; however, 5 miles below the outfall the numbers increased steeply to over 33,000/mm² and further downstream, more than 30 miles beyond the outfall, numbers in excess of 50,000/mm² were noted. Such enormous numbers were composed of not just *Chlorococcum* and *Achnathes*, but also of the original algae although *Cocconeis* did not reappear for 30 miles.

2. HALOGENS. The possibility that chlorination of heated water could result in a decrease in the algae population was indicated by the data collected in the Delaware River during a thermal study (53). Protozoan cysts are also subject to damage or death from chlorination. It has been noted that 1-hour exposure to 8 mg/liter of chlorine at 10°C killed cysts of *Entamoeba histolytica* (82). While the destruction of this human parasitic form is of great importance, the significance of the destruction of encysted free living forms must also be considered.

Experiments have demonstrated the potential of halogens in influencing algal populations under different ecological conditions. The addition of chlorine and bromine in various or equal concentrations to suspensions of *Chlorella sorokiniana* caused increasing mortality, while combined halogens did not accelerate the rate of death. Bromine caused a decrease in the numbers of *Chlorella* in the light but not in the dark. Chlorine killed algae both in the light and dark (83). In a study of the Vellar Estuary in India, molybdenum levels were found to be correlated with the chlorinity of the water; that is, the lowest concentrations of molybdenum were found in low chlorinity waters (84).

3. PESTICIDES. Information concerning the effects of pesticides upon algae and protozoa is limited. The ability of submerged plants to concentrate pesticides to several times their original levels (85) and the implications of further passage and magnification through food webs are of great significance. Also of importance are the toxic effects of the pesticide on the organism throughout its life cycle. Information from laboratory and field studies can assist us in better understanding the potential effects of pesticide residues in aquatic systems. For example, the effects of dieldrin

on populations of cultured *Navicula seminulum* var. *Hustedtii Patr.*, a moderately sensitive freshwater diatom common to many unpolluted streams, was examined in axenic culture. Concentrations of dieldrin at 1.8 ppm resulted in only a 10.6 percent reduction in cell numbers, while dieldrin at 12.8 ppm caused a reduction to half the control number at the end of a 5-day growth period. No increase in the number of cells occurred at concentrations of 32 ppm (86). In addition to the reproductive damage noted in this study, the possibility of accumulation of dieldrin in the cells should be considered. Indeed, recent studies dealing with the biological concentrations of pesticides and their effects on metabolism demonstrated that certain Cyanophyta (*Microcystis aeruginosa* and *Anabaena cylindrica*) and Chlorophyta (*Scenedesmus quadricauda* and *Oedogonium* sp.) proved to be, for the most part, highly resistant to dieldrin, endrin, aldrin, and *p,p'*-DDT at concentrations of 5, 10, and 20 µg/ml. (*M. aeruginosa*, however, was killed by less than 5 µg/ml of dieldrin, aldrin, or endrin.) The concentrations of pesticides in the algae were at least 120 times those initially occurring in the natural waters, and their concentration by algae was evident 30 minutes after exposure. In terms of function, none of the pesticides examined at the concentration of 1 ppm produced any significant effect upon the respiration of the freshwater algae tested in this study (87). However, lower levels under different conditions could induce subtle damage resulting in functional changes which could have vital consequences in community structure. A few ppb of DDT in seawater, for example, has the potential to considerably reduce the rate of photosynthesis of certain components of the phytoplankton community (88). Also, photosynthesis and growth in cultured marine algal plytoplankton isolates from various environments were affected by DDT, dieldrin, and endrin. Different tested species (*Dunaliella*, *Cyclotella*, *Coccolithus*, *Skeletonema*) exhibited different sensitivities ranging from complete insensitivity up to 1000 ppb to toxicity at 0.1–1.0 ppb of the pesticides (89). Numerous other pesticides have the potential to inflict severe damage on aquatic ecosystems, and environmental problems arising in part from their increased use may often exceed the benefits of the proposed use.

4. SURFACE-ACTIVE AGENTS. *Paramecium aurelia*, when treated with anionic surface-active agents, lost the ability to respond to a difference in electrical potential (90). Treatment with cationic and anionic surface-active agents did not affect the protozoan's normal movement towards the cathode. A nonionic surface-active agent at concentrations of 1.0, 5.0, and 10.0 mg/liter stimulated the growth of Chlorophyta for the first few days but later caused a retardation in growth. Initial growth retardation occurred at concentrations of 50 mg/liter. A cationic surface-active agent

applied at 5.0 mg/liter hindered growth of Chlorophyta, while an anionic surface-active agent applied at 10.0 mg/liter proved to be slightly toxic for 25 days, at which time growth commenced (91). In another study, the blue-green alga, *Oscillatoria*, exhibited growth inhibition at concentrations of 5.0 mg/liter of either of two ammonium chloride cationic complexes (Arquad C and Arquad D) (92). The effects of surface-active agents on kelp were also investigated and sodium dodecyl sulfate at concentrations of 10 mg/liter and zephiran chloride at concentrations of 1.0 mg/liter produced marked inhibiting effects (93). In the same investigation, *Macrocystis pyrifera* displayed 50% inactivation of photosynthesis in the bottom fronds during 4 days exposure at concentrations of zephiran chloride less than 1.0 mg/liter while concentrations of 5–10 mg/liter of sodium dodecyl sulfate were needed in order to produce similar effects. Alkyl benzene sulphonate produced complete inactivation in 48 and 96 hours at 2 and 5 mg/liters, respectively, but 0.5 mg/liter produced only slightly detrimental effects (72, 73, 94).

5. REDUCED INORGANIC MATERIALS. Paper plant effluents are potential sources of large quantities of sulfide-reduced compounds (95). While the specific effects of these substances upon algae and protozoa are not well defined, an investigation has been conducted on H_2S toxicity which provides some interesting insights on spatial relationships (96). In systems where anaerobic bottom conditions exist with H_2S concentrated below and limited dissolved oxygen concentrations above, protozoa capable of tolerating low oxygen concentrations but incapable of H_2S tolerance are restricted to the upper limited oxygen layer. Examples of this type are *Spirostomum ambiguum*, *Halteria grandilenna*, and *Paramecium caudatum*. In another investigation the influence of both H_2S and oxygen was noted with reference to movement of *Loxodes rostrum* in a pond (97). The protozoa moved out of the H_2S containing bottom layer, but also avoided the zone of higher oxygen content, and as a result of such layer restriction were found in a narrow stratum in sufficient numbers to be visible to the naked eye as a white layer. In general, with the exception of sapropelic species, H_2S is more toxic to protozoa than other gases such as H_2, CO_2, or CH_4 (98).

D. Suspended Solids. The range of variation of suspended solids concentrations in natural systems is enormous. For example, McCarthy and Keighton in 1964 (99) reported that sediment concentrations for the Delaware River at Trenton, New Jersey, ranged from 1 to 4,100 ppm. A variety of both direct and indirect effects of suspended solids upon aquatic organisms has been discussed (100). These are: (*a*) mechanical or abrasive action; (*b*) blanketing action or sedimentation; (*c*) reduction of light penetration; (d) availability as a surface for growth of bacteria, fungi, etc.; (e) adsorption and/or absorption of various chemicals; and (f) reduction of temperature fluctuations. An example of pollution by excess suspended solids was provided by a study of the influence of highway construction on a stream. Here, drastically reduced production at all levels of the biotic community due to sediments occurred. During the period of heavy sedimentation there was an apparent reduction of 61% in primary producers. Accompanying this effect was a 58% reduction in the energy required by the heterotrophic community members (101).

E. Radioactive Materials. Consideration of radioactive wastes and/or extraneous materials as pollutants affecting, directly and indirectly, the community structure of algae and protozoa has recently received increasing attention because of plans to utilize atomic power plants as energy sources, and increased use of isotopes in hospitals and industry. Algae, other lower forms of life, and higher aquatic vegetation accumulate isotopes by both adsorption and absorption. Generally, lower organisms are more resistant to ionizing radiations than higher forms and the acute dosage causing 50% mortality in algae ranged between 8,000–100,000 Roentgens while the range which produced 100% mortality was between 250,000–600,000 Roentgens. Protozoan susceptibility to radiation was noted within a range of 10,000–300,000 Roentgens which induced 50% mortality and 18,000–1,250,000 Roentgens causing 100% mortality (102). These ranges are not likely to occur in natural systems.

Accumulation and concentration of radionuclides in aquatic communities is dependent upon many interrelated phenomena including the component organisms, metabolic pathways and the complexity of the aquatic ecosystem. The uptake of a radionuclide depends upon the ratio of the radionuclide to the nonradioactive form of the same element (103).

The recent studies of the uptake and accumulation of radioactive materials by both living and dead plankton are of interest (104). The accumulation of ^{32}P and ^{65}Zn by living and formalin-killed plankton was studied in an experiment where the plankton were collected above a nuclear reactor, enclosed in plankton nets, and permitted to float through the reactor fallout areas. Accumulation of ^{65}Zn by living plankton was significantly lower than that of the dead plankton. On the other hand, the accumulation of ^{32}P by living plankton appeared to be higher than that of dead plankton. It was noted that the ability of algae to accumulate phosphorus in excess of their immediate metabolic needs may be responsible for this difference.

The following generalizations are of obvious ecological significance in dealing with communities (105):

1. Radioactive materials are taken into the body of an organism either through physiological processes and incorporated directly into the tissues, or they are absorbed on the surfaces of the organism. In general, adsorption and absorption are governing mechanisms for the lower forms of life while ingestion is the principle route for predators.

2. The concentration of certain radioelements reaches a higher level in the lower plant and animal forms, such as bacteria, protozoa, and phytoplankton, than in higher forms such as vertebrates. In such instances there is an inverse correlation between the complexity of body structure and the concentration of the radioelement in question.

3. Certain plants and animals have a predilection for concentrating specific radionuclides in certain tissues or organs. Iodine, for example, is concentrated in the thyroid, silicon in the tests of diatoms, calcium in the shells of mussels, calcium and phosphorus in the boney skeleton of vertebrates, and cesium in soft tissues.

4. Although certain radioelements may occur in amounts acceptable for drinking water, many freshwater organisms have the ability to concentrate them to levels that might be harmful.

10-3. Conclusions

We have tried to show that algal and protozoan species frequently have complex requirements that may rival those of the higher organisms. Aggregations of these species (communities) in nature are constantly undergoing a replacement of species as environmental conditions change and competitive advantage shifts. Although this is no different from successional processes of higher organisms, there has been a tendency to view these changes as chaotic and fortuitous. This is probably due to the short time span involved and the difficulty of determining spatial relationships in microbial communities. It is now becoming increasingly evident that the same ecological principals which apply to higher plants and animals are also valid for algae and protozoa. Microbial communities have structure which is maintained despite succession, and other homeostatic mechanisms are operative. We must stop regarding microbial communities as random aggregations of species thrown together by chance and regard them as structural communities produced by the same ecological determinants as communities of higher organisms.

Although many types of pollution exist, the response of algal and protozoan communities is generally rather simple, a reduction in complexity usually evidenced by a reduction in the number of species present. If one regards the universal microbial species pool as an information system about the natural environment, then species will always be available to replace those eliminated by normal environmental change. Any change outside the normal range will cause a disequilibrium situation not easily rectified by the usual homeostatic mechanisms. In short, the information system of the universal species pool does not fit. This results in a simplification of the system with the usual ecological consequences (*e.g.*, increased oscillation). In an earlier paper, one of us (Cairns) defined pollution as "any environmental change which alters the species diversity more than 20 percent from the empirically-determined level for that particular locale" (106). Although one might dispute the percentage, the general definition still seems applicable.

With a finite ecological base and increased pressures on this base, we must learn to use it wisely. This means (*a*) developing standards to protect the system, and (*b*) learning enough about its operational prerequisites to manage it well. At present we are sadly deficient in both areas.

Acknowledgments

A substantial portion of this manuscript was based on information gathered under a grant provided by the U.S. Department of the Interior, Office of Water Resources Research, as authorized by the Water Resources Research Center Project B-017-VA. We are deeply indebted to Bruce C. Parker and Robert A. Paterson for reviewing a draft of this manuscript and offering many helpful comments and suggestions.

Addendum

Both authors and both reviewers realize that there is no clearcut distinction between algae and protozoans. For example, *Chlamydomonas* is included in both groups. This taxonomic confusion may produce some uncertainty in the generalizations—a regrettable but unavoidable situation in the present circumstances.

REFERENCES

1. R. Patrick, M. H. Hohn, and J. H. Wallace, *Notulae Naturae Acad. Nat. Sci. Phila.* **259**, 1 (1954).
2. R. Patrick, *Proc. Acad. Nat. Sci. Phila.* **101**, 277 (1949).
3. J. Cairns, Jr., *Notulae Naturae Acad. Nat. Sci. Phila.* **375**, 1 (1965).

4. J. Cairns, Jr., in *Biological Aspects of Thermal Pollution* (P. A. Krenkel and F. L. Parker, Eds.), Vanderbilt Univ. Press, Nashville, Tenn., 1969.
5. J. L. Brooks and S. I. Dodson, *Science* **150**, 28 (1965).
6. J. Hrbacek, M. Dvorakova, V. Korinek, and L. Prochazkova, *Verhandl. Int. Verien. Limnol.* **14**, 192 (1962).
7. J. Hrbacek, *Rozprovy Cesk. Akad. Ved. Rada Mat. Prirod. Ved.* **72**, 116 (1962).
8. W. Pennington, *J. Ecol.* **29**, 204 (1941).
9. M. H. Hohn, *Western Petrol. Refineries Assoc.* **PC 56**, 5 (1956).
10a. J. Cairns, Jr., D. W. Albaugh, F. Busey, and M. D. Chaney, *J. Water Pollution Control Fed.* **40**(9), 1607 (1968).
10b. J. Cairns, Jr. and K. L. Dickson, *J. Water Poll. Control Fed.* **43**, 755 (1971).
11. J. Cairns, Jr., K. L. Dickson, and W. H. Yongue, *Trans. Amer. Micros. Soc.* **90**, 71 (1971).
12. R. Patrick, J. Cairns, Jr., and S. S. Roback, *Proc. Acad. Nat. Sci. Phila.* **118**(5), 109 (1967).
13. G. R. Lanza, J. Cairns, Jr., and K. L. Dickson, *Assoc. Southeastern Biol.* **17**, 52 (1970), Abstract.
14. L. E. Noland and M. Gojdics, *Research in Protozoology* (T. Chen, Ed.), Permagon, New York, 1967, pp. 215.
15. K. S. Warren, *Nature* **195**(4836), 47 (1962).
16. K. S. Warren and S. Schenker, *Amer. J. Physiol.* **199**(6), 1105 (1960).
17. G. K. Reid, *Ecology of Inland Waters and Estuaries*, Van Nostrand Reinhold, New York, 1961.
18. B. H. Ketchum, *Ann. Rev. Plant Physiol.* **5**, 55 (1954).
19. R. J. Wright and J. E. Hobbie, *Ecology* **47**(3), 447 (1966).
20. E. G. Bozniak, "Laboratory and Field Studies of Phytoplankton Communities." Ph.D. Thesis, Washington Univ., St. Louis, 1969.
21. P. J. Wangersky, *Am. Sci.* **53**(3), 358 (1965).
22. A. Collier, *North Amer. Wildlife Conf. Trans.* **18**, 463 (1953).
23. G. E. Fogg, *Algal Cultures and Phytoplankton Ecology,* Univ. Wisconsin Press, Madison, Wisc., 1966.
24. B. Parker, *Science* **133**(3455), 761 (1961).
25. B. Parker, *Ecology* **42**(2), 381 (1961).
26. L. Provasoli, *Eutrophication, Causes. Consequences, Correctives,* National Academy of Sciences, Washington, D.C., 1969.
27. B. C. Parker, *Nature* **219**(5154), 617 (1968).
28. R. R. L. Guillard and J. H. Ryther, *Can. J. Microbiol.* **8**, 229 (1962).
29. R. R. L. Guillard and V. Cassie, *Limnol. Oceanog.* **8**, 161 (1963).
30. H. B. N. Hynes, *The Biology of Polluted Waters,* Liverpool Univ. Press, Liverpool, 1960.
31. S. P. Chu, *J. Ecol.* **31**, 109 (1942).
32. S. P. Chu, *J. Ecol.* **31**, 109 (1943).
33. C. E. ZoBell, *Proc. Nat. Acad. Sci.* **21**, 517 (1935).
34. W. Rodhe, *Symbolae Botan. Upsalienses* **10**, 1 (1945).
35. W. T. Edmondson, in *Eutrophication, Causes, Consequences, Correctives,* National Academy of Sciences, Washington, D.C., 1969.
36. A. M. Beeton, in *Eutrophication, Causes, Consequences, Correctives,* National Academy of Sciences, Washington, D.C., 1969.
37. C. C. David, *Limnol. Oceanog.* **9**, 275 (1964).

38. J. W. G. Lund, *J. Ecol.* **38**, 1, 15 (1950).
39. R. F. Legge and D. Dingelein, *Can. Res. Develop.* 19 (1970).
40. A. A. Schaeffer, *J. Exp. Zool.* **8**, 75 (1910).
41. E. Bosset, *Monatsbull. Schweiz. Ver. Gas Wasserfachm.* **45**, 146 (1965).
42. E. A. Thomas, in *Eutrophication, Causes, Consequences, Correctives,* National Academy of Sciences, Washington, D.C., 1969.
43. R. Reid, *P. Protozool.* **16**, 103 (1969).
44. B. G. Sweeney, *Amer. J. Bot.* **38**, 669 (1951).
45. W. J. Robbins, A. Hervey, and M. E. Stevens, *Bull. Torrey Botan. Club* **77**, 423 (1950).
46. S. S. Bamforth, *Limnol. Oceanog.* **3**(4), 398 (1958).
47. T. Kolflat, *Thermal pollution—1968,* Hearings before the Subcommittee on Air and Water Pollution of the Committee on Public Works, United States Senate, 90 Cong. 2nd Sess., February 1968. U.S. Govt. Printing Office, Washington, D.C., p. 63.
48. E. Marre, *Physiology and Biochemistry of Algae* (R. A. Lewin, Ed.), Academic Press, New York, 1962.
49. R. Margalef, *Hydrobiologia* **6**, 83 (1954).
50. R. Anderson and R. Lommasson, *Butler Univ. Botan. Studies* **13**, 113 (1958).
51. J. Cairns, Jr., *Ind. Wastes* **1**, 150 (1956).
52. J. R. Krezoski, *Benton Harbor Power Plant Limnological Studies. Part III. Some Effects of Power Plant Waste Heat Discharge on the Ecology of Lake Michigan. Special Report on the Ecology of Lake Michigan*, Special Report No. 144 of the Great Lakes Research Division, The University of Michigan, Ann Arbor, Mich., 1969.
53. F. J. Trembley, *Research Project on Effects on Condenser Discharge Water on Aquatic Life. Progress Report 1956–1959*, Lehigh University Institute of Research, Bethlehem, Pa., 1960.
54. L. P. Beer and W. O. Pipes, *A Practical Approach: Environmental Effects of Condenser Water Discharge in Southwest Lake Michigan*, Staff Report, Commonwealth Edison Company, 1968.
55. S. H. Hutner, S. Baker, S. Aaronson, H. A. Nathan, E. Rodriguez, S. Lockwood, M. Sanders, and R. A. Petersen, *J. Protozool.* **4**, 259 (1957).
56. N. Rosenbaum, J. Erwin, D. Beach, and G. G. Holz, Jr., *J. Protozool.* **13**, 535 (1966).
57. J. A. Gross and T. L. Jahn, *J. Protozool.* **9**, 340 (1962).
58. J. A. Gross, *J. Protozool.* **9**, 415 (1962).
59. A. C. Giese and B. McCaw, *J. Protozool.* **10**, 173 (1963).
60. K. Hashimoto, *J. Protozool.* **9**, 161 (1962).
61. J. Cairns, Jr., *Univ. Kansas Sci. Bull.* **158**, 209 (1969).
62. J. Cairns, Jr., and K. L. Dickson, *Trans. Kansas Acad. Sci.* **73**, 1 (1970).
63. J. Cairns, Jr., *Notulae Naturae Acad. Sci. Phila.*, **387**, 1 (1966), plus 43 pp. supporting data deposited as document No. 8902 with the AID Aux. Pub. Proj. Photodupl. Serv., Library of Congress, 1966.
64. J. Cairns, Jr. and R. L. Kaesler, *Hydrobiol.* **34**, 414 (1969).
65. J. Cairns, Jr., R. Patrick, and R. L. Kaesler, *Notulae Naturae Acad. Nat. Sci. Phila.* No. 436, 1970.
66. S. S. Roback, J. Cairns, Jr., and R. L. Kaesler, *Hydrobiol.* **34**, 484 (1969).
67. J. Cairns, Jr. and R. L. Kaesler (Submitted for Publication). *Trans. Amer. Fisheries Soc.* (In Press)

68. R. Patrick, in *Water Quality Criteria*, Report of the National Technical Advisory Committee to the Secretary of the Interior. Section III, Fish, Other Aquatic Life, Wildlife, 1968, p. 234.

69. G. Bringmann and R. Kuhn, *Gesundh. Ingr.* **80**, 115 (1959).

70. G. Bringmann and R. Kuhn, *Gesundh. Ingr.* **80**, 239 (1959).

71. K. A. Clendenning and W. J. North, *Quart. Prog. Rept. Inst. Marine Resources,* Univ. California, La Jolla, IMR Ref. 58-6 (1958).

72. K. A. Clendenning and W. J. North, in *Proceedings of the 1st International Conference on Waste Disposal in the Marine Environment*, Pergamon, New York, 1960.

73. W. J. North and K. A. Clendenning, *Ann. Prog. Rept. Inst. Marine Resources,* Univ. California, La Jolla, IMR Ref. 58-11 (1958).

74. W. J. North and K. A. Clendenning, *Ann. Rept. Inst. Marine Resources,* California (1959).

75. G. P. Fitzgerald, G. C. Gerloff, and F. S. Borg, *Sewage Ind. Wastes,* **24**(7), 888 (1952).

76. T. E. Maloney and C. M. Palmer, *Water Sewage Works* **103**, 509 (1956).

77. E. N. Willmer, *J. Exp. Biol.* **33**, 583 (1956).

78. M. Suhama, *J. Sci. Hiroshima Univ., Ser. B* **20**, 33 (1961).

79. V. Tartar, *Exp. Cell Res.* **13**, 317 (1957).

80. L. G. Williams and D. I. Mount, *Amer. J. Botany* **52**, 26 (1965).

81. R. J. Benoit, J. Cairns, Jr., and C. W. Reimer, in *Reservoir Fishery Resources Symposium,* American Fisheries Society, 1968, Allen Press, Lawrence Kansas.

82. H. Kott and Y. Kott, *J. Protozool., Suppl.* **14**, 44 (1967) (Abstract).

83. Y. Kott, *J. Protozool. Suppl.* **15**, 44 (1968) (Abstract).

84. J. D. Burton, *J. Cons. Perma. Explor. Mer. Copenhague* **33**, 103 (1969).

85. C. H. Hoffman and A. T. Drooz, *Amer. Midl. Nat.* **50**, 172 (1953).

86. J. Cairns, Jr., *Mosquito News* **28**, 177 (1968).

87. B. D. Vance and W. Drummond, *J. Amer. Water Works Assoc.* **61**, 360 (1969).

88. C. F. Wurster, *Science* **159**, 1474 (1968).

89. D. W. Menzel, J. A. Anderson, and A. Randtke, *Science* **167**, 1724 (1970).

90. H. M. Butzel, Jr., L. H. Brown, and W. B. Martin, Jr., *Physiol. Zool.* **33**, 39 (1960).

91. A. J. Wurtz, *J. Water Pollution Control Fed.* **33**, 687 (1961).

92. O. B. Williams, C. R. Groninger, and N. F. Albritton, *Producers Monthly* **16**, 14 (1952).

93. E. A. Pearson, R. D. Pomeroy, and J. E. McKee, "Summary of marine waste disposal research program in California," State Water Pollution Control Board Publication No. 22 (1960).

94. W. J. North, K. A. Clendenning, and H. L. Scotten, *Quart. Prog. Rept. Inst. Marine Resources,* Univ. California, La Jolla, IMR Reference 60-10 (1959).

95. J. E. McKee and H. W. Wolf, in J. E. McKee and H. W. Wolf, *Water Quality Criteria,* The Resources Agency of California, State Water Quality Control Board, Publication 3-A, 1963.

96. H. Bick, *Arch. Hydrobiol.* **54**, 506 (1958).

97. V. W. Rylov, *Int. Rev. Ges Hydrobiol. Hydrogr.* **11**, 179 (1923).

98. J. Nikitinky and F. K. Mudrezowa-Wyse, *Centralbl. Bakt.* (abt. 2) **81**, 167 (1930).

99. L. T. McCarthy, Jr., and W. B. Keighton, Geological Survey Water-Supply Paper 1779-X. U.S. Gov. Printing Office, Washington, D.C., 1964.

100. J. Cairns, Jr., *Purdue Univ. Eng. Bull. Ext. Ser.* **129**, 16 (1967).

101. K. L. King and R. C. Ball, "The influence of highway construction on a stream," Research Report No. 19, Michigan State University Agricultural Experiment Station, East Lansing, Mich, 1964.

102. L. R. Donaldson and R. F. Foster, *Nat. Acad. Sci.–Nat. Res. Council* **551**, 96 (1957).

103. J. E. McKee and H. W. Wolf, in J. E. McKee, and H. W. Wolf, *Water Quality Criteria,* The Resources Agency of California, State Water Quality Control Board, Publication 3-A, 1963.

104. C. E. Cushing and D. G. Watson, *Oikos* **19**, 143 (1963).

105. L. A. Krumholz and R. F. Foster, *Nat. Acad. Sci.–Nat. Res. Council* **551**, 88 (1957).

106. J. Cairns, Jr., *Water Resources Bull.* **3**, 47 (1967).

11 Ecological Control of Microbial Imbalances

Ralph Mitchell, Laboratory of Applied Microbiology, Division of Engineering and Applied Physics, Harvard University, Cambridge, Mass.

11-1. Introduction

The stability of aquatic ecosystems is dependent on the diversity of the native population. In most natural waters a balance between the different communities of microorganisms has been achieved. The microbial population is extremely heterogeneous and well balanced. When there is a large input of foreign microorganisms into these systems, the population becomes unbalanced. A portion of the native microflora reacts to these perturbations in a homeostatic manner in order to restore stability to the community.

Intermicrobial predators are present in all aquatic ecosystems. There are a group of microorganisms capable of preying on algae, fungi, bacteria, and possibly on viruses. These organisms normally comprise a very small portion of the total population. Under normal conditions they survive by consuming susceptible native microorganisms. However, when a foreign microorganism enters the ecosystem in large numbers or when a native microorganism, such as an alga, is put under stress, the predator populations temporarily respond and attempt to return the biological equilibrium by consuming the pollutant. These processes are discussed in some detail in this chapter.

11-2.   Predator–Prey Relationships

A.   Bacteria as Prey.   The quantity of enteric bacteria entering natural waters has increased dramatically in recent years. Each year more sewage, frequently untreated, is dumped into rivers, lakes, and the sea. Because the sea is a natural dumping ground for untreated domestic wastes, the processes controlling destruction of enteric bacteria there have been intensively studied.

The implication of a biological agent in the destruction of enteric bacteria in natural seawater was suggested as long ago as 1885, when DeGiaxa (1) reported that the microorganisms which caused typhoid and cholera survived longer in heat-sterilized than in filter-sterilized seawater. Further support for this hypothesis came from the data of Vaccaro *et al.* (2) who showed that in North Atlantic waters *Escherichia coli* survived ten times longer in winter than in summer. A typical killing curve of *E. coli* in seawater displays a lag phase, one of rapid decline, followed by the development of resistant microorganisms. Bactericidal activity increases at higher temperatures, which would explain the more rapid die-away curves observed in the summer months. All of these studies point to a biological mechanism of destruction of enteric bacteria in the sea. This hypothesis is strengthened by the work of Saz *et al.* (3) who showed that the active factor in the destruction of *Staphylococcus aureus* in seawater is a thermolabile macromolecule, which they suggest may be associated with algal blooms.

Thermolabile antibacterial materials have been extracted from plankton (4, 5). Acrylic acid is present in extracts of the alga *Phaeocystis* (6). This acid inhibits a number of pathogenic bacteria. It is apparent that the frequently high nutrient level of water containing sewage will stimulate algae, some of which produce antibacterial materials which could be partially responsible for the destruction of enteric bacteria.

Many attempts have been made to implicate antibiotic production in the bactericidal action of seawater. However, Krassilnikova (7) reported that, despite the large number of antibiotic producing bacteria in the ocean, these microorganisms displayed insignificant antimicrobial activity when cultivated in seawater. Similarly, Carlucci and Pramer (8) concluded that antibiosis was not a significant factor in the destruction of *E. coli* in seawater. These conclusions are not altogether surprising in view of the

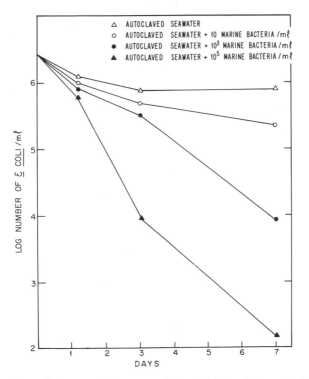

Figure 11.1 Plot of the survival curves of *Escherichia coli* in autoclaved seawater after the addition of different concentrations of marine microorganisms (9).

elaborate nutritional requirements for antibiotic production and the comparative scarcity of nutrients in the sea.

Figure 11.1 shows that there is a direct proportionality between the size of the native marine microbial population and the rate and extent of decline of the population of *E. coli* (9). When *E. coli* is added to filter-sterilized seawater there is a minimal decrease in the coliform count in the absence of marine microorganisms. In the presence of 10^5/ml marine bacteria the *E. coli* count typically declines in 5 days from 10^7 to 10^2/ml. This effect is also observed in field studies (9). Seawater sampled close to shore where there is a large native microbial population is strongly antagonistic to *E. coli*. In contrast, seawater sampled from the deep Atlantic Ocean, where the microbial count is very low, displays very weak activity against *E. coli*. It is obvious that the native marine microflora plays an important role in the destruction of intestinal bacteria in the sea.

The intestinal bacteria are utilized as a food source for a number of

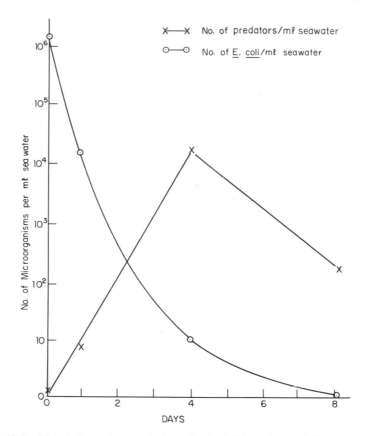

Figure 11.2 Stimulation of a population of microbial predators following the addition of *Escherichia coli* to seawater (12).

different microbial predators living in the sea (10). These antagonists can be estimated by counting clearing zones on solid media containing high concentrations of *E. coli*. Data in Figure 11.2 show that in unpolluted seawater the concentration of predators is very low. However, in the presence of large numbers of intestinal bacteria this group of microorganisms develops rapidly (11). This response is not confined to seawater and occurs in both freshwater and soil. Nor is the phenomenon specific for *E. coli*, as will be shown later in this chapter.

Different groups of predators are associated with this phenomenon. Marine bacteria capable of preying on *E. coli* by enzymatically digesting the cell walls develop in response to the addition of intestinal bacteria (9).

These predators are pseudomonads which can utilize either inert materials or cell walls of a wide range of Gram-negative bacteria as a sole carbon source.

The most interesting group of microorganisms associated with the decline of *E. coli* belongs to the genus *Bdellovibrio* (11). Halophilic forms of these tiny bacteria, which are obligately parasitic on Gram-negative bacteria, are present in the sea. The *Bdellovibrio* population in uncontaminated seawater rarely exceeds 10/ml. However, in the presence of *E. coli* it rapidly increases to 10^3–10^4/ml, causing extensive destruction of the *E. coli* population. These parasites grow well on old and physiologically incapacitated bacteria. The inability of *E. coli* to compete with marine bacteria for nutrients in the sea makes them an easy target for these parasites. Bdellovibrios presumably maintain themselves in unpolluted seawater by preying on inactive cells of Gram-negative marine bacteria.

Protozoa are the most voracious predators of *E. coli* (12). An amoeba of the genus *Vexillifera* usually develops in response to inoculation of the intestinal bacteria to seawater. This organism, when inoculated to sterile seawater at a concentration of 100 cells/ml, reduces the *E. coli* population from 10^9 to 10^4/ml in 4 days. Marine ciliates yield similar results. It is apparent that a wide range of predators and parasites respond to the addition of *E. coli* to seawater. All of these microorganisms compete for the intestinal bacteria as a nutrient source and cause rapid and effective removal of the intestinal bacteria from the sea.

Pramer *et al.* showed that artificial seawater displays strong bactericidal activity against *E. coli* (13). They expressed the opinion that the role of biological antagonists has been overemphasized and that more attention should be paid to the effect of chemical components of seawater on intestinal bacteria. However, filtered natural seawater exerts a very weak bactericidal action compared to Pramer's artificial seawater, and it seems likely that artificially formulated seawater is more bactericidal than filtered natural seawater.

Enteric bacteria die at the same rate in freshwater as in seawater. The rate of destruction is proportional to the concentration of native microorganisms in the habitat. In Charles River water, where the indigenous population is very high because of high concentrations of organic matter, the rate of destruction of intestinal bacteria is rapid. Complete eradication occurs in 2 to 5 days. A contrasting situation is found in the very clean Walden Pond water, which is almost free of bacteria. In this water the intestinal bacteria remain viable for periods as long as 2 weeks. Thus we have a paradoxical situation. Waters high in organic matter, which may be polluted, have a greater capacity for self-purification from microbial pollutants than very clean water with low concentrations of organic matter, which have little or no ability to destroy intestinal pathogens, at least initially.

The microorganisms involved in destruction of enteric bacteria in freshwater appear to be similar to those identified in seawater. Cell wall lytic bacteria and Bdellovibrios play a minor role. The ameboid and ciliate protozoa actively destroy coliforms. The observation that the difference in chemical composition between seawater and freshwater does not have any significant effect on either the rate or mechanisms of kill of *E. coli* provides further evidence to support the hypothesis that the destruction of bacterial pollutants in natural waters is caused primarily by a group of predators among the native microbial population, which develop in response to the immigration of a foreign microorganism. This predatory population consumes the foreign bacteria and, in conformity with predator–prey theory, dies back to a minimal level when the source of food is depleted.

B.   Yeasts and Fungi as Prey.   The prime source of fungal pollution in natural water is domestic sewage. Yeasts and fungi pathogenic to man have been found in coastal water off Southern California (14), and intestinal yeasts are found in the guts of seagulls living in warm tropical waters (15). Intestinal yeasts could not be isolated from temperate estuarine waters of Portuguese rivers (16). Apparently temperature is an important factor in the survival of these yeasts in estuaries, because of their inability to grow at low temperatures.

Microorganisms antagonistic to intestinal yeasts and fungi are common in seawater. Approximately 8% of all bacteria isolated from amphipods inhibit nonmarine yeasts including *Candida albicans* (17). These bacteria may be ecologically significant in destroying yeasts entering the sea from nonmarine sources. Preparations of marine algae have been tested against pathogenic yeasts and fungi (18). Extracts of 11 of 35 algae tested were strongly antagonistic to the pathogens *Candida albicans* and *Histoplasma capsulatum*. These algal extracts may be important as new sources of antifungal chemotherapeutic agents. However, one would not expect organisms which require homogenization to display antimicrobial activity to be ecologically significant in the destruction of other microorganisms in the ecosystem.

Bacteria capable of destroying nonmarine fungi by enzymatically lysing their cell walls are common in the sea (19). When mycelial mats of the terrestial fungus *Pythium deBarianum* are placed in sterile seawater, they remain intact indefinitely. Mats placed in fresh nonsterile seawater and incubated at 22° on the shaker disintegrate completely in 10 to 20 days.

Disintegration in standing culture occurs in 25 to 35 days. Repeated inoculation of the fungus into seawater presumably enriches the culture specifically for antifungal microorganisms and reduces the disintegration time to 6 days (19). Thus it is clear that an antagonistic marine microflora is responsible for the destruction of this nonmarine fungus. The predominant predator in this system is an agar decomposing pseudomonad. These predatory bacteria act by lysing the cell walls of the fungal prey in a similar manner to that observed in the natural system.

There is no apparent difference between destruction of nonindigenous fungi in seawater and in freshwater. When marine fungi are added to freshwater the mycelia degrade rapidly and predators similar to those discussed above are isolated. It appears that the predatory group of microorganisms antagonistic to foreign fungi act by lysing the fungal cell walls in both marine and freshwater ecosystems.

C. Algae as Prey.   The algal population inevitably becomes unbalanced in eutrophic waters, and one genus ultimately predominates. The populations in eutrophic waters are notoriously unstable. At the height of productivity some essential nutrient for the predominant alga frequently becomes limiting and productivity stops, leaving a massive biomass of physiologically incapacitated cells. These inactive algae are highly sensitive to predation by the native microflora.

Canter and Lund (20) have followed the periodicity of populations of the diatom *Asterionella* in a number of English lakes during a period of several years. This diatom has a spring and fall maximum. The end of the spring increase was correlated with the decline in silica concentration. The decline of the autumn growth is of interest because of the marked correlation of the decline with growth of a parasitic fungus, *Rhizophidium*. Epidemics of this parasite occur in the late autumn. Ultimately most of the diatom cells become infected and die. Figure 11.3 shows that at the beginning of the bloom 10% of the *Asterionella* culture was infected with the parasite. At the peak of the bloom infection rose to 26% of the algal cells causing a rapid decline in the algal population.

Other diatoms and the blue-green alga *Oscillatoria* become infected by this parasitic fungus (21). The normal spring increase of *Asterionella* is delayed, both by growth of *Oscillatoria* and by epidemics of *Rhizophidium*. *Oscillatoria* also becomes infected with a species of *Rhizophidium* which causes a drastic decline in the population of the blue-green alga.

A group of predators among the native microflora is associated with the decline of the dioflagellate *Chlamydomonas* in lake water and of the diatom *Skeletonema* in seawater (10). Table 11.1 shows the effect of the native microflora on the two algae. When the *Chlamydomonas* is grown in

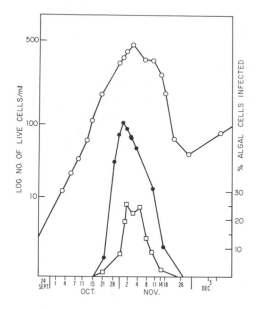

Figure 11.3 Relationship between the numbers of *Asterionella formosa* and its parasite, *Rhizophidium planktonicum,* during an autumn bloom in Esthwaite water (20). (○) Number of live *Asterionella* cells/ml; (●) number of live *Rhizophidium* cells/ml; (□) percentage of *Asterionella* cells infected by *Rhizophidium*.

sterile lake water and then placed in the dark, the death rate is insignificant in 2 weeks. However, in fresh natural lake water a 90% death rate is observed within 10 days. The effect of the native microflora on *Skeletonema* is even more dramatic. The diatom population remains intact for 2 weeks in sterile seawater, but in fresh natural seawater a 90% decline occurs in 4 days. Microorganisms responsible for destruction of both of

Table 11.1   Microbial Destruction of the Freshwater Alga *Chlamydomonas* in Lake Water and of the Marine Alga *Skeletonema* in Seawater (10)

Time, Days	*Chlamydomonas*		*Skeletonema*	
	Sterile Lake Water, No./ml	Nonsterile Lake Water, No./ml	Seawater, No./ml	Nonsterile Seawater, No./ml
0	4×10^6	9×10^5	1×10^5	2×10^5
4	4×10^6	7×10^5	1×10^5	4×10^3
7	2×10^6	2×10^5	2×10^5	—
11	2×10^6	3×10^3	—	—

these algae belong to the genus *Pseudomonas*. The predator for each alga is genus specific, and may act by degrading a portion of the cell wall, although this has not yet been confirmed.

These results and those of Canter and Lund provide an excellent example of the capacity of the native microflora to exert a stabilizing effect in natural waters, counteracting the perturbations caused by eutrophication. Unfortunately, this natural self-purification process often results in the development of anaerobic conditions and in the release of toxic or malodorous chemical products in the waters. It should be possible to make use of these predatory microorganisms to prevent the domination of the ecosystem by one alga and to prevent excessive productivity.

Safferman and Morris (22) have isolated a virus which lyses blue-green algae, and have suggested that these viruses might be used to control algal blooms. Table 11.2 shows the activity of this virus against *Plec-*

Table 11.2   Effect of a Virus Against Blue-Green Algae on an Algal Bloom[a]

Days	Algae Filament Length/ml, μ		Virus Count (Plaque Forming), Units/ml
	0–100	100–300	
0	115,000	20,000	6×10^5
3	165,000	55,000	4×10^4
6	—	—	1×10^7
7	5,000[b]	0	2×10^9

[a] From Safferman and Morris (22).
[b] None of the filaments measured more than 50 μ.

tonema boxyanum in an artificial pond. The virus concentration increased 3000-fold and the algal population was completely lysed in 7 days. The virus is active against a narrow range of algae and effective control would depend on the development of a number of viruses encompassing the whole range of blue-green algae.

Shilo (23) has isolated bacteria which are capable of lysing *Nostoc, Plectonema,* and *Anacystis,* and has discussed the possibility of using antagonistic bacteria for biological control of blue-green algae. However, it does not seem reasonable to attempt biological control by adding microorganisms to a system in which they are normally present but not dominant. The native population of microorganisms will inevitably cause the inoculated population to die back to the level normally present in that ecosystem. A more reasonable approach might involve the temporary alteration of the environment to stimulate native antagonists.

11-3. Viruses

A. Survival in Seawater. Enteric viruses apparently survive well in natural waters and pose a serious public health hazard, either by ingestion in drinking waters or by concentration in shellfish. Large quantities of viruses are present in fecal material and more than 100,000 polio particles may be present in a gram of feces (24). Neither primary nor secondary sewage treatment significantly reduces the number of enteric viruses in sewage samples, and chlorine treatment is frequently ineffective (25). The storage of moderately polluted waters at warm surface water temperatures can inactivate 99% of coxsackie virus in approximately 7 days. At colder temperatures, when the water is grossly polluted, or in the absence of the microflora, the survival time is much longer (26).

Figure 11.4 shows the effect of water temperature on different enteric viruses (27). Most viruses survive for longer than 56 days in winter and for less than 32 days in summer in estuarine waters. When an estuary becomes polluted, polio virus survives 35 days compared to 14 days in clean water. The normal marine microflora apparently is responsible for the destruction of the virus, and pollution by raw sewage tends to protect the enteric virus (27).

There is good circumstantial evidence that infectious hepatitis virus

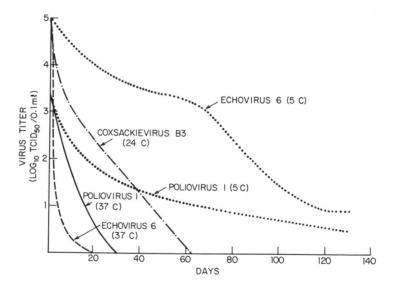

Figure 11.4 Survival of enteric viruses in seawater (27).

survives well in estuarine waters. Enteric viruses are commonly isolated at distances of more than 4 miles from sewage outfalls and in shellfish. Liu *et al.* (28) measured the uptake of polio virus into hard clams, and showed that the virus was rapidly concentrated in the shellfish with maximal viral contamination occurring within a few hours. This plateau was maintained as long as there was sufficient virus in the surrounding seawater. The virus in the digestive system was not absorbed onto, nor did it penetrate cells, and when the clams were placed in clean water the virus population declined to a nondetectable level within 48 to 96 hours.

B. Inactivation Processes. Most research on virus inactivation in natural habitats deals with seawater, probably because of the ability of shellfish to concentrate enteric viruses. Animal viruses are rapidly inactivated in fresh natural seawater. The inactivating agent is thermolabile, suggesting that it is of a biological origin (29). Antiviral activity is inhibited by proteins and amino acids. A marine bacterium, a strain of *Vibrio marinus,* has been associated with antiviral activity (30). When this bacterium is added to autoclaved seawater, the antiviral activity is restored, indicating that the bacterium excretes some as yet undefined antiviral agent.

These results are interesting in the light of Waksman's (31) earlier pessimistic discussion of the inability of his group to isolate antiviral agents from natural sources. Viruses are surprisingly resistant to inactivation. Stanley (32), emphasized the resistance of viruses to proteolytic enzymes. However, Chang (33) showed the specificity of this resistance in his observation that Group B *Arthropod*-borne viruses are more easily digested by proteases than are Group A viruses. The results indicate the danger of generalizing about the degree of resistance of viruses to inactivation by enzymes. Nonenzymatic inactivation agents may also be of ecological significance. Vavra and Dietz (34) described an antiviral agent produced by an actinomycete. This agent was active against both DNA and RNA viruses.

It is apparent that the factors controlling virus survival in seawater are very complex. Recent studies have shown that there are two components, one biological and the other chemical, controlling inactivation of the bacterial virus ϕx-174 in seawater (35). The effect of the microflora on the virus was shown in a series of experiments in which native microorganisms were returned to sterile seawater where there was virtually 100% survival during the test period. In the presence of 10^3/ml native bacteria the virus declined from 10^{12} to 10^{10}/ml in 6 days. When the native bacterial population was increased to 10^6/ml, the virus population declined from 10^{13} to 10^6/ml in 6 days. There is a direct correlation between the size of the native microbial population and the rate of inactivation of the virus. The responsible microbial antagonists have not been identified and it remains to be determined if the inactivation is caused by toxic metabolites or by antiviral enzymes. The abiotic factor can be demonstrated in filter sterilized seawater. Inactivation is even more rapid than in seawater containing microorganisms. This chemical inactivation does not seem to be associated with salinity, since the virus survives well in 2.5% NaCl. It is assumed that inactivation is caused by heavy metals in the water. The observation that chemical inactivation is even stronger than inactivation caused by components of natural seawater gives the impression that the presence of microorganisms lowers the antiviral activity of seawater. This is not surprising if one accepts the theory that the nonantagonistic portion of the marine microflora acts as a protective colloid for the virus, preventing inactivation.

This information is of the utmost importance in explaining how viruses survive in the sea. It is quite probable that virus particles coming into the sea adsorb to bacteria or algae or other debris and are carried to shellfish in this manner. The presence of gross pollution in the sea increases the chances of survival by providing a greater adsorptive surface.

The processes controlling inactivation in freshwater differ from those in seawater only in the absence of a chemical inactivating agent. Studies in freshwater have shown that the rate of inactivation is proportional to the size of the native microbial population (Figure 11.5). The virus can be protected by adsorption to living or dead cells in the same manner as in seawater. The nature of the microbial inactivating agent in fresh water remains, as in seawater, a mystery.

11-4. Chemical Communication between Predator and Prey

Communication between organisms living in aquatic habitats is frequently achieved by the transmission of signals by chemicals excreted by one organism and detected by another organism. Typical chemical communication systems involve mate recognition, aggregating and alarm materials, and detection of prey by predators (36). Table 11.3 shows the range of communication within species. These chemicals are known as pheromones. Chemical communication also occurs between unrelated organisms. The attraction of the predator starfish to their prey, the oyster, is an example of nonpheromonal communication between different species.

Microorganisms, because of their rapid growth and the ease with which metabolic products are produced, would seem to be natural candidates for chemical communication. Adler (37) has shown that *Escherichia coli* is

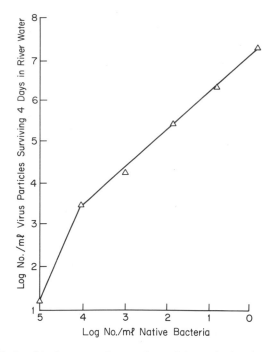

Figure 11.5 Relationship between the number of bacteria in river water and the survival of the bacteriophage ϕx-174 in that water (10).

attracted by low concentrations of glucose and other monosaccharides. Not all of the attractants are necessarily metabolized by the bacterium. Motile bacterial predators on the diatom *Skeletonema* and on the fungus *Pythium* are strongly attracted to exudates of their algal and fungal prey (38). This attraction is highly specific.

Inhibition of communication between microbial predator and prey would have a detrimental effect on self-purification processes. Materials which have a petrochemical origin are becoming increasingly abundant in our natural waters today. These include crude oil and synthetic chemicals. Blumer (39) has shown that crude oil inhibits the communication between starfish and their oyster prey.

The ability of motile bacteria to detect food sources is seriously impaired by low concentrations of chemical pollutants. Table 11.4 shows that large numbers of marine bacteria are attracted to a nonspecific carbon source, nutrient broth. Addition of hydrocarbons including toluene and phenol apparently mask the chemoreceptors and prevent detection of the nutrient source by the bacteria. An interesting characteristic of this inhibition is

Table 11.3   The Occurrence of Chemical Communication Systems in Protistan and Animal Phyla[a]

Taxa	Activity of Pheromone	Chemical Nature of Pheromone
Protista		
Volvox sp.	Female substance induces gonidia to develop into sperm packets	High molecular weight, over 200,000; probably a protein
Paramecium bursaria	Mate recognition, by cilial contact	Apparently a protein
Aschelminthes		
Brachionus spp. (rotifer)	Recognition of females by males, followed by breeding	Not a protein; otherwise unknown
Annelida		
Lumbricus terrestris (earthworm)	Alarm and evasion; secreted in mucus	Unknown
Mollusca		
Helisoma spp. and some other aquatic snails	Alarm: self-burying or escape from water	Polypeptides from tissue; mol. wt. about 10,000
Arthropoda		
Decapoda		
Portunus sanguinolentus (crab)	Sex attractant from female urine	Unknown
Cirripedia		
Balanus balanoides and *Elminius modestus* (barnacles)	Aggregation and settlement of larvae, by contact with pheromone on substratum	Protein
Arachnida		
Lycosidae (wolf spiders)	Female sex attractant	Unknown
Salticidae (jumping spiders)	Female sex attractant	Unknown

[a] From Wilson (36).

the failure of the hydrocarbons to inhibit motility of the organism and an apparent lack of enzymatic inhibition. The damage is not permanent and the bacteria regain their ability to detect nutrients when they are washed free of hydrocarbons in fresh seawater.

Sublethal concentrations of these materials in natural waters may affect chemical communication between higher organisms, inhibiting such essential activities as mate detection aggregation and detection of prey by predators. They also inhibit microbial communication and may result in increased accumulation of enteric bacteria, algae, and viruses in the water.

Table 11.4   Inhibition by Hydrocarbons of the Ability of Motile Marine
Bacteria to Detect a Food Source

Material in Capillary	No. of Bacteria Attracted to a Capillary
Artificial seawater (ASW)	2,000
ASW + nutrient broth (NB)	100,000
ASW + NB + 0.6% phenol	2,000
ASW + NB + 0.6% toluene	2,000

11-5. Conclusions

Natural waters have a strong inherent capacity to purify themselves and to prevent gross pollution by microbial pollutants. The rate of inactivation of foreign bacteria, fungi, and viruses, and of abnormally high concentrations of algae in natural waters is proportional to the size of the native microbial population. Some portion of the native microflora is responsible for the destruction of these pollutants. Specific native predators respond to high concentrations of these bacterial, fungal, and algal pollutants, and rapidly inactivate them. Destruction of viruses is a great deal more complex. It involves unknown biological antagonists, chemical antagonists in seawater, and the protective action of biological colloids.

This ability of natural waters to rid themselves of microbial contamination is a function of the extremely diverse native microflora of natural ecosystems. The development of an abnormally high concentration of one microorganism is prevented, both by competition for nutrients and by the rapid development of specific antagonists. This population of antagonists is only temporary and declines rapidly as soon as the microbial imbalance disappears.

This type of information may prove useful in the development of unconventional approaches to water pollution problems. It may be possible to treat microbial pollutants in natural waters directly *in situ* as an alternative to conventional waste treatment. This approach may require artificial reinforcement or stimulation of native predatory microorganisms in some manner not yet determined.

REFERENCES

1. V. DeGiaxa, *Z. Hyg. Infektionskrankh.* **6**, 162 (1889).
2. R. F. Vaccaro, M. P. Briggs, C. L. Carey, and B. H. Ketchum, *Amer. J. Public Health* **40**, 1257 (1950).

3. A. K. Saz, S. Watson, S. R. Brown, and D. L. Lowery, *Limnol. Oceanog.* **8**, 63 (1963).
4. M. Aubert, H. Lebout, and J. Aubert, *Ann. Inst. Pasteur* **106**, 147 (1964).
5. E. G. Jorgensen, *Physiol. Plant.* **15**, 530 (1962).
6. J. M. Sieburth, *Science* **132**, 676 (1960).
7. E. N. Krassilnikova, *Mikrobiologiya* **30**, 545 (1962).
8. A. F. Carlucci and D. Pramer, *Appl. Microbiol.* **8**, 251 (1960).
9. R. Mitchell and Z. Nevo, *Nature* **205**, 1007 (1965).
10. R. Mitchell, *Nature* **230**, 257 (1971).
11. R. Mitchell, S. Yankofsky, and H. W. Jannasch, Nature **211**, 891 (1967).
12. R. Mitchell and S. Yankofsky, *Environ. Sci. Technol.* **3**, 574 (1969).
13. D. Pramer, A. F. Carlucci and P. V. Scarpino, in *Marine Microbiology* (C. H. Oppenheimer, Ed.), Thomas, Springfield, Ill., 1963.
14. N. Dabrowa, J. W. Landau, V. D. Newcomer, and O. A. Plunkett, *Mycopathol. Mycol. Appl.* **24**, 137 (1964).
15. N. van Uden and R. C. Branco, *Limnol. Oceanog.* **8**, 323 (1963).
16. I. Taysi and N. van Uden, *Limnol. Oceanog.* **9**, 42 (1964).
17. J. D. Buck and S. P. Meyers, *Limnol. Oceanog.* **10**, 385 (1965).
18. A. M. Welch, *J. Bacteriol.* **83**, 97 (1962).
19. R. Mitchell and C. Wirsen, *J. Gen. Microbiol.* **52**, 335 (1968).
20. H. M. Canter and J. W. C. Lund, *New Phytologist* **47**, 238 (1943).
21. H. M. Canter and J. W. C. Lund, *Ann. Botany* (London) [N.S.] **15**, 360 (1951).
22. R. S. Safferman and M. E. Morris, *J. Amer. Water Works Assoc.* **56**, 1217 (1964).
23. M. Shilo, *Bacteriol. Rev.* **31**, 180 (1967).
24. A. B. Sabin, *Amer. J. Med. Sci.* **230**, 1 (1955).
25. S. Kelly and W. W. Sanderson, *Sewage Ind. Wastes* **31**, 683 (1959).
26. N. A. Clark and S. L. Chang, *J. Amer. Water Works Assoc.* **51**, 1299 (1959).
27. T. C. Metcalf and W. C. Stiles, in *Transmission of Viruses by the Water Route* (G. Berg, Ed.), Wiley-Interscience, New York, 1967.
28. O. C. Liu, H. R. Seraichekas, and B. L. Murphy, in *Transmission of Viruses by the Water Route* (G. Berg, Ed.), Wiley-Interscience, New York, 1967.
29. E. Lycke, S. Magnusson, and E. Lund, *Arch. Ges. Virusforsch.* **17**, 409 (1965).
30. S. Magnusson, K. Gundersen, A. Brandberg, and E. Lycke, *Acta Pathol. Microbiol. Scand.* **71**, 274 (1967).
31. S. Waksman, *Science* **29**, 665 (1945).
32. W. M. Stanley, *Science* **80**, 339 (1940).
33. S. L. Chang, *Amer. J. Hyg.* **52**, 194 (1950).
34. J. J. Vavra and A. Dietz, in *Antimicrobial Agents and Chemotherapy,* American Society for Microbiology, 1965.
35. R. Mitchell and H. W. Jannasch, *Environ. Sci. Technol.* **3**, 941 (1969).
36. E. O. Wilson, in *Chemical Ecology* (E. Sondheimer and J. B. Simeone, Eds.), Academic Press, New York, 1970.
37. J. Adler, *Science* **166**, 1588 (1969).
38. I. Chet, S. Fogel, and R. Mitchell, *J. Bacteriol.* **106**, 863 (1971).
39. M. Blumer, in *Oil on the Sea* (D. P. Hoult, Ed.), Plenum, 1969.

Part V
Microbial Parameters of Pollution

12 New Approaches to Assessment of Microbial Activity in Polluted Waters

Holger W. Jannasch, Woods
Hole Oceanographic Institution
Woods Hole, Mass.

It was an early observation that certain aquatic organisms may serve as excellent indicators for degrees of pollution in rivers, lakes, and seawater. In 1909, Kolkwitz and Marsson started to compile lists of aquatic plants and animals that were recommended for practical use in water analysis (1). It is surprising that microorganisms were hardly considered in these empirical studies, the more so as Winogradsky and Beijerinck had shown by that time that bacteria react quickly to changes of the environmental conditions *in vitro* as well as *in situ*. This apparent lack of communication between ecologically orientated microbiology and sanitary engineering is analogous to a similar situation between general and medical microbiology and still persists to a degree in research on applied problems in the two

291

fields. In recent years, this incompatibility of the biochemical and the sanitary aspects in microbiological studies of pollution is being resolved by the creation of interdisciplinary discussions, meetings, and joint research projects.

As a corollary of this history, it becomes understandable that microbiological tests of medical importance have developed into widely accepted routine techniques, while research in the general microbiology of freshwater and seawater has had little effect on water quality studies. The well-standardized bacteriological tests in existence are capable of, but also limited to, obtaining specific information on the origin and source of certain pollutants and on potential health hazards. The most important applications of indicator organisms are dealt with in chapter 14.

La Rivière (2) deplores the fact that the "microbiology of pollution" has always been approached in a strictly defensive manner, being primarily concerned with tracing and removing pathogenic microorganisms, although the predominant role of microorganisms in the "self-purification" process is well known. New impulses in microbial ecology in recent years are apt to change this situation. A better understanding of microbial activities in complex waste materials may lead to more controlled conversion processes aiming at an efficient microbial recycling of valuable mineral or organic materials. Many successful developments in industrial microbiology have demonstrated advantages in using microorganisms as inexpensive catalysts in various unspecific and specific transformations.

This chapter presents a critical discussion of various approaches to the determination of biochemical activities in water caused by whole microbial populations or by population segments of special metabolic types. In the absence of workable routine techniques, it seems worthwhile to look at a number of recent efforts at measuring microbial activities in water with the objective of determining their potential as pollution indicators. This paper does not include work on algae and on methods used in connection with studies on waste water or sludge.

12-1. Biomass Determinations

The assessment of microbial activity is possible to a degree by a determination of the microbial biomass. It appears advantageous, of course, to limit the latter to the mass of "viable" microorganisms, excluding the metabolically inactive cells. Viability, however, is determined by growth resulting in visible colonies on agar plates or by turbidity in liquid media. This demonstrates the difficulty in (*a*) defining "bio-"mass and (*b*) interpreting biomass determinations with regard to the actual microbial activity in the water sampled.

A. Microbial Counts. Ever since Koch (3) devised the technique of counting bacterial colonies on, or in, solidified transparent media, "plate counts" thus determined have been used as a principal criterion of pollution in natural waters. The so-called "total count" of the microbial population implies an indirect measure of *in situ* activity in contrast to numbers of a specific indicator organism. It was realized early that such counts are far from total due to a variety of reasons, mainly the limited choice of media in growing an unknown variety of organisms of different nutritional requirements, and the variable degree of clumping of microbial cells or their attachment to detritus particles. The same difficulties apply when colonies are counted on membrane filters or when the "most probable number" (MPN) of microorganisms is determined statistically from a series of dilution tubes.

For that reason, the *direct* microscopic counting of microbial cells either on membrane filters or on glass slides has been employed. Although the directly determined counts tend to be considerably higher, it is even more difficult to standardize this technique than the choice of media in the indirect counting techniques (4). Tests which distinguish living from dead cells are based on using fluorescent dyes (*e.g.,* acridine orange) or antibody labeling techniques. The former is hampered by the difficult problem of maintaining the critical dye concentration in the presence of unknown dye-adsorbing materials in the water samples. Tracing active microbial cells by radioactive-labeled antibodies is restricted to a few suitable strains of bacteria.

A differentiated direct microscopic enumeration technique (5) used in lake waters involves four separate countings on membrane filters, including the number of bacteria, after incubating separate samples for 8 hours with and without prior removal of zooplankton by filtration. An equation is used for calculation of the "bacterial production" from the various parallel counts. In a more recent approach, direct counts are combined with determinations of heterotrophic CO_2-fixation (see Section 12-5.A).

B. Determination of Cell Constituents. Next to the determination of the living cell as a whole, the measurement of a cell constituent would theoretically suffice if the two following requirements were met: (*a*) The concentration of the particular material must be constant in relation to the rest of the biomass or metabolic activity, and (*b*) it must be unstable and rapidly degraded outside of the living cell.

This is not true, unfortunately, for the bulk materials of microbial cells (protein, carbohydrates, lipids, etc.) that can be conveniently used for most growth determination in pure cultures under defined conditions. Other cell constituents are present in much smaller quantities, and the analysis lacks the necessary sensitivity.

In 1964, Lewin *et al.* (6) suggested that adenosine triphosphate (ATP) might represent an exception to that rule. It is one of the most widely distributed biochemically essential materials of low molecular weight in living matter. The analysis of ATP is based on the firefly reaction. Light is emitted when ATP activates a mixture of luciferine and luciferase. This can be measured by scintillation photometry or directly with sensitive photomultipliers. The technique has been developed for waste water studies (7) and in marine biology (8).

From pure culture work, it is not entirely clear to what degree the two requirements given above are met. The ATP content of a bacterial cell is closely related to the temporary metabolic activity. A large number of cells in the mixed microbial flora of unpolluted or polluted water will be present in dormant or in other forms of inactive stages, while a few organisms, including flagellates or protozoa, may contribute the major share of ATP measured. The practiced reconversion of ATP into microbial counts, therefore, seems neither appropriate nor necessary.

From practical experience it appears that ATP is lost rapidly from cells on death (8). At the same time, it was found that the ratio of organic carbon to ATP remained relatively constant at a value of about 250 (by weight) in most populations of aquatic microorganisms (9). It must be remembered that organic carbon is not a suitable parameter of living cell material. The degree of significance and interpretability of ATP measurements in mixed microbial populations is currently being evaluated in many laboratories.

Desoxyribonucleic acid (DNA) has been suggested as another parameter of biomass determinations. Its analysis is far less sensitive than that of ATP, and according to the function of DNA, the amount per cell is affected by the physiological state of the population (10, 11). Little is known about the survival of DNA on death or outside of the living cell in a natural mixed population.

12-2. Respirometry

The fact that even complete determinations of microbial biomass do not necessarily reflect the biochemical activity of the particular population makes direct measurements of these activities desirable. Most microbiological studies on polluted waters are concerned with heterotrophic activities. In aerobic metabolism, the oxidation of available organic matter is directly determined by following the oxygen uptake in a closed system. The measurement of oxygen as the terminal electron acceptor has definite advantages for other analytical procedures that involve breaking up the electron transport chain for the determination of an accumulated intermediate product.

A. Oxygen Uptake. In principle, any respirometer, including Warburg and Gilson apparatuses, can be used to measure oxygen uptake of water samples. The choice of the method will depend on the sensitivity required. Rates of oxygen uptake in polluted waters are relatively low compared to heavy suspensions of microbial cells. Oversized reaction vessels have been used in some cases. Dodson and Thomas (12) suggested a concentration of the microplankton prior to respirometric measurements. This technique was used in a recent study (13) comparing rates of oxygen uptake in offshore seawater with biomass determinations. The concentration procedure retains single cells and detritus particles and permits modifications with regard to expected rates of oxygen uptake in measuring pollution effects in natural waters.

The "biological oxygen demand" (BOD), as routinely performed, represents only an assessment of the amount of degradable organic materials in a water sample. A detailed discussion on the BOD as a tool in pollution studies is given in the next chapter.

B. Tetrazolium Salts. Next to measuring the terminal hydrogen acceptor, the metabolic activity of mixed microbial populations may be assessed by determining the rate of any other reaction of the electron transport system. This offers, theoretically at least, the advantage of including anaerobic processes.

The rate of electron transport is reflected by the activity of the coenzymes nicotinamide-adenine-dinucleotide (NAD) and flavin-adenine-dinucleotide (FAD) as intermediate electron acceptors. If suitable electron acceptor dyes are present, the activity of NAD and FAD can be measured by the visible color change of those dyes. In the late 1940s, 2,3,5-triphenyltetrazolium chloride (TTC) was introduced for distinguishing active and inactive bacterial colonies on agar plates in genetic studies. The dye is reduced to the red formazan. During later developments of the technique, the reduced dye was extracted from microbial cells in organic solvents for spectrophotometric measurements.

Inherently, the reaction interferes with the normal electron transport process, competing with ubiquinone as the regular successive hydrogen acceptor and forming an insoluble product at the reduction site. The incubation time is, therefore, more critical in this technique than in the measurement of oxygen uptake.

Other tetrazolium salts vary in their solubility and color in the oxidized or reduced state and interact at different sites of activity within the electron transport chain (7). In the interpretation by Packard (14), 2-(p-iodophenyl)-3-(p-nitrophenyl)-5-phenyltetrazolium chloride (INT) is suitable for measuring the "potential oxygen utilization rate." In a recent study on offshore seawater, this approach was compared to simultaneous measurements of ATP and respirometry of concentrated samples of microplankton (15). Because of the simplicity of using indicator dyes, this approach will certainly be of continued interest in pollution research.

12-3. Special Metabolic Groups

While a variety of microbial species can be isolated from mixed aquatic populations, their ecological significance and their quality as indicators for specific microbial *in situ* transformation processes are largely unknown. The present discussion is restricted to some qualitative and quantitative tests that have proven to be useful in microbiological pollution studies.

A. Qualitative Tests.   Most techniques designed to yield information about a microbial process in a water sample belong in the category of enrichment cultures in which specific substrates and/or selective chemophysical conditions of incubation produce a predominant population of the particular metabolic type, the presence of which is tested. The initiated microbial transformation or growth can be measured chemically or observed directly with the microscope.

In the case of nitrifying or nitrate-reducing bacteria, the accumulation of nitrite under defined conditions provides a very sensitive test of detection. Blackening of iron-containing agar stab cultures indicates the presence of sulfate-reducing bacteria in the inoculum. Gas production combined with a simple test for CO_2 and combustible gases has been employed for the demonstration of denitrifying and fermenting bacteria. Under defined conditions, a strong rise of the pH can be used as an indication of urea decomposition. The disappearance of sulfur and thiosulfate accompanied by a pH decrease shows the presence of sulfur-oxidizing bacteria. The decrease of dry weight of certain solid materials like chitin or cellulose has been used as an index of their degradation.

In earlier studies, rates of microbial reactions induced by the addition of a test substrate have often been related to population size in the initial inoculum. The rate of ammonia production from peptides added to a water sample has been given as a relative but quantitative figure for the actual rate of ammonification in the natural environment. Similarly, nitrate uptake in a bottled water sample has often been measured as a quantitative index of the original activity in the unsampled water. Such interpretations disregarded the limited significance of the tests. Below saturation values of the substrate concentration, microbial growth responds in an almost proportional manner to the amount added. This fact constitutes the general problem in measuring *in situ* rates of substrate-related activities if the natural concentration of the rate-limiting substrate is unknown.

B.   Quantitative Determinations.   In contrast to the problem discussed above, microbial activities measured in any supplemented sample or bacteriological medium may well be used for a quantitative characterization of specific environmental conditions. For instance, rates of various biochemical processes in bottles of media inoculated with specific bacterial strains and exposed to the deep sea indicated a strong inhibitory effect of high pressure/low temperature conditions (16). The results do not indicate actual microbial activities in the natural environment, but demonstrate on a quantitative basis that biodegradation of organic materials by certain organisms will be far slower in the deep sea than in shallow marine waters.

Some of the microbial enrichment procedures have been perfected to such a degree that the presence of a single cell per sample can be detected. If combined with the serial dilution method, "most probable numbers" of particular metabolic types of bacteria can be obtained. Such counts may be interpreted as specific biomass determinations, as discussed earlier, but do not indicate *in situ* activities.

An example for such a technique is the counting of sulfate-reducing bacteria using Senez' enrichment medium (17). Similarly, nitrate production has been measured in the case of ammonia-oxidizing and nitrate-reducing bacteria, and Durham tubes have been used in enumerating gas-producing microorganisms. The serial dilution technique has recently been applied in estimating the number of methane-producing bacteria (18).

Two further approaches come close to yielding information about the actual *in situ* activity. If labeled thymidine was added as a tracer material to a macroscopically visible cluster of *Leucothrix mucor* collected in a small vial, growth could be measured after a certain incubation period by counting the newly formed cells autoradiographically (19). A recently developed enzymatic technique of measuring the nitrogen-fixing capacity of a mixed microbial population has led to a number of applications in microbial ecology (20). This method is of special interest in pollution studies because of the characteristic nitrogen-fixing activity during blooms of blue-green algae.

Purple sulfur bacteria commonly occur in polluted ponds and lagoons and are indicative of conditions favorable for anaerobic photosynthesis and the possible presence of a number of organic compounds (21, 22). The extent of their ecological role in polluted waters is largely unknown.

12-4. Growth Determinations

The general statement that "Growth is the expression *par excellence* of the dynamic nature of living organisms" (23) may be extended toward the applicability of measuring growth as an integrated response of microorganisms to environmental conditions.

Specific growth rate usually is measured as the increase of a growth product (cell number, dry weight, etc.) in time and is often mistakenly identified with biomass. The latter, however, represents the result of growth and does not reflect the conditions that determine the rate of growth at any given time. In a closed system, a batch culture, the biomass is still low when the growth rate attains its maximum value; in a later stage, when the limiting substrate nears exhaustion, growth approaches zero in the stationary phase while the biomass is at its maximum.

In an open system, a continuous culture, new medium is constantly added and growth may proceed indefinitely. In the chemostat as a well-defined continuous culture system (*i.e.,* complete mixing; pure culture; constancy of medium composition, medium flow, and culture volume) the concentration of the growth limiting substrate in the culture vessel is a direct measure of the growth rate (24).

Neither of the two artificial systems, batch culture or chemostat, really reflect the situation of the mixed population in natural water. Their value lies in their applicability as tools in studying growth as a most sensitive index of complex nutritional conditions (25).

A. Batch Culture. Any water sample containing microorganisms confined in a bottle turns into a batch culture by definition. The microorganisms present will continue to metabolize, exhausting the growth limiting substrate and, in some cases, accumulating an intermediate product. This, in turn, will eventually lead to changes in species predominance. In general, bottle experiments are affected by the imposed change of growth conditions.

This obvious disadvantage does not apply if the natural population is replaced by an organism of known growth characteristics and if cell numbers, dry weight, or other parameters of growth are measured as relative indices for the complex nutritional composition of the water sample. In such approaches, the natural population is removed by filtration. It is an inherent disadvantage of most crude bioassay procedures of this kind that the particulate phase of a natural water is excluded from the test.

True bioassay methods are used when chemical or physical analyses of a known substrate are neither sensitive nor specific enough, and they are based on a proportional relationship between growth, mostly given in yield of dry weight, of a particular assay organism and the concentration of the known substrate (26). In contrast to bioassay techniques, simple growth experiments using any suitable test strain estimate the potential microbial activity in a water sample and do not provide more detailed information on specific limiting, inhibitory, or stimulatory factors.

B. Continuous Culture. If tests of specific growth rates are conducted in continuous culture systems, the sterile-filtered water is metered into the chemostat from large reservoirs. The establishment of a steady state of the test organism, indicated, for example, by a constant population density, shows that the growth rate is equal to the experimental dilution rate (flow rate per volume of the culture vessel) at this point.

In experiments with natural seawater, steady states could only be reached at extremely low dilution rates (0.08 to 0.16 hours^{-1}), or at retention times of 6 to 12 hours, with water from highly contaminated inshore basins (27). In other cases, steady states of the test culture were established rapidly in seawater supplemented with various substrates, mainly carbohydrates as carbon and energy sources. In unsupplemented offshore seawater, growth rates of test organisms were too slow to result in steady-state populations.

It was found, however, that growth rates could be calculated during transient state of the population in the chemostat if the washout rate of the test organism reached a constant value (24). During transient state, the population density (x) changes with time (t):

$$x = x_0 e^{(\mu - D)t}$$

where x_0 is the initial population density in cell counts or dry weight/ml), μ is the growth rate, and D the dilution rate of the system (both in hours^{-1}). From this equation the growth rate can be calculated according to:

$$\mu = D + \frac{1}{t} \ln \frac{x}{x_0}$$

By using this technique, generation times (defined as the reciprocal of the growth rate) of more than 100 hours have been found in offshore seawater (27).

Natural water was metered directly into the chemostat without prior filtration and storage aboard research vessels, and it was possible to measure the growth rate of a test organism in the presence of the natural population of microorganisms. This approach led to studies on competition for the unknown growth limiting factor or for a supplemented substrate (27). On this basis, studies on characterizing different types of pollution, including physical effects and toxic materials, may also consider aspects of population dynamics.

12-5. Determination of Substrate Uptake

The use of radioactive-labeled material was adopted for ecological studies soon after its introduction to physiology and biochemistry. Since the rates of substrate conversion in natural systems are mostly below saturation values, the concentration of the added substrate is critical. The high sensitivity of analytical techniques using radioactive tracers is therefore of the utmost importance. Another advantage of using labeled substrates is the specificity of the reaction measured in complex media.

The photosynthetic fixation of carbon dioxide by phytoplankton offered itself as an ideal expression of "primary productivity" in natural waters and became a routine technique. The only interference that represents a serious handicap under certain conditions is the nonphotosynthetic, that is, respiratory and chemolithotrophic, uptake of carbon dioxide.

In contrast to measuring photosynthesis, it is far more difficult to find a unique substrate, the conversion of which would be indicative of the reverse reaction, the degradation of organic materials. There is a multitude of substrates to choose from, and the specific rate of decomposition of every single one cannot be expected to reflect the rate of the total degradation process. Dissolved oxygen may be the most easily and accurately measured substrate in such studies.

A. Heterotrophic Uptake of CO_2. It was originally shown by the discovery of the Wood-Werkman reaction that small amounts of CO_2 are incorporated into organic matter or cell material during heterotrophic aerobic growth. Russian workers have attempted to make use of this fact in ecological studies by measuring the dark uptake of $^{14}CO_2$ in water samples. The obvious difficulty of this approach is that the ratio between the utilization of inorganic and organic carbon is extremely small and, probably, highly variable. Sorokin (28) showed that in pure cultures of a variety of isolates during growth on various organic substrates, the amount of CO_2 assimilated was 2 to 3% of the organic carbon in the biomass.

When $Na_2^{14}CO_3$ was added to samples of lake water incubated for 24 hours in the dark, 6% of the organic carbon produced was present in labeled form (29). In a water reservoir where chemosynthesis was assumed to be absent, Kusnetzov and Romanenko (30) found the ratio between heterotrophic and photosynthetic fixation of CO_2 to be 1.17.

The approach is certainly unique with regard to the fact that CO_2 is indeed utilized in practically all heterotrophic assimilatory conversions. The obvious problem is to separate the small fraction of heterotrophic CO_2 uptake from the probably larger fraction of chemolithotrophic CO_2 uptake. There seems to be more information necessary on the constancy of heterotrophic CO_2 uptake relative to conversion rates of particular organic substrates.

B. Uptake of Organic Substrates and Production of CO_2. In contrast to measuring the heterotrophic uptake of CO_2, the determination of incorporated carbon from labeled organic substrates into cell material is far less susceptible to the sources of error mentioned above. The obvious disadvantage is the substrate specificity. A difficulty that applies to all carbon-uptake measurements is the unknown rate of carbon turnover in mixed populations and nonreproducible situations. A substantial amount of incorporated carbon might be respired within the set incubation time. One way of coping with this problem is to keep incubation time to a feasible minimum, usually about 3 hours. Another way is to supplement the data with a determination of the CO_2 produced.

The measurement of the "relative heterotrophic potential" by the incorporation of carbon from ^{14}C-labeled acetate was first suggested in 1961 (31). Other approaches were based on the sole determination of CO_2 produced from uniformly labeled glucose (32). Turnover times of 1 to 60 days were calculated for glucose in seawater, the fluctuations being dependent on seasonal changes of nutrient concentrations and temperature (33). Glucose was also used in studies where the relative biomass of heterotrophic microorganisms attached to suspended matter in seawater was calculated on the basis of the known uptake rate per cell of a marine vibrio strain (34).

All of these studies have been done in more or less unpolluted fresh- and seawater. Stumm-Zollinger and Harris (35) have suggested that higher nutrient levels and population densities of microorganisms in polluted waters might well facilitate the application of this general approach.

C. Kinetic Approaches. When the uptake rate of acetate in a water sample was related to the concentration of substrate, a function similar to that of enzymatic reactions described by the Michaelis-Menten equation (31) was found. The reciprocal plot of the data makes it possible to determine graphically the kinetic constants that describe the turnover time of the substrate and its concentration, K_s, at half of the maximum uptake rate. This K_s value cannot be compared to the corresponding Michaelis constant because it also contains the unknown portion of the substrate already present in the water sample. Exploitation of this fact in an assay technique has been suggested (36). If a culture of an organism with a known K_s is used, the natural concentration of the substrate (S) can be calculated from the graphically obtained value for $(K_s + S)$. It must be noted, however, that the K_s of a pure culture relative to a specific substrate is also affected by the varying composition of the water from sample to sample.

The technique of using the kinetic uptake of added substrates by the microbial population of whole water samples to estimate heterotrophic activity did not originally consider the possible loss of labeled carbon due to CO_2 production even during short incubation times. It was found that 8 to 60% of the incorporated labeled carbon may be lost by respiration during 3-hour incubation times, the percentage being mainly dependent on the type of substrate used (37).

From the available data it appears that this particular uptake behavior has been observed only in situations where a predominance of one particular species within the mixed microbial population was most likely. In oligotrophic marine waters the uptake of increasing concentrations of an added substrate did not follow the saturation curve necessary for a kinetic analysis (38). If, however, the samples were enriched with small amounts of the particular substrate about 24 hours prior to the addition of the labeled portion, a predominant population of one species developed and the originally irregular relation between uptake rate and substrate concentration changed into a saturation curve amenable to kinetic analysis. This finding suggests that the inherent theoretical problems of the kinetic approach in measuring heterotrophic activity do not apply in nutrient-rich or polluted waters.

While the need for broad applications of measuring microbial activities in polluted waters excludes an undue sophistication of methods, a number of approaches developed in aquatic microbiology may well evolve into practical techniques for microbiological pollution studies.

Acknowledgment

Contribution number 2599 of the Woods Hole Oceanographic Institution. Supported by the National Science Foundation, grant BO 20956.

REFERENCES

1. R. Kolkwitz, *Oekologie der Saprobien,* Piscator, Stuttgart, 1950.
2. J. W. M. la Rivière, in *Global Impacts of Applied Microbiology, 3rd Intl. Conf.,* Bombay, 1969.
3. R. Koch, *Mitt. Kais. Ges. Amt.* **1**, 1 (1881).
4. H. W. Jannasch and G. E. Jones, *Limnol. Oceanog.* **4**, 128 (1959).
5. S. I. Kusnetzov, *Verhandl. Int. Ver. Limnol.* **13**, 156 (1958).
6. G. V. Lewin, J. R. Clendenning, E. W. Chappelle, A. H. Heim, and E. Rocek, *Bioscience* **14**, 37 (1964).
7. J. W. Patterson, P. L. Brezonik, and H. D. Putnam, *Proceedings 24th Purdue Industrial Waste Conference,* 1, 1969.
8. O. Holm-Hansen, *Limnol. Oceanog.* **14**, 740 (1969).
9. J. D. H. Strickland, *Symp. Soc. Gen. Microbiol.* **21**, 231 (1971).
10. D. Herbert, *Symp. Soc. Gen. Microbiol.* **11**, 391 (1961).
11. O. Maaloe and N. O. Kjeldgard, *Control of Macromolecular Synthesis,* Benjamin, New York, 1966.
12. A. N. Dodson and W. H. Thomas, *Limnol. Oceanog.* **9**, 455 (1964).
13. L. R. Pomeroy and R. E. Johannes, *Deep-Sea Res.* **15**, 381 (1968).
14. T. T. Packard, Thesis, Univ. of Washington, 1969.
15. O. Holm-Hansen, L. R. Pomeroy, and T. T. Packard, *Limnol. Oceanog.* (In Press).
16. H. W. Jannasch, K. Eimhjellen, C. O. Wirsen, and A. Farmanfarmaian, *Science* **171**, 672 (1971).
17. Y. Abd-el-Malek and S. G. Rizk, *Nature* **182**, 538 (1958).
18. M. L. Siebert, D. F. Toerien, and W. H. J. Hattingh, *Water Res.* **2**, 545 (1968).
19. T. D. Brock, *Science* **155**, 81 (1967).
20. W. D. P. Steward, G. P. Fitzgerald, and R. H. Burris, *Proc. Nat. Acad. Sci. U.S.* **58**, 2071 (1967).
21. N. Pfennig, *Ann. Rev. Microbiol.* **21**, 211 (1967).
22. H. W. Holm and J. W. Vennes, *Appl. Microbiol.* **19**, 988 (1970).
23. C. B. van Niel, in *Chemistry and Physiology of Growth of Microorganisms,* A. K. Parpart (ed.), Princeton Univ. Press, Princeton, N.J., 1949.
24. D. Herbert, R. Elsworth, and R. C. Telling, *J. Gen. Microbiol.* **14**, 601 (1956).
25. H. W. Jannasch, *Verhandl. Int. Ver. Limnol.* **17**, 25 (1969).
26. S. H. Hutner, A. Cury, and H. Baker, *Anal. Chem.* **30**, 849 (1958).
27. H. W. Jannasch, *J. Bacteriol.* **99**, 156 (1969).
28. J. I. Sorokin, *Zh. Obscch. Biol.* **22**, 265 (1961).
29. W. I. Romanenko, *Mikrobiologija* **33**, 679 (1964).
30. S. I. Kusnetzov and W. I. Romanenko, *Verhandl. Int. Ver. Limnol.* **16**, 1493 (1966).
31. T. R. Parsons and J. D. H. Strickland, *Deep-Sea Res.* **8**, 211 (1961).
32. H. Kadota, U. Hata, and H. Hyoshi, *Mem. Res. Inst. Food Sci. Kyoto Univ.* **27**, 28 (1966).
33. P. J. le B. Williams and C. Askew, *Deep-Sea Res.* **15**, 365 (1968).
34. H. Seki, *Appl. Microbiol.* **19**, 960 (1970).
35. E. Stumm-Zollinger and R. H. Harris, *Fifth Rudolp Res. Conf.* (In Press).
36. R. T. Wright and J. E. Hobbie, *Ecology* **47**, 447 (1966).
37. J. E. Hobbie and C. C. Crawford, *Limnol. Oceanog.* **14**, 528 (1969).
38. R. F. Vaccaro and H. W. Jannasch, *Limnol. Oceanog.* **12**, 540 (1967).

13 Biochemical Oxygen Demand

Anthony F. Gaudy, Jr., Bioen-
vironmental Engineering and
Water Resources, School of Civil
Engineering, Oklahoma State
University, Stillwater, Okla-
homa

13-1. Introduction

In considering the topic of biochemical oxygen demand (BOD), it is essential to distinguish between the concept of BOD and the BOD test itself. The idea, or concept, of assessing the pollutional potential of a waste water which contains an available organic carbon source for aerobic organotrophic microorganisms by measuring the amount of oxygen utilized during growth of the organisms on a sample of the waste water, is certainly defensible. One of the primary reasons for purifying or treating a polluted water before its recycle to the water resource is to deplete this oxygen demand on shore so that the organic matter which remains in the waste will not place a drain on the dissolved oxygen resources of the receiving body. Water is truly a multipurpose resource, and most of the uses for which it may be optimally employed are negated if dissolved oxygen levels required to maintain the desired ecological balance in the stream are not maintained. Therefore, measuring the pollutional potential of an aqueous waste in accordance with its potential drain on the dissolved oxygen resource is a reasonable and logical concept which provides a broad-scale approach to assessing the "strength" of a waste.

305

Pollution control scientists and engineers find it advantageous to employ a colligative property of organic wastes, and when that colligative property has a direct effect on the quality of the environment, as does the BOD, there should be no doubt about the value of the concept. Today, however, many seem willing to discard the concept because they see inadequacies in the BOD test, which supposedly employs the concept. Possible replacements for the BOD test which have been considered as means of assessing the pollutional strength of a waste are the COD (chemical oxygen demand) test and direct measurement of organic carbon. A point to be delineated is that both of these methods, which are valuable tools, are employed as measures of the amount of organic material in the sample. Unfortunately, many workers in the water pollution control field have come to consider the BOD test also to be a measure of organic material or even of organic carbon in the sample. As a measure of organic carbon, it is at best a very indirect method. It is far better to consider the BOD of a waste to be precisely what it is, that is, an assessment of the amount of oxygen used for the respiratory activities of microorganisms which utilize organic matter in the waste for growth, and for further metabolism by these and other organisms of the cellular components synthesized from the waste. When considered in this way, it is apparent that the concept goes right to the heart of the pollutional problem; that is, it deals with the amount of oxygen which will be required for all of the microbial metabolic reactions resulting from the introduction into a water resource of a particular amount of carbon source. As a measure of organic matter, the BOD test cannot compare with the other two analyses mentioned above, and it should not be used for the same purposes as are COD and carbon analyses.

Manometric procedures which permit use of less dilute samples than can be permitted in the BOD bottle retain the basic concept, that is, measurement of biological oxygen demand. Such techniques can be employed to assess the oxygen requirements of the microorganisms in utilizing an organic waste substrate, but they provide little or no information about the behavior of the sample in the receiving stream. That is, they do not provide data which relate to the kinetics of oxygen utilization in the stream; whereas the BOD test has been supposed to depict kinetically the mode of deoxygenation in the receiving stream. The validity of the latter assumption will be discussed at length in this chapter.

The BOD test has been employed by regulatory agencies and by designers and operators of treatment plants as a measure of the pollutional potential of a certain amount of organic waste. The mechanism and kinetics of BOD exertion were considered to be so thoroughly defined that data obtained from the BOD test could be used, in conjunction with a reaeration coefficient, in a differential equation designed to predict the course of depletion and recovery of the dissolved oxygen in the receiving water. Perhaps it is too much to hope that the complex ecosystem which exists in a water resource can ever be approximated closely enough in the laboratory to allow adequate or accurate predictive formulations for occurrences in the natural environment. On the other hand, perhaps there are ways in which this can be accomplished or approximated if we do not insist that the concept of BOD be subordinated to the ideas and conventions which have developed concerning the standard dilution technique or standard test. Perhaps the fruitful use of this important concept can be enhanced by altering the testing technique and the ways in which we interpret and use the test results.

The statement made by Hoover, Jacewicz, and Porges (1) concerning the BOD test is an interesting and important one, worthy of repeating: "The BOD test is paradoxical. It is the basis of all regulatory actions, and is used routinely in almost all control and research studies on sewage and industrial waste treatment. It has been the subject of a tremendous amount of research, yet no one appears to consider it adequately understood or well adapted to his own work." This statement was made in 1953, and it is in large measure applicable today even though we now have a more adequate understanding of the biological events which can take place in the BOD bottle. The author contends that in the main the BOD test is still rightfully the subject of research rather than a reliable analytical parameter for use in research. Many might argue with this view. One often hears a view expressed which in essence states that the mechanism is rather simple: oxygen is used by the organism for growth, and the kinetics of this respiration are such that its course can be approximated by the kinetics of a first-order decreasing rate ("monomolecular law") reaction. Any attempts to delve deeper are sometimes spoken of as merely "fine tuning" of a necessarily coarsely selected parameter. The author would, however, remind such critics of BOD research that recognition of the biological nature of the O_2 uptake dates back to the nineteenth century (1884) when Dupre concluded that the oxygen depletion he observed in a stored sample of polluted water was due to the presence of growing organisms, that is, "microphytes" (2), and that during balanced growth of microorganisms, O_2 uptake does not follow "monomolecular law."

We are now approaching the twenty-first century, and if we are to enjoy fully the affluence provided by our advanced technology, we will of necessity live in a more "finely tuned" environment. If our technology is to permit us to continue advancing, it would seem that a more finely tuned understanding and usage of BOD on the part of the applied scientists responsible for environmental control is warranted.

13-2.   Development of the BOD Test

Much of the developmental history of BOD up to approximately 1943 has been expertly presented by Phelps (2). Prior to Dupre's observation on the gross mechanism of oxygen depletion (or uptake), the use of oxygen as a yardstick for assessment of pollutional potential of a waste was investigated by Frankland, who showed that the amount of oxygen consumed upon storage of a sample of water containing organic matter was dependent upon the time of storage. The two essential principles of the present BOD test were thus established more than 80 years ago; if one is to employ oxygen utilization as the yardstick for measuring the potential drain on the dissolved oxygen resources of a receiving body resulting from discharge of organic-containing wastes, he must be reconciled to the fact that the mechanism is biological in nature, and he must be concerned not only with the amount of O_2 used but with the rate at which it is used. Thus it was necessary to adopt a standard incubation time and to define the rate of reaction mathematically.

The early studies of the Royal Commission on sewage disposal had made use of a 5-day incubation period. Some studies were made to determine whether this time could be reduced, but it was generally concluded that the 5-day incubation period allowed time for the "reaction" to develop sufficiently.

The concept of BOD aroused the interest of investigators in the United States, and much work was undertaken to improve and standardize the technique and to investigate the kinetic course of O_2 uptake. Studies of Streeter and Phelps (3) led in 1925 to the following conclusions or generalizations concerning the course of biochemical oxidation of organic matter: "The rate of the biochemical oxidation of organic matter is proportional to the remaining concentration of unoxidized substance, measured in terms of oxidizability." This law was expressed mathematically as follows:

$$-\frac{dL}{dt} = K_1 L \tag{1}$$

In integrated form:

$$L_t = L_0 e^{-K_1 t} \qquad \text{or} \qquad L_t = L_0 10^{-k_1 t} \tag{2}$$

The symbol L_0 represents the total oxidizability (BOD) of the organic matter initially present in the sample, that is, at zero time. This term has become known as the ultimate BOD. L_t is the amount of oxidizability (BOD) remaining to be expressed at the corresponding time, t. K_1 (nat-

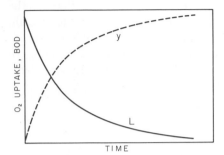

Figure 13.1 Exertion of biochemical oxygen demand in accordance with first-order decreasing rate ("monomolecular law") kinetics.

ural logarithms) or k_1 (common logarithms) is the proportionality or velocity constant for the "reaction." It is sometimes referred to as the specific rate constant.

Events which conform to this type of kinetic equation can be described by first-order kinetics. This formulation is also the kinetic mode of a monomolecular reaction. It states that the rate at which the amount of oxidizability or oxidizable material remaining decreases becomes increasingly lower as the material is oxidized, eventually becoming asymptotic to some lower limit, as seen in Figure 13.1. The dotted line represents the same kinetic relationship in terms of oxidizability, or BOD, expressed or exerted; that is, $y = (L_0 - L_t)$ at time t. Substituting for L_t in Equation 2, the amount of BOD expressed, y, is given as:

$$y_t = L_0(1 - e^{-K_1 t}) \quad \text{or} \quad y_t = L_0(1 - 10^{-k_1 t}) \qquad (3)$$

It is this first-order decreasing rate mode of kinetics which requires further discussion and elucidation as this text is developed. It is interesting to note here that this type of kinetic relationship not only could be fitted to the data reported by Streeter and Phelps, but it was also in line with the early work (1909) of Phelps in his studies leading to the development of the relative stability test (2).

Theriault, in his classical report in 1927, also observed that the course of first stage, or carbonaceous, exertion of BOD for Ohio River water conformed to the monomolecular law (4). Thus the dissolved oxygen (DO) sag curve equation of Streeter and Phelps (3), which has become a widely used expression for predicting or estimating the assimilating capacity of a receiving stream, is one in which two opposing first-order decreasing rate kinetic processes are combined. This equation is usually written as:

$$\frac{dD}{dt} = K_1 L - K_2 D \qquad (4)$$

D represents the deficit of dissolved oxygen in the stream, that is, the difference between the DO saturation and the actual DO in the stream. K_2 represents the velocity or proportionality constant of the first-order decreasing rate expression depicting the course of physical reaeration of the stream. Integration of this equation with proportionality constants converted to base 10 leads to the well-known "sag equation":

$$D_t = \frac{k_1 L_0}{k_2 - k_1} (10^{-k_1 t} - 10^{-k_2 t}) + D_0 10^{-k_2 t} \qquad (5)$$

This equation has become one of the most widely quoted kinetic expressions in the field of pollution control. The assimilating capacity of the stream is of importance, although perhaps indirect interest in this chapter; however, the equation is introduced primarily because an application of it in modified form is employed later in this chapter to determine the BOD exertion curve. Also, it should be emphasized that the calculation of assimilating capacity based on prediction of the course of oxygen depletion (BOD exertion) and replenishment is one of the primary uses to which the BOD dilution technique is put, and that estimates of BOD and of assimilating capacity are inseparable.

The wide adoption of the sag curve, coupled with the growing need to develop a more reliable technique for BOD measurements, led to a concentration of research on the dilution technique for BOD determinations. It is important to emphasize here that the work of Phelps (2), Streeter and Phelps (3), and Theriault (4), which led to the establishment of the monomolecular equation for exertion of carbonaceous BOD (and the sag equation), was accomplished prior to most of the work which was done later to establish "optimum" procedures and conditions for the BOD test. Much research has gone into development of these procedures and establishment of the need for environmental control of the test conditions. The early work was concerned, for the most part, with sewage and river water, whereas later interest centered around industrial wastes and involved standardization of dilution waters to which were added complements of inorganic nutrients essential for microbial growth. Much of this research was aimed at establishing a procedure which would give not only reliable but more realistic evaluation of the pollutional potential of the waste sample rather than predicting the course of O_2 uptake in a specific receiving water. In effect, the technique and application of the BOD test changed somewhat after development of the first-order decreasing rate kinetic "law" had been concluded.

Today the standard conditions for the BOD test, the medium, pH, and so on are akin to a minimal growth medium which might be employed by a microbiologist in culturing microorganisms. The growth medium is a very dilute one, and often the nature of the "carbon source," that is, the waste, is unknown. Indeed the waste sample itself may contain many essential nutrients, but the inorganic constituents which are added in the medium (dilution water) resemble the usual minimal medium which is designed to ensure that the carbon source is the limiting nutrient. This is, of course, as it should be in view of the fact that the main usage of the BOD test today is in rating a waste water sample in terms of its maximum pollutional potential or, as some would say, as a measure of the biologically available organic carbon in the sample. It seems possible that in the early studies the carbon source may not necessarily have been the limiting nutrient.

13-3. Mechanism and Kinetics

A. Some Effects of Autotrophic Metabolism. In the discussion which follows it is important to note that attention is turned toward the mechanistic and kinetic occurrences in the laboratory test or in the receiving stream during the course of exertion of carbonaceous BOD. It is necessary to distinguish between O_2 uptake due to aerobic organotrophic metabolism of the carbon sources in the waste (carbonaceous BOD) and O_2 uptake caused by the aerobic autotrophic organisms, for example, *Nitrosomonas* and *Nitrobacter*. These nitrifying bacteria, which may be present in certain well-aerated seed material and in highly treated waste water treatment plant effluents, can cause an O_2 uptake which is sometimes termed the "second-stage uptake" or the nitrogenous BOD. In this case oxygen utilization is brought about by the oxidation of ammonia to nitrite (*Nitrosomonas*) and then to nitrate (*Nitrobacter*). Some of the chemical energy released upon this progressive removal of electrons is trapped as biologically usable and highly transferable chemical energy, for example, ATP. A reduced inorganic compound such as ammonia may also serve as an electron donor in the reductive (synthetic) fixation of inorganic carbon, for example, CO_2, for which energy is supplied by ATP. The overall oxidation reaction can be written as:

$$2NH_3 + 4O_2 \rightarrow 2NO_3^- + 2H^+ + 2H_2O$$

It can be seen that for each milligram/liter of NH_3, nearly 4 mg/liter O_2 can be consumed. This can amount to a significant "BOD" in wastes containing large amounts of ammonia or nitrogenous organic matter, for example, protein from which ammonia can be released in deaminative reactions accompanying the aerobic assimilation of the organic carbon.

Under the usual conditions, exertion of this autotrophic BOD is not a factor during the 5-day incubation period since it normally does not occur until after this time. Usually the number of nitrifying bacteria in the seed or sample is low, their growth is relatively slow, and their proliferation depends upon adequate supplies of both reduced inorganic nitrogen compounds and inorganic carbon. This stage of O_2 uptake can occur in the receiving stream, but cannot be expected to be a determinative factor in producing the low point of a DO profile in a receiving stream.

One of the chief causes for concern with noncarbonaceous BOD arises from the estimation of the so-called "ultimate BOD," that is, L_0. This term has been used synonymously with the "20-day BOD" which is a holdover from the early work on the BOD of municipal sewage wherein the velocity constant, k_1, was generally believed to be approximately 0.1 day^{-1}. For this velocity constant, in accordance with "monomolecular law," 99% of the carbonaceous BOD would be exerted by the twentieth day; hence, 20-day BOD's are sometimes equated with the ultimate BOD, L_0. Since O_2 uptake due to nitrification usually occurs during the second week of incubation, the nitrogenous exertion affects the values of L_0 or y_{20}, but not, in the usual case, the 5-day BOD. A lack of agreement may thus be observed between actual measurement of 20-day BOD, or ultimate BOD, and calculation of L_0 based on 5-day BOD data, and the amount of the discrepancy is at least partially dependent upon the amount of noncarbonaceous BOD exerted. Therefore, L_0 may not be, in actuality, the total carbonaceous BOD if L_0 is based on 20-day BOD determination.

Phelps (2) expressed the view that, since the oxygen in NO_2^- and NO_3^- is available as a source of oxygen to some bacteria, the DO used in effecting nitrification does not really represent a depletion of the total oxygen resource. There is some validity in this reasoning if one is interested in the total oxygen resource, and this opinion enjoys some current acceptance. The author disagrees with this viewpoint on the basis that it is not really the total O_2 resource which is of importance in assessing pollution. It is important to realize that the bound oxygen in nitrite and nitrate is not available as an electron acceptor to most of the forms of life in the receiving stream, and that those microorganisms which can use it do not draw upon this resource until the DO level is nil or very low. Thus the bound O_2 resource would not be used until the stream was in dire stress, that is, until little or no dissolved oxygen remained. If the stream is to maintain a healthy ecosystem, dissolved oxygen, not bound oxygen, is required, and any agency which depletes it, such as nitrification, should not be idly dismissed when assessing effects of pollutants on the stream.

It is important to point out that the nitrifying bacteria are not the only chemoautotrophic organisms which can exert BOD. The hydrogen bacteria (H_2 as electron donor), the colorless sulfur bacteria (H_2S, S, and $S_2O_3{}^{2-}$ as electric donors) and the iron bacteria (Fe^{3+} as electron donor) also possess this ability. However, these sources of energy are usually not present in sufficient quantities to cause serious O_2 depletion. One of the products of the oxidation of sulfur is $SO_4{}^{2-}$ which, like $NO_2{}^-$ and $NO_3{}^-$, can be construed as a source of bound oxygen. However, the primary user of this oxygen resource is the strict anaerobe, *Desulfovibrio,* and a product of sulfate reduction is H_2S. Therefore, use of sulfate as an oxygen resource is to be avoided, because at least the bottom of the receiving body must be in an anaerobic condition for this to occur. It is emphasized, therefore, that aerobic autotrophic metabolism leading to the production of bound oxygen (*e.g.,* by nitrifiers and colorless sulfur bacteria) must be construed as a depletion of the available oxygen resources in the receiving stream.

Another factor which can complicate observations during the course of exertion of carbonaceous BOD is the photosynthetic production of oxygen by algae and their subsequent oxygen utilization in the absence of light. Such complications can be easily controlled in the laboratory by preventing light from entering the sphere of the reaction.

In highly enriched streams algae represent a deleterious factor because they contribute to the organic loading (both suspended and benthal). In dark periods a large algal population can consume large amounts of oxygen. Concerted efforts are now being made to control algal nutrients. In sum, all of these factors contribute to the author's conclusion that any attempt by pollution control workers to consider photosynthetic oxygen production as an asset in calculating the DO resource in the stream is an erroneous approach. In healthy streams, however, algae are present in relatively small numbers and are an asset, occupying a useful niche in the ecosystem.

B. Carbonaceous BOD. It is now appropriate to examine, as best we can, the course of biochemical and ecological reactions which occur during metabolism of organic carbon sources in the rather dilute aerobic environments extant in the stream and in the BOD bottle. In the BOD bottle and, to a large measure, in the receiving stream (depending upon the reach of stream examined, entry of new organic carbon sources, etc.), it seems essentially correct to consider the ecosystem as a closed system. In this sense, the BOD bottle is akin to the microbial growth tube with two important differences. The concentration of food supply is rather low in the BOD bottle because the aim of the technique is to make the carbon source, not the O_2 supply or any other nutrient source, the limiting factor, and, most importantly, the microbial population is a heterogeneous or natural one.

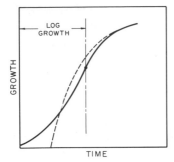

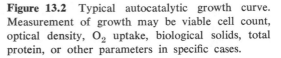

Figure 13.2 Typical autocatalytic growth curve. Measurement of growth may be viable cell count, optical density, O_2 uptake, biological solids, total protein, or other parameters in specific cases.

If we can agree on the similarity between the events in the BOD reaction vessel and the microbial batch growth tube, it seems reasonable to look to the field of microbiology for some initial insights. If one plots data generated by sampling at various times a system inoculated with a relatively small number of cells in relation to the carbon supply, the familiar curve of microbial growth is developed (solid line, Figure 13.2).

A curve of this shape is usually observed, whether viable count, total count, biological solids concentration, or optical density is used to measure growth. The general shape of the O_2 uptake curve is similar. It is not implied that growth and oxygen uptake curves for any system would be superimposable. The shapes of both curves, however, would be those of an autocatalytic process; the inflection point would separate the first-order increasing rate portion (logarithmic growth) and the decreasing rate portion which sometimes may approximate first-order decreasing kinetics. The relative position of the inflection point which separates the two parts of the curve may be expected to vary, depending on the parameter of measurement employed and on the microbial population whose growth is being assessed. The important point is that we would expect to observe both a logarithmic and a declining phase, regardless of whether growth or oxygen uptake is chosen as the parameter of measurement.

The initial lag period is not represented in the figure. If either a metabolic acclimation or an adaptation (population selection) is required before the log phase is entered, its extent (with regard to time) can be determined upon examination of a semilogarithmic plot of the data. At the termination of the curve, one would expect that the population had exhausted the available carbon source.

As our discussion progresses we will attempt to build upon this figure and fill in some of the other parameters of concern, such as the curve for removal of the carbon source which permitted the growth curve to be generated, events which occur after the removal of the carbon source, etc. But first, the curve above should be compared with the one produced by monomolecular "law," which was shown in Figure 13.1. That curve disallowed a log phase of growth and O_2 uptake.

In attempts to "force" BOD curves into the monomolecular mold, the log phase is sometimes treated as a lag phase (dotted line, Figure 13.2). If one does employ this procedure, he should be aware that the phase he treats as a lag phase may actually be a log phase, and that he may be discarding that portion of his data which is most critical from the standpoint of its use in making a credible assessment of the critical DO in the receiving stream. On the other hand, it should be realized that the inflection point shown on the curve occurs at an early time, much before the end of the standard 5-day incubation period. Therefore, for the curve shown, if the investigator is interested in obtaining a number for the purpose of keeping an historical record of the performance of a treatment plant or in assessing the BOD loading as a design criterion to size a trickling filter or activated sludge tank, it might make relatively little difference to him whether this early portion of the curve were treated as a lag or a log phase.

In the early days of investigation on the exertion of BOD, the data indicated that the O_2 uptake curve was produced with an ever-declining rate (Figure 13.1), that is, "monomolecular law." The present standard BOD techniques were developed with the goal of providing optimum growth conditions, and it can be expected that a log growth phase will occur and will produce a log phase of O_2 uptake.

There are conditions under which respiration (O_2 uptake) can proceed without a log phase. For example, under conditions of severe nitrogen limitation, O_2 uptake may be initiated with either linear or decreasing rate kinetics; the latter may approximate a first-order decreasing rate. Also endogenous respiration, that is, biological O_2 uptake due to oxidation of organic matter which has already been taken up by the organisms from the external medium, can proceed with either linear or first-order decreasing kinetics. Often when one examines accumulated oxygen uptake curves under either of the conditions cited above, the early stages appear linear but, after tracing the O_2 utilization for a longer time, the decreasing rate of O_2 uptake becomes apparent. It is not unlikely that first-order decreasing rate kinetics could have been observed in the early studies on BOD because of nutrient deficiencies or because, in some of the studies on river water, most of the O_2 uptake measured resulted from rather dense microbial populations respiring organic materials which had already been incorporated into the cells.

It is important to note that in the long term overall curve of O_2 uptake, the logarithmic phase of uptake may represent a relatively small portion (perhaps no more than 25%) of the total amount of oxygen ultimately used by the ecosystem in the BOD bottle. In experiments in which samples were taken daily for measurement of O_2 utilization over a period of 20 days, the existence of a log phase might not be detected.

Up to this point, a curve of the general shape expected for measurements of growth, O_2 uptake, or BOD exertion has been shown (Figure 13.2), and it has been stated that at the point at which the curve terminates all of the substrate or original carbon source added to the system has been used for growth, that is, it has been assimilated and used for cell replication. The reader is cautioned, however, that there are exceptions to the general rule that the leveling off of the S-curve marks the time of removal of the carbon source (5, 6).

In the general case, if organisms could now be removed from the bottle, the liquid remaining would contain very little organic matter, that is, its BOD would be very low.

If BOD exertion is measured, with a known substrate, it can be shown that at this point only 35 to 45% of the "theoretical oxygen demand" has been exerted. The theoretical oxygen demand is defined as the amount of oxygen which would be required to oxidize the substrate to CO_2 and water. For an actual waste sample the theoretical COD cannot be calculated, since the qualitative and quantitative compositions are unknown, but it can be approximated by measurement of the COD. In neither case is theoretical oxygen demand equivalent to "ultimate" BOD, or L_0, since this equality would imply total biological oxidation of the waste. It is extremely unlikely that total oxidation in the BOD bottle would occur during any reasonable incubation time. An even greater discrepancy between theoretical oxygen demand, or COD, and L_0 arises when a waste contains organic material which is relatively inert to biological oxidation. For example, the significantly large proportion of lignin in pulp mill wastes leads to a very large difference in the COD and BOD of the waste. The term "theoretical oxygen demand" is therefore applied only to pure, readily metabolizable organic compounds and has been used primarily in research with such compounds.

Turning attention again to the termination point of the curves shown in Figure 13.2, it should be pointed out that, at this point, although the BOD has been removed from solution, a considerable portion of the "ultimate" BOD remains to be expressed. The organic substrates in the waste sample have been converted to cells. The O_2 uptake curve roughly parallels the growth curve up to the point at which the curves level off (termination of Figure 13.2). After this point the viable count in the system will usually be found to decrease, and the continued rise in O_2 uptake can not be attributed solely to endogenous metabolism (7, 8).

The decrease in numbers of viable bacteria can occur for various reasons, all of which can lead to increased O_2 uptake. After attaining maximum numbers, some species of bacteria undergo death and autolysis. In addition, after the external carbon source is depleted the starvation conditions which exist in the ecosystem enhance competitive interactions, including predation. Thus, a considerable amount of metabolism can occur after the initial carbon source in the waste sample has been removed from the system, and the respiration (O_2 uptake) associated with it is registered as carbonaceous BOD.

The portion of the developing BOD (accumulated O_2 uptake) curve shown in Figure 13.2 represents a period which is usually less than 2 days' incubation. A portion of the BOD which was formerly soluble is now suspended BOD, and it is this portion which now remains to be expressed. In many instances there is a discernible pause in O_2 uptake before the second stage of carbonaceous BOD exertion, which may take the form of a second S-curve in sequence with the one shown in Figure 13.2, proceeds. It is emphasized that the general ecological situation has undergone a rather severe change from that which existed at the start of the incubation period. At the start there was a relatively large external food supply in relation to the number of seeding organisms, whereas now the microbial population is large and the concentration of external food is very low. It might be expected that a period of time might elapse before some cells in the population could synthesize the enzyme systems required to solubilize and utilize the organic material of other cells for growth, or before that portion of the predatory population which could ingest certain bacterial species would grow up to sufficient numbers to exert a discernible O_2 uptake. The pause which often occurs has become known as the "plateau" in BOD exertion. Busch ascribed oxygen uptake which occurred after the plateau to a combination of endogenous bacterial respiration and growth of protozoa. He believed that a large initial population of protozoa in the BOD bottle could mask or blur the plateau (9).

Thus far, in this description of the course of BOD exertion, an attempt has been made to present the most generalized concept possible in analyzing occurrences in heterogeneous populations. Thus the sequential O_2 uptake curves have been attributed to sequential growth cycles, first of bacteria which remove the original external carbon source in the waste sample, and second, of either protozoa or other bacteria using the cells which grew during the first cycle as a carbon source. However, it should be recognized that sequential growth (hence O_2 uptake) curves can occur on two or more fractions of the original carbon sources (5). Thus the development of a plateau in the O_2 uptake curve does not necessarily signal the completion of removal of the original carbon source of the waste sample. Also, it should be recognized that even if the original waste sample consisted of a single organic compound, for example, glucose or fructose, the microbial population can, during the process of growing on this one compound, produce and cause to be accumulated in the medium a variety of other organic compounds (metabolic products) which can be subject to sequential metabolism after the original compound has been utilized (6). Thus the picture is not quite as simple and straightforward as the one herein given with respect to the position of the point of substrate removal along the O_2 uptake curve. Also, as another caution against overgeneralization or simplification of the course of events, it should be noted that a pause or plateau in the course of O_2 uptake can be shown to exist in systems in which the seed material consists of a pure culture of bacteria (6, 10). Thus the development of such a discontinuity in the course of exertion of carbonaceous BOD is not uniquely dependent upon the presence of a mixed microbial population consisting of bacteria and bacterial predators. However, it should be realized that in most ideal seeding materials, protozoa will be present and can be expected to exert an effect on the overall kinetics of the system.

The mechanism by which protozoa contribute to O_2 uptake may still be the subject of some debate. Nearly 40 years ago Butterfield et al. (11) interpreted their experimental results in the form of a theory which ascribed to the protozoa a somewhat passive or indirect role in the exertion of BOD. According to these workers, the chief function of the protozoa was to "keep the bacterial population reduced below a saturation point and thus to provide conditions suitable for continuous bacterial multiplication, this in turn resulting in a more complete oxidation." Such a theory would be credible if, at the time the protozoa reduced the bacterial population, external carbon sources were still present in the medium so that the bacteria could continue to grow and thus more completely oxidize the waste. The theory of Butterfield et al. was based largely on the surmise that some of the original carbon source, albeit a decreasing amount, was present in the medium throughout the incubation period. It has since been shown that, in the general case, the original external carbon source in the waste sample has been removed by the time the "plateau" is attained, and before the protozoan population has increased to any great extent (7, 12, 13). Consequently, there is little if any external carbon source for the reduced numbers of bacteria to metabolize.

Various aspects of the comparative evidence for the indirect and the direct role of the protozoa in the exertion of BOD have been presented in some detail by Bhatla and Gaudy (14). They obtained experimental evidence for the direct contribution to oxygen uptake by the protozoa. In addition to experimental evidence attesting to the fact that a second major stage of carbonaceous BOD exertion occurred during growth of protozoa in the absence of any original carbon source in the waste, it was also shown that a second stage of oxygen uptake could occur in the presence of chloramphenicol, an antibiotic which prevents protein synthesis in bacteria but not in protozoa. In the face of this direct evidence in support of a major O_2 uptake by the predators, we are forced to conclude that the promulgation of the oxygen uptake curve during the exertion of carbonaceous O_2 demand beyond the plateau region is caused mainly by the ingestion and utilization for growth, by the protozoa, of the bacteria which were produced prior to the plateau.

Since the BOD bottle is a closed system, the types of organisms in the inoculum as well as the competitive events which occur after the bottle is sealed determine the mechanistic and kinetic patterns of BOD exertion during the incubation period. If one employs an inoculum which has been previously exposed to the carbon source in the waste for some time, natural selection processes tend to establish the requisite food chain in the ecosystem. Thus the inoculum would be expected to be poised to perform in the sequential fashion depicted above, and promulgation of the BOD curve beyond the time of removal of the carbon sources is indeed usually brought about by the predatory growth of the protozoa. One cannot, however, always depend upon the ecological events being manifested sequentially in the oxygen uptake (BOD) curve. Only when there is a significant time differential between the peak in bacterial numbers and the period of rapidly increasing protozoan numbers is this pause or plateau discernible.

With due regard for the need for caution concerning oversimplification or generalization, attention is turned to Figure 13.3. Figure 13.3*A* represents an idealized version of the complete sequence of events by which a discernible plateau is usually generated in the exertion of BOD. The portion of the figure to the left of the plateau shows the relationship between bacterial growth, O_2 uptake, and removal of substrate. The identity of this portion of the curve with Figure 13.2 will be readily recognized. When the plateau is reached, the maximum concentration of bacteria has been attained and the soluble substrate has been essentially exhausted. The growth of protozoa during the period of substrate removal has been slow. Their initial numbers were rather low, their growth rate is slower than that of bacteria, and their food supply (bacteria) is low during the early period of substrate removal. Perhaps the bacterial cells which have grown up in the early stages consisted of some species which were not readily available to the protozoan species present in the seed. The bacteria have gone through logarithmic and declining growth. The termination of this period of growth and substrate removal is accompanied by a discernible retardation of the accumulated O_2 uptake curve. The dotted line in the O_2 uptake curve extending to the right of the plateau is intended to depict the probable course of O_2 uptake in a system in which no predators were

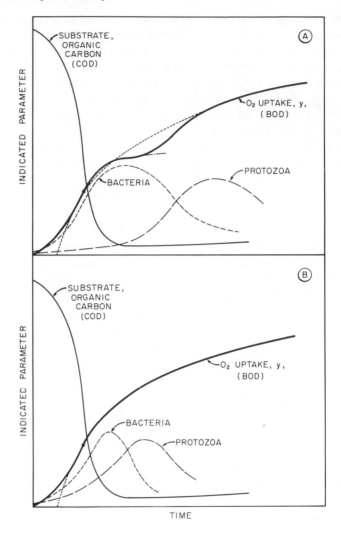

Figure 13.3 Relationship between O₂ uptake, organic carbon removal, and microbial growth. In (A), growth of the protozoa lags bacterial growth sufficiently to cause development of a plateau in O₂ uptake, whereas in (B), predator growth does not lag sufficiently to allow development of a plateau.

present. This uptake might be termed "endogenous respiration," if this is defined as respiration after the original external carbon source has been assimilated.

With regard to the protozoa, the ecological situation has undergone a drastic change. They are now existing in an environment rich in carbon source (the bacteria) and they experience a growth cycle which is naturally accompanied by oxygen utilization. The time differential between growth cycles has caused the development of a plateau in O_2 uptake. The plateau can be considered as a second stage of O_2 uptake which would continue along the endogenous curve (dotted line) if there were no protozoa in the system. The O_2 uptake resulting from metabolism of the protozoa represents a third stage, but the endogenous stage is of sufficiently low magnitude that it may be neglected and the O_2 uptake by the protozoa is referred to as the second carbonaceous stage. In the stream the protozoa in their turn become food for higher organisms in the ecosystem. This aerobic metabolism also exerts an O_2 demand; one of the reasons wastes are treated is to allow this O_2 demand to proceed in an unrestricted manner at DO levels suitable to the higher organisms.

The dotted line on the BOD curve is intended to show how, by neglecting a logarithmic growth phase for the bacteria (which is sometimes considered as a lag phase in O_2 uptake) and by neglecting the pause or plateau between the bacterial and protozoan phases of O_2 uptake, one might approximate the curve of BOD exertion during this incubation period by a curve approximating first-order decreasing rate kinetics. This procedure is often employed by workers in the field, although it is hoped that by the termination of this chapter the reader will be convinced that such an expedient is neither acceptable nor is it necessary in assessing the kinetics of BOD exertion.

In Figure 13.3*B* the course of events is depicted in a system similar to that of Figure 13.3*A*, except that the initial protozoan population is somewhat larger and perhaps more diverse, and the ecological relationships are such that the bacteria developed early in the growth cycle are readily available to the predominant protozoan species. The end result of this situation is one which closes the time differential between bacterial and protozoan life cycles, and the closing of this differential eliminates the pause, or plateau, in O_2 uptake. In such a case it is much easier for the investigator to reconcile the kinetics of BOD exertion to a single first-order decreasing kinetic mode, as seen by the dotted line on the O_2 uptake curve. However, in both parts of Figure 13.3 it is seen that in using this expedient, one is forced to neglect the log phase of growth, that is, the log phase of O_2 uptake. This completely reverses the kinetics of the O_2 uptake curve, changing a first-order increasing rate to a first-order decreasing rate. Adoption of the first-order concept, or recommendation of its use on the basis of its being "all right to use for practical purposes," seems to the author unwise and impractical for reasons which are developed below.

Very few investigators would argue against the fact that in the BOD bottle the ratio of carbon source to initial viable microorganisms in the inoculum is sufficiently high in the usual case to permit the development of a logarithmic growth phase. Also, few would dispute the fact that the oxygen uptake curve during growth is of the same general shape as the growth curve. There may be cases in which the material (waste water) being tested for BOD consists essentially of a bacterial suspension, that is, the external carbon source had already been taken up or assimilated by the microorganisms before the waste was put into the BOD bottle. In such a case a bacterial logarithmic phase would not be expected during the incubation period. This situation has been suggested by Hoover *et al.* (1). In such cases the apparent decreasing rate first-order kinetics could actually prevail. However, in the majority of cases, BOD exertion results from utilization by the microorganisms of an external, soluble food supply. Thus, during the early portion of the O_2 uptake curve, the kinetics of BOD exertion should be first-order increasing (logarithmic) rather than first-order decreasing.

One may wonder what effect this occurrence would have on the so-called dissolved oxygen sag curve in a receiving stream. Isaacs and Gaudy (15) made a study of some of the factors governing the process of physical reaeration using a 670-liter simulated receiving stream. In these laboratory investigations it was possible to define the reaeration constant very closely and thus to calculate the amount of oxygen transferred to the water at any time so long as the existing dissolved oxygen concentration and the DO deficit were known. Organic matter was added to this "stream," and the resultant DO profile (sag curve) was determined (16). The oxygen utilization (BOD exertion) which produced the DO sag could then be calculated, using a numerical integration procedure.

It is significant to note that the O_2 uptake (BOD curve) observed in this open system consisted of two sequential autocatalytic curves of the type shown in Figure 13.3A. The maximum DO deficit in the sag curve occurred immediately before the plateau in the BOD exertion curve, and the downward leg of the DO profile formed an S-curve. The plateau in BOD exertion corresponded to a period of rapidly rising DO concentration, which was followed by a deceleration of the DO recovery corresponding to a second major stage of O_2 uptake. Uptake curves were also developed, using the standard BOD dilution technique, and it was found that the DO sag calculated using these data and the sag equation (Equation 5) were in serious disagreement with the observed DO profile. In fact, the maximum DO deficit calculated in the standard fashion was only about one-third the actual maximum deficit. On this basis, the assumption of first-order decreasing rate kinetics for BOD exertion would not only be incorrect for scientific reasons but would not serve even for practical engineering purposes. It should be clear that logarithmic growth in the system, with a specific growth rate constant of any reasonable magnitude, can lead to a rapidly increasing O_2 uptake and a concomitantly rapidly decreasing downward leg of the DO profile. The large difference between the actual first-order increasing kinetics of O_2 uptake and the assumed first-order decreasing kinetics therefore can lead to a serious discrepancy between the actual DO profile and that calculated by the usual procedure.

Further work by Jennelle and Gaudy (17) was designed to examine the course of BOD exertion in two systems open to the atmosphere, the simulated stream mentioned above and an open stirred reactor, and in closed systems (both stirred and quiescent) consisting of 2.4 liter and standard 300 ml BOD bottles. Growth and shifts in the ecological balance were assessed by counts of viable bacteria and protozoa and by measurements of the optical density of the water. Substrate removal was assessed by COD and carbohydrate analyses. Thus a variety of analyses were available for correlation with the DO sag data and O_2 uptake curves for these systems. It was found that the dissolved oxygen profiles and the BOD exertion curves calculated from them exhibited two sequential S-curves, the first associated with growth of bacteria and removal of carbon source from the waste, and the second with growth of the protozoa on the bacteria synthesized during the substrate removal phase. In the closed systems the standard dilution technique was employed. The BOD curves for these systems (corrected for dilution) were not comparable to those obtained in the open systems. Thus, information obtained in the standard manner could not have been used to predict DO sags in the open systems. Both the magnitude and the rate of O_2 uptake were lower in the closed systems (corrected for dilution) than in the open systems simulating stream conditions. In the closed bottles the quiescent and stirred systems yielded comparable O_2 uptake curves, indicating that mixing or stirring at the degree of mixing employed did not affect the O_2 uptake curves. Indeed, the only factor to which the difference in BOD curves could be attributed was that the closed systems were necessarily more dilute than the open systems simulating the receiving stream. Only when the waste concentration in the BOD bottle was the same as that in the open system were the BOD curves relatively comparable. To run such an experiment it was necessary to supersaturate the dilution water for the closed systems with dissolved oxygen in order to avoid the need for decreasing the concentration of substrate.

The relationship between logarithmic growth rate constant and initial substrate concentration, the "Monod equation," is usually given as

$$\mu = \frac{\mu_m S}{k_s + S} \tag{6}$$

μ_m represents the maximum growth rate constant which can be developed under the particular operational conditions employed when the initial substrate concentration, S, is so high that μ is independent of S. k_s is a constant which determines the curvature of a plot of μ *vs.* S, and is numerically equal to the substrate concentration at which μ is half the value of μ_m. The equation describes a hyperbolic plot similar in general shape to a Michaelis-Menten plot of the velocity of an enzymic reaction, v, versus substrate concentration. The equation relating μ and S was originated by Monod on the basis of analysis of bacterial growth curves.

This relationship has also been tested using heterogeneous populations of sewage origin and found to be applicable to natural populations (18, 19). Thus we can say that the proportionality factor, μ, which determines the logarithmic rate of bacterial growth, changes with the initial substrate concentration in the system. The first-order increasing rate equation for logarithmic growth can be stated as

$$\frac{dN}{dt} = \mu N \qquad (7)$$

where $N =$ the number of cells per unit volume. As stated previously, the O_2 uptake curve has the same shape as the population growth curve during the logarithmic phase of growth. Therefore, the proportionality constant, or rate constant, for O_2 uptake, which we may call k_i, might be expected to be affected by substrate concentration in the same way that μ is affected. Thus the rate constant for BOD exertion should be affected by the dilution factor, since the substrate concentration in the standard BOD incubation technique is necessarily so low as to be rate-limiting. However, the fundamental concept which has been adopted, inherently, in the standard BOD dilution technique may be stated as follows: ". . . the rate of biochemical degradation of organic matter is directly proportional to the amount of unoxidized material existing at the time . . ." For example, ". . . a 10 per cent dilution uses oxygen at 1/10th the rate of a 100 per cent sample" (20).

Jennelle and Gaudy (17) have reported results of experiments in which various concentrations of standard BOD substrate (glucose–glutamic acid mixture) were placed in BOD bottles under identical seeding conditions, and O_2 utilization was measured. In all cases a logarithmic phase (first-order increasing rate) O_2 uptake was developed. The values of k_i increased with increasing concentration of substrate. Thus the velocity constant was affected by the dilution factor. In general, a plot of the values of k_i *vs.* S yielded a curve similar in shape to the Monod curve for μ *vs.* S. Thus the rate constant for the process (as well as the rate of the process) has been shown to be dependent upon the substrate concentration.

We are now faced with two important discrepancies which point out the need for revision of the classical application of the concept of biochemical oxygen demand to assess the effect of organic waste waters on the receiving stream. First, it has been shown that a logarithmic growth phase is developed in response to the food source of which the waste is comprised. This fact negates the use of the sag equation on a theoretical basis. Furthermore, because of the increasing first-order rate (rather than a decreasing first-order rate) and the rapidity of approach to the plateau region after the inflection point in O_2 uptake, the maximum deficit (or low point on the DO sag curve) attained is lower than that which would be approximated by forcing the data into a "monomolecular" configuration, thus militating against the utility of the sag equation from a practical basis. Second, the magnitude of the BOD exertion rate constant (usually obtained in BOD bottles in the laboratory) is affected by the dilution factor (*i.e.,* the substrate concentration S or L). But, in the usual case, the waste sample being examined must be diluted to avoid exhaustion of the dissolved oxygen in the closed system. Hence, any estimate of "the" rate constant obtained in systems more dilute than the system expected in the receiving stream also leads to a higher estimate of assimilation capacity than actually would exist in the stream.

Incorporation of these newer concepts into the fundamental view of biochemical oxygen demand would help provide unifying mechanistic and kinetic principles. It is often said that there is really no good reason for expecting O_2 uptake in the BOD bottle to conform with the "monomolecular law" when some organic food is present under optimum growth conditions. In point of fact, it does not; it occurs in the predictable autocatalytic fashion. The only excuse for forcing the data into a single first-order decreasing rate mode is the ease of dealing with the kinetic expression of the data. This is not a concept but a convenience, which unfortunately has militated against conceptual progress regarding the exertion of BOD.

At this point the reader may or may not agree that the writer has successfully clarified some former concepts and added new conceptual factors for consideration in the matter of biochemical oxygen demand. In any event, it is not sufficient nor in keeping with the aims of this chapter to show why the prevalent monomolecular concepts of BOD exertion are inadequate or why the dilution technique does not permit a true representation of the BOD exertion. Early in this chapter the BOD concept, not necessarily the BOD test, was defended. Thus far, clarification and new inputs to the concept have in the main provided arguments against the way the concept is presently applied to assess the drain on the DO resources in the receiving stream. A successful conclusion would seem to demand some suggestion or recommendation concerning the use of BOD.

13-4.   Suggested Improvements in BOD Measurement

If one were to use a series of BOD bottles to determine in the laboratory the course of O_2 uptake due to a sample of waste, and if the dilution
factor were the same as that expected in the stream (one would probably
have to use supersaturated dilution water), the BOD exertion curve which
was generated would be similar to that expected in the stream (17). The
expected similarity could be enhanced if the river water rather than the
standard dilution water were employed. If one could define the physical
reaeration constant, k_2, for a reach of stream of interest, numerical integration of these two processes [discontinuous autocatalytic curve(s) of
O_2 uptake and a continuous single phase monomolecular curve for reaera-
tion] might yield more realistic estimates of the DO profile in the receiving
water. Thus one would have used the BOD bottle, but not the dilution
technique nor the standard sag equation. A further improvement in the
technique could be achieved by using an open stirred reactor, for example,
a large battery jar, rather than BOD bottles (17). In this case the investigator would first determine the physical reaeration characteristics of the
system at the temperature, volumes of water, and stirring rate employed.

The stirring rate could be then set to yield k_2 values in a range which
might be reasonably expected in the receiving stream under investigation.
Experimental runs could be made at various ratios of waste water to river
water (dilution factors), and the DO sag can thus be measured directly
at the assumed rate of reaeration. Delineation of the DO profile during
each experimental run would yield the necessary data to combine numerically with the calculated reaeration input, thus permitting computation of
the BOD exertion curves for each dilution rate. Ideally, the dilutions employed would bracket those expected (or permitted) in the receiving
stream. Experiments run at three or four dilutions might suffice to bracket
the various flows of waste water and river discharges expected. It might
be possible to interpolate, graphically, the BOD curves for other dilutions.

Having determined the BOD curves, they could now be combined with
various k_2 values expected in successive reaches of the river for prediction
of the DO profile in the receiving stream. This application of the concept
of BOD has advantages over the more standard or accepted practice, since
it allows direct use of the O_2 uptake curves which are actually generated.
These curves may not fit any standard mathematical representation of the
kinetics, but it is not required for this method that they do, whereas it is
necessary in the sag equation. The procedure recommended above also
permits use of a single large volume reactor rather than a series of BOD
bottles, and permits generation of BOD exertion curves under conditions

which more closely simulate stream conditions than do those in the BOD bottle.

The method recommended allows the worker to make an estimate of the DO profile, regardless of the shape or kinetics of the BOD curve. The BOD curve may be S-shaped up to the plateau or not; the curve may or may not exhibit a plateau; the curve may or may not be approximated by a single first-order decreasing rate reaction. All types of curves are possible, depending upon the state of the effluent entering the stream. When the effluent consists largely of microbial substrates rather than microorganisms, a curve of the type shown in Figure 13.2 would be expected; when the substrate has been largely assimilated before the effluent is tested for O_2 uptake, the curve generated may be one which can be justifiably approximated by an ever-decreasing curvature, that is, rate of O_2 uptake (Figure 13.1). In this case the O_2 demand would be one which is being expressed largely by the bacterial predators, that is, mechanistically the curve would be beginning beyond the plateau. Even here a logarithmic growth phase for the predators (and thus a log increase in O_2 uptake) might be expected.

In conclusion, concerning the mechanistic concept of BOD (and its inseparable kinetics), it has been shown that the kinetics are not, and in the general case would not be expected to be, those of a first-order decreasing rate reaction. Logarithmic growth occurs, and O_2 uptake therefore increases according to first-order kinetics; furthermore, the rate constant for this portion of the BOD curve is dependent upon substrate concentration, that is, the dilution factor. The inflection point and discontinuities in the BOD curve make it difficult to devise a mathematical expression which could be used to modify the sag equation. If a new sag equation could be developed, taking into account a deoxygenation process involving two A-curves separated by a plateau or pause of variable duration, the expression would be extremely unwieldy. Therefore, the approximate procedure using a BOD curve developed as recommended in this chapter seems the most immediate way of using the BOD concept for predictive evaluation of the biological demand of organic waste on the O_2 resources of an aqueous environment. It offers a practical application of a more refined BOD concept.

13-5. BOD in Design and Operation of Biological Treatment

Concerning the standard BOD test, the author is somewhat more reluctant to offer recommendations concerning its future status or ways of improving it which are as specific as those made above regarding the application of the concept to stream control. Perhaps if users of the BOD test could revise their tendency to consider the 5-day value as one which lies along a well-defined "monomolecular" curve, allow that the dilution factor can affect the rate constant(s) for the "reaction," and accept the fact that one cannot expect the reproducibility and precision for the natural microbial world to be that which might be expected from a formula involving a physical process, for example, a quantity of flow in an open channel, we might in many respects improve the test, not so much by changes in the procedure, but by reassessment of the interpretation of the results.

In essence, the BOD test is not the all-purpose measurement we have made of it.

There is no real need to employ the 5-day BOD test as a functional loading parameter in the design of a biological treatment plant. In obtaining information upon which to base decisions concerning the sizing of an activated sludge tank, for example, if one wishes to know the amount of biochemical oxygen demanding material which can be removed in a reasonable aeration time, he can, in a relatively simple laboratory study, determine the course of purification of the waste using the COD test as a parameter. The difference between initial and residual COD in a batch system, or between influent and effluent COD in a continuous flow system, represents the organic matter which has been removed from the waste in terms of its oxidizability (21, 22). This removal has been brought about by the microorganisms and is therefore interpretable as a usable design criterion. The information obtained using such a procedure provides a measure of the O_2-demanding substrate in the waste which has been removed via biological processes in a reasonable aeration period, that is, it is a parameter consistent with the aims of the treatment plant. It is important, however, not to interpret the difference between the initial and residual COD as an amount of biochemical oxygen demand exerted. The BOD which is exerted, that is, the O_2 uptake during the time between measuring the initial filtrate COD and the residual filtrate COD, is considerably less than the ΔCOD so calculated, since a portion of the COD which has been removed is incorporated in newly synthesized organic matter (i.e., it has been used to make some new cells). Since COD is a measure of chemically oxidizable material, the ΔCOD may be precisely defined as the amount of oxygen required to oxidize chemically the organic material which has been removed biologically during the aeration period which intervened between samplings for the COD determinations. In practice, even if the cells can be completely separated from the waste, ΔCOD could only be equated to ultimate BOD removal if one assumes total oxidation of the organic matter in the waste.

However, there is no technologically justifiable need to relate the plant design criteria to any sort of standard BOD. The use of a ΔCOD (i.e., COD of the influent minus COD of the residual) rather than a ΔBOD seems highly recommendable because it obviates the need to use a test which is subject to variable results, interpretations, and much criticism. The ΔCOD relates directly to the purpose of treatment, that is, removal of biochemically available organic matter, which is indeed the way in which the BOD test has been applied to design. It is not the principle of the use of BOD for this purpose which is objectionable; rather it is the use of a standard technique which is inadequate for the purpose for which it is needed.

If one used both filtered samples and settled supernatant for COD determinations, he could also assess the overall efficiency which could be expected, that is, the efficiency measurement would include the settleability of the sludge as well as the removal of soluble organic matter.

Concerning daily assessment of operational efficiency of the plant once it is constructed, the operator might best employ the ΔCOD procedure as well. It is more rapid than the BOD, and provides the operator with direct assessment of the efficiency of removal of organic matter. If the ΔCOD procedure were performed on filtrate and on supernatant, both the biochemical and the overall efficiency of the plant could be easily assessed on the day on which that efficiency was obtained, and the immediate availability of this information should aid the operator in planning operational procedures for the next day.

13-6. BOD as an Aid in Regulation of Water Quality

Thus far, replacing the BOD test with other procedures for three of its major applications (DO sag analysis, design, and operation of treatment plants) has been recommended. At this point the reader may ask if, according to the author, there is really any place for a standard BOD dilution technique as we know it today.

Looking to the future, it is becoming more and more apparent that all used water containing significant amounts of organic matter will be subjected to some sort of treatment before reentry into the water resource. Thus the standard BOD test will become less and less significant for raw wastes.

It is, however, important to keep some record of estimates concerning the pollutional oxygen demand on the receiving stream of the effluent from the waste water treatment plant. While the daily assessment of ΔCOD provides information on the daily performance of the plant in removing biochemically available organic matter, it provides little direct information concerning the effect of the residual or effluent biochemical oxygen demand on the DO resource in the receiving body. The estimate of the DO profile using the procedures outlined previously, which was made prior to building the plant, is usually based upon the least favorable stream conditions. The estimate or prediction should be checked from time to time by monitoring the dissolved oxygen in the receiving body under all conditions, but especially during the worst predicted conditions of flow, temperature, etc. However, a continuing stream monitoring program for dissolved oxygen, while it is important in assessing the overall situation in the receiving body, is really not satisfactory for regulatory needs, because many changing industrial and commercial activities may pollute the stream, and regulation of each demands an assessment of the effect of each regarding its individual contribution to the total DO situation in the receiving body.

The open stirred reactor procedure suggested in this chapter could be used for this purpose but running such experiments daily would consume a great amount of time and effort. A BOD bottle technique could be usefully applied here. If BOD determinations were set up on the effluent from each treatment plant, preferably using water from the receiving body as diluent and a dilution factor approximating that which exists in the receiving body, the resultant O_2 uptake, after 5 days or some other specified interval, would provide, along with the ΔCOD filtrate and ΔCOD supernatant data from the plant, a way of assessing that effluent's contribution to the overall dissolved oxygen situation in the receiving stream due to carbonaceous BOD. This use of the BOD test would help the regulatory agency and the treatment plant manager to pin down the cause(s) for the DO situation in the receiving body and could serve as a basis for regulatory fines or charges and for future remedial actions or corrective operational procedures.

13-7.   Summary

It is concluded that the concept of carbonaceous BOD is vitally important for future progress in the control of water pollution. Through research on exertion of BOD, the concept has been refined and the kinetics and mechanisms of BOD exertion are much better understood today than they were in the past. Important applications of the concept do not need to involve "the BOD test."

Methods for utilizing the concept of BOD in other procedures which do not utilize the standard BOD technique have been suggested in this chapter. These procedures would replace the standard BOD test in estimating assimilatory capacity and in determining design criteria and treatment efficiency. The present BOD test, with some minor modifications as to application, can be useful in the long-term regulation of treatment plants designed to remove organic matter when the results of such a test are used along with other data on plant efficiency and data from a stream monitoring program. Monitoring of the aqueous environment is an important undertaking which has been inadequately accomplished to date because of the expense involved and a general lassitude on the part of those who must bear such costs. The recent general awakening to the critical need to protect water as a vital and reusable resource is doing much to correct this situation and to point up the need for continual monitoring and management of natural bodies of water. The use of a BOD test can help regulatory agencies and plant personnel in accomplishing this major task.

In the early stages of development of the BOD test the apparent aim was to devise some sort of test which would assist in the regulatory process. Accomplishment of this aim led to the development of the concept of BOD which was only partially understood and developed before major attention was focused upon standardization of the test. It would seem that the concept was then manipulated in order to accommodate the various applications of the test which were devised. This has led to a new series of investigations of the concept and the test itself, and the mechanistic and kinetic understanding gained by these studies will hopefully provide new insight into restrictions which should be placed upon applications of the test.

REFERENCES

1. S. R. Hoover, L. Jasewicz, and N. Porges, *J. Water Pollution Control Fed.* **25**, 1163 (1953).
2. E. B. Phelps, *Stream Sanitation*, Wiley, New York, 1944.
3. H. W. Streeter and E. B. Phelps. *Public Health Service Bull.* **146** (1925).
4. E. J. Theriault, *Public Health Service Bull.* **173** (1927).
5. M. N. Bhatla and A. F. Gaudy, Jr., *Appl. Microbiol.* **13**, 345 (1965).
6. M. N. Bhatla and A. F. Gaudy, Jr., *Biotechnol. Bioeng.* **7**, 387 (1965).
7. M. N. Bhatla and A. F. Gaudy, Jr., *J. Water Pollution Control Fed.* **38**, 1441 (1966).
8. A. F. Gaudy, Jr., M. N. Bhatla, R. H. Follett, and F. Abu-Niaaj, *J. Water Pollution Control Fed.* **37**, 444 (1965).
9. A. W. Busch, *J. Water Pollution Control Fed.* **30**, 1336 (1958).
10. I. S. Wilson and M. E. Harrison, *J. Inst. Sewage Purification* **3**, 261 (1960).
11. C. T. Butterfield, W. C. Purdy, and E. J. Theriault, *Public Health Rept.* **46**, 393 (1931).
12. T. R. McWhorter and H. Heukelekian, in *Proceedings 1st International Conference on Water Pollution Research,* **2** Pergamon Press London, 1962.

13. C. P. L. Grady, Jr., and A. W. Busch, *Proceedings 14th Oklahoma Industrial Waste Conference, Stilwater, Okla.,* 1963.

14. M. N. Bhatla and A. F. Gaudy, Jr., *J. Sanit. Eng. Div. Amer. Soc. Civil Eng.* **91**, SA3, 63 (1965).

15. W. P. Isaacs and A. F. Gaudy, Jr., *J. Sanit. Eng. Div. Amer. Soc. Civil Eng.* **94**, SA2, 319 (1968).

16. W. P. Isaacs and A. F. Gaudy, Jr., *Purdue Univ. Eng. Bull. Ext. Ser.* **52**, 165 (1967).

17. E. M. Jennelle and A. F. Gaudy, Jr., *Biotechnol. Bioeng.* **12**, 519 (1970).

18. A. F. Gaudy, Jr., M. Ramanathan, and B. S. Rao, *Biotechnol. Bioeng.* **9**, 387 (1967).

19. M. Ramanathan and A. F. Gaudy, Jr., *Biotechnol. Bioeng.* **11**, 207 (1969).

20. C. N. Sawyer and P. L. McCarty, *Chemistry for Sanitary Engineers,* McGraw-Hill, New York, 1967.

21. J. M. Symons, R. E. McKinney, and H. H. Hassis, *J. Water Pollution Control Fed.* **32**, 841 (1960).

22. L. L. Hiser and A. W. Busch, *J. Water Pollution Control Fed.* **36**, 505 (1964).

14 The Coliform Count as a Measure of Water Quality

Harold W. Wolf, Division of
Criteria & Standards, Bureau
of Water Hygiene, U.S. Public
Health Service, Rockville, Md.

14-1. Origin of the Coliform Count

A. Early Bacteriology. The science of sanitary water bacteriology began in 1880 when Von Fritsch described *Klebsiella pneumonia* and *K. rhinoscleromatis* as organisms characteristic of human fecal contamination. A short time later, Escherich identified *Bacillus coli* as an indicator of fecal pollution. Both observers considered human feces as a dangerous source of pollution while the feces of other warm-blooded animals were not considered a health hazard. From this origin, the current coliform group developed to include numerous microorganisms of diverse biochemical and serologic characteristics (1).

In the years that have passed since the development of the coliform group as an indicator of fecal pollution, a tremendous amount of information has accumulated on the physical and biochemical characteristics of this group. Attempts have been repeatedly made to correlate physical and biochemical characteristics of certain members of the group more closely with human fecal contamination in an attempt to sharpen the sanitary significance of such organisms. Parr studied the biochemical in-

333

formation available on coliform strains from fecal and nonfecal sources, and he selected the indole, methyl red, Voges-Proskauer, and citrate tests as the combination of reactions that would best render a separation into the two sources. This combination of four procedures has been designated the "IMViC test" (1).

In this classification, IMViC types $++--$, $+---$, and $-+--$ are considered to be of fecal origin, and types $--++$, $--+-$, and $---+$ are considered to be of soil origin. The remaining 10 possible IMViC types fall into an intermediate group (1).

Standard Methods (2) defines the coliform group as including "all of the aerobic and facultative anaerobic Gram-negative, nonspore-forming, rod-shaped bacteria which ferment lactose with gas formation within 48 hours at 35°C." The coliform group as defined is equivalent to the "B. coli group" as used in the 3rd, 4th, and 5th editions of *Standard Methods*, and to the "coli-aerogenes group" as used in the 6th, 7th, and 8th editions. The use of the membrane filter technique has led to the establishment of another definition for coliforms which is described in *Standard Methods* as follows: "All organisms which produce a dark colony (generally purplish-green) with a metallic sheen within 24 hours of incubation (on a specified medium and at a specified temperature) are considered members of the coliform group." The coliform group thus defined is not necessarily the same as that defined previously, but it probably has the same sanitary significance, particularly if suitable studies have been made to establish the relationship between the results obtained by the different laboratory procedures (2).

B.   Manifestations in Early Standards.   At the close of the nineteenth century an increasing amount of diversity was being practiced among a number of sanitary bacteriologists. American Public Health Association (A.P.H.A.) members appointed a committee which submitted a report in 1897 of recommended procedures which became widely accepted. The enthusiastic reception of the recommended procedures led to the publication by the A.P.H.A. in 1909 of the 1st edition of *Standard Methods*. Jackson (3) proposed further restrictions which appeared in the 2nd edition of *Standard Methods* in 1912. The U.S. Public Health Service's first bacteriological standards appeared shortly thereafter in 1914.

The involvement of the U.S. Public Health Service in the problem of water supply (and subsequent federal promulgation of coliform criteria) originated with the enactment by Congress in 1893 of the Interstate Quarantine Act. These statutes authorized (*a*) research investigations and studies related to the cause and prevention of diseases; (*b*) public health regulations designed to prevent the spread of communicable diseases from one state to another; and (c) the acceptance of assistance and the right to assist states in obtaining compliance with regulations. But it was not until 1912 that a regulation was adopted which prohibited the use on interstate carriers of the common drinking cup.

It was quickly realized that even a most sanitary drinking cup would be of no value if the water placed in it were originally foul. Hence, in 1914 the first Public Health Service Drinking Water Standard adopted a bacteriological standard which was applicable to any water supply provided to an interstate common carrier. Such a water supply is called an Interstate Carrier Water Supply. At one time there were over 2000 such supplies. Today, the list has dwindled to some 709; however, it is estimated that these supplies serve 82 million people.

The first PHS Drinking Water Standard for coliform organisms read as follows:

Not more than 1 out of 5 ten cc portions of any sample examined shall show the presence of the bacillus coli group when tested as follows:

a) Five 10 cc portions of each sample tested shall be planted, each in a fermentation tube containing not less than 30 cc of lactose peptone broth. These shall be incubated 48 hours at 37°C and observed to note gas formation.

b) From each tube showing gas, in more than 5% of the closed arm of fermentation tubes, plates shall be made after 48 hours' incubation, upon lactose litmus agar or Endo's medium.

c) When plate colonies resembling B. coli develop upon either of these plate media within 24 hours, a well-isolated characteristic colony shall be fished and transplanted into a lactose broth fermentation tube, which shall be incubated at 37°C for 48 hours.

For the purposes of enforcing any regulations which may be based upon these recommendations the following may be considered sufficient evidence of the presence of organisms of the Bacillus coli group.

Formation of gas in fermentation tube containing original sample of water (a).

Development of acid-forming colonies on lactose litmus agar plates or bright red colonies on Endo's medium plates, when plates are prepared as directed above under (b).

The formation of gas, occupying 10% or more of closed arm of fermentation tube, in lactose peptone broth fermentation tube inoculated with colony fished from 24-hour lactose litmus agar or Endo's medium plate.

These steps are selected with reference to demonstrating the presence in the samples examined of aerobic lactose-fermenting organisms.

The 1914 Standard also recommended the additional plating of smaller portions for more polluted waters and also adherence to the 1912 edition of *Standard Methods* for media and methods.

A discussion by the committee that recommended the first drinking water standards is also included in the published report. The discussion notes that in any consideration of "standards of purity," purity is an absolute term, and that, therefore, the recommendations are more properly "limits of permissible impurity." In this context, the report emphasizes that these limits are recommended only for the specific application in mind: the control of the sanitary quality of the water supplies of common carriers. The committee pleaded, "It is a fact so well established as to need no further discussion that the results of bacteriological and chemical examination of a sample of water ought always to be correlated with a knowledge of the source, treatment, and storage of the supply in order to enable a just estimate of the sanitary quality of such supply."

Another notable comment appears later in the text: "Bacteria of the *Bacillus coli* group are normally inhabitants of the intestinal tract of warm-blooded animals and it is believed that under ordinary conditions they do not multiply, in nature, outside of the animal body; that in drinking water supplies they tend, on the contrary, to die out rather rapidly. The presence of such bacteria in water may accordingly be considered valid evidence that the water has been polluted with the intestinal discharge of some of the higher animals and the numbers present may be considered a fair index of the extent of such pollution. Since practically all of the diseases which are known to be commonly transmitted through water supplies are due to germs which are discharged from the intestines of infected persons, pollution with intestinal discharge is not only the most offensive but by far the most dangerous kind of pollution to which water supplies are exposed."

A concluding comment again recognized the importance of the sanitary survey: "It is requested that the recommendation of these hard-and-fast limits of bacteriological impurity be not interpreted as minimizing in any way the importance of field surveys in estimating the sanitary quality of water supplies in general. It is always desirable to obtain information from as many angles as possible, and this is, indeed, necessary in order to form an altogether fair estimate of an individual supply."

The suggestions of the committee apparently went largely unheeded because in 1919, J. O. Cobb, a Public Health Service medical officer stated, "Water supplies heretofore have been passed upon by bacteriological standards. Very little attention has been paid to (sanitary) survey standards which engineers would likely set. I think the time has come to adopt the engineering point of view. The bacteriologist will only be a checker-up and I think I can convince you . . . that we should adopt the (sanitary) survey method of accepting a water supply rather than the bacteriological" (4).

Recognition of statistical aspects of the distribution of bacteria in a system was indicated by the 2nd revision of the PHS Drinking Water Standards (1925) which read as follows:

1) Of all the standard (10 cc) *portions* examined in accordance with the procedure specified below, not more than 10% shall show the presence of organisms of the *B. coli* group.

2) Occasionally three or more of the 5 equal (10 cc) *portions* constituting a single standard *sample* may show the presence of *B. coli*. This shall be allowable if it occurs in more than—

a) Five percent of the standard samples when 20 or more samples have been examined;

b) One standard sample when less than 20 samples have been examined.

Note,—It is to be understood that in the examination of any water supply the series of samples must conform to *both the above requirements*, (1) and (2). For example, where the total number of samples is less than 6, the occurrence of positive tests in 3 or more of the 5 portions of any single sample, although it would be permitted under requirement (2), would constitute a failure to meet requirement (1). . . . The *Standard portion* of water for this test shall be 10 cc. The *Standard Sample* for this test shall consist of 5 standard *portions* of 10 cc each.

The use of the coliform organism as an indicator of contamination received a jolt from the Illinois Supreme Court. In *People* v. *Bowen* (1941) 376 Ill. 317, 33 N.E. (2d) 587. Bowen, the Director of the Department of Public Welfare of the State of Illinois, was indicted and found guilty by the trial court of palpable omission of duty for failing to take proper measures to render safe a drinking water at a state hospital. As a result, an epidemic of typhoid fever resulted with much illness and many deaths. On appeal, the Supreme Court reversed the decision:

. . . the most that can be said for any of the 158 exhibits, is that it showed the water to be either positive or negative as to coli aerogenes. It appears from the record that coli aerogenes or colon bacillus may be friendly or inimical, and that the mere presence of the colon bacillus in water proves exactly nothing as far as typhoid fever is concerned. The tests seem to have been made by a method of broth fermentation, and determined nothing more than the presence or absence of some kind of colon bacillus. It further appears that this type of bacillus is present in the air one breathes, in milk, on fruits and practically everywhere.

. . . It is further apparent that colon bacillus may be of the fecal or nonfecal types and that so far as typhoid is concerned, it is only the fecal type from man alone (not from animals) that can spread the disease. The typhoid bacillus could not possibly have been identified by the laboratory means used in any of these reports and none of them is of any value to the People in an attempt to prove the guilt of the defendant. . . .

. . . Even if these reports were of any probative value, they would necessarily tend to disprove, rather than prove, a case against the defendant. The water from these wells was consumed by all inhabitants of Manteco for more than 8 years prior to this epidemic. This test over a period of 8 years, if looked upon as a laboratory experiment, would go a long way toward proving that the water actually was safe for human consumption, because there is no evidence of any abnormal condition as to typhoid occurring during that period of time. . . . (5).

C.   Transition to Fecal Coliforms.   The increasing reports of recoveries of coliform organisms from nonfecally contaminated environments together with reports of the multiplication of coliforms in some natural environments served to stimulate researchers in the separation of fecal from nonfecal coliforms. Added to this stream of activity must be the increasing use of standards for the enforcement of water pollution control measures as well as the continuing problem of the sanitary interpretation of a few coliform cultures. Sanitary bacteriologists therefore reverted to a study of the elevated-temperature test as a means of differentiating fecal from nonfecal coliforms, a test originally proposed by Eijkman in 1904. Eijkman had observed that coliform bacteria derived from the gut of warm-blooded animals produced gas from glucose at 47°C, while coliform strains from nonfecal sources failed to grow at this temperature. A number of workers followed through on Eijkman's observation with the following conclusions:

1. The most acceptable temperature of incubation for the separation of fecal from the nonfecal coliform group was 44.5°C in a water bath.
2. A small percentage of the fecal coliform strains did not grow when incubated at this temperature, but an equal percentage of nonfecal coliforms did grow.
3. The EC medium, described by Perry and Hajna, gave the most rapid results as it required only 24 hours' incubation.
4. The test should be used only as a confirmatory procedure on coliform cultures obtained from the presumptive test.
5. In the evaluation of results, all coliforms from the feces of warm-blooded animals must be considered as fecal coliforms, and all cultures isolated from unpolluted soils must be considered as nonfecal coliforms (1).

Applying the concepts above as guides, Geldreich and associates studied the coliform organisms isolated from the feces of several warm-blooded animals, including humans, cows, sheep, pigs, chickens, ducks, and turkeys; from 223 soil samples with no known fecal pollution, collected in 26 states; from 28 fecal-polluted soil samples from feed lots or locations recently flooded with domestic waste water; and from 15 species of plants and 40 species of insects. Coliform strains isolated from these samples were purified, and their reactions to Standard Method's Completed Test, to the IMViC test, and to 44.5°C temperature tests were determined. From the results of these studies it appears that separation of fecal coliforms from nonfecal coliforms can be obtained using the elevated-temperature test which is superior to biochemical procedures in simplicity of technical procedure and in time required to complete. Most importantly, the Completed Test need not be applied to a tube that is positive by the elevated-temperature test, because complete correlation exists between the positive reaction in both procedures. Coliform strains not derived from feces, however, are negative in the elevated-temperature test and positive in the Completed Test (1).

Fecal coliform organisms may be considered indicators of recent fecal pollution by warm-blooded animals. Because no satisfactory method is currently available for differentiating human fecal coliform organisms from those that derive from animal origin, it is necessary to consider all fecal coliform organisms as indicative of dangerous contamination. The presence of any type of coliform organism in treated drinking water suggests either inadequate treatment or access of undesirable material to the water after treatment. Insofar as bacterial pathogens are concerned, the coliform group is considered to be a reliable indicator of treatment inadequacy. Whether these considerations can be extended to include rickettsial and viral organisms has not yet been definitely determined (6).

Some sanitary bacteriologists are proposing the use of fecal coliforms as a better means of evaluating the sanitary significance of raw waters. No suggestion is being made to substitute fecal coliforms for the traditional coliforms when used in the evaluation of treated waters. As pointed out by a State Health Department water supply official recently, a properly treated, properly operated, soundly constructed and maintained public water supply system should provide essentially no coliform organisms to the consumer's tap. Hence, what would be the point in changing from total coliforms to fecal coliforms for evaluating treated water quality? Another factor that must be considered is that as the quality of source water diminishes, the treatment provided must be increasingly more dependable.

The drinking water standard presently in use (approximately one coliform per 100 ml) is, in a sense, a standard of expedience. It does not exclude entirely the possibility of acquiring an intestinal infection. It is attainable by the economic development of available water supplies, their disinfection, and, if need be, treatment in purification works by economically feasible methods. It is not a standard of perfection. Fair and Geyer note that the application of the standard to areas of high endemicity may well be challenged (7). Similarly, one may question its relevance for application to renovated wastewaters for potable uses.

14-2.   Current Water Quality Usage

A.   Drinking.   The standards of quality applied to drinking waters in the United States are recorded in the various revisions of the PHS Drinking Water Standards. We have already taken a look at the first (1914) and second (1925) standards. Subsequent revisions appeared, the next being in 1942 in which it was required that bacteriological samples be "collected at representative points throughout the distribution system," and also specifying a minimum requirement for the number of samples to be taken—a requirement that varies with the population served. The 1942 revisions were momentous in that the emphasis of regulatory agencies shifted from concern solely for the adequacy of treatment to concern also that the distribution system might not be as unreactive nor as tight as desired. Quality deterioration in the distribution system is today widely recognized. Yet to be recognized is that the routine surveillance for coliforms as practiced today is a part of a municipality's cross connection control program. If routine surveillance for coliforms is supposed to be a part of such a control program, then one must ask if enough samples are being taken to demonstrate lapses with any degree of efficiency? Is there not a better indicator for this use than a coliform? What about a chemical or physical indicator? The next revision of the PHS Drinking Water Standards will have to answer these questions, as well as others. If quality in the distribution system is of interest, it would be well for the standards to specify that a sample shall also be taken where the water enters that system. Similarly, the next PHS Standards should specify raw water samples so that efficiency of treatment will be a part of the operating record.

The effectiveness of the present indicator system in contributing to the control of the classical communicable diseases is now legendary. But what about its effectiveness in controlling virus infection? Ineffectiveness is perhaps indicated by the occasional occurrences of gastroenteritis attacks by unknown (suspected viral) etiologic agents, by known cases of waterborne infectious hepatitis, and by the suspicion of a viral etiology in certain cancers. The recognized waterborne virus, infectious hepatitis, is known to be more resistant to chlorine procedures than is the coliform indicator. What about the other viruses?

The Committee on Environmental Quality Management of the American Society of Civil Engineers, Sanitary Engineering Division, concludes that "Prudence requires . . . an assumption of waterborne viral gastroenteritis as a disease entity. On the other hand, an association of viruses in water with production of tumors is in the realm of conjecture at this point" (8).

If the source water for a potable supply is to be a reclaimed water, the sanitary survey loses its significance. Health protecting criteria for this use do not exist in the United States, although such a practice has been reported from South Africa. As increasing water shortages forces the issue in the United States, we shall likely see the development of additional indicator systems to be used along with the coliform. For example, a bacterium of special importance in any reclaimed water application for intimate human use is *Pseudomonas.* Some species of this ubiquitous organism cause obstinate ear, urinary, and other type infections; some also are among the prominent denitrifiers, a use that is likely to be exploited on an increasing scale in the future; some are of sufficient importance to the health of fish to be of concern to commercial fisheries; and some grow prodigiously in and on tertiary treatment devices such as reverse osmosis and electrodialysis membranes and in sand and carbon filtration and adsorption beds. At this point in the development of reclaimed water technology, it appears that some means of assessing strains and/or quantities of certain types of *Pseudomonas* would be essential as an adjunct to the coliform tests. In developing any uses for reclaimed waters, specific quantitative tests for a host of pathogens, notably *Salmonella* and *Shigella,* will always be an important requirement.

B. Shellfishing. Control of water quality in shellfish growing areas utilizes both the coliform organism and the sanitary survey. Recent research on the sanitary significance of the fecal coliform is likely to lead to its adoption in water as, indeed, it already is used to judge the quality of shellfish in marketing (9). The current *USPHS Manual of Recommended Practice* (10) classifies shellfish growing waters into four categories: approved, conditionally approved, restricted, and prohibited.

The approved area classification states that the coliform median MPN of the water does not exceed 70/100 ml, and not more than 10% of the samples ordinarily exceed an MPN of 230/100 ml in those portions of the area most probably exposed to fecal contamination during the most unfavorable hydrographic and pollution conditions. These limits need not apply if it can be shown by detailed study that the coliforms are not of direct fecal origin and do not indicate a public health hazard. This MPN value is based on a "typical" ratio of coliforms to pathogens and would not be applicable to any situation in which an abnormally large number of pathogens might be present. Consideration must also be given to the possible presence of industrial or agricultural wastes in which there is an atypical coliform-to-pathogen ratio.

The coliform criteria for a conditionally approved area are the same as for an approved area but the sanitary survey aspects differ. Criteria for the restricted area state that the coliform median MPN of the water does not exceed 700/100 ml and not more than 10% of the samples exceed an MPN of 2300/100 ml. The prohibited area criteria exceed these figures.

C. Recreation and Bathing. The clear rationale of the coliform criteria applied to drinking and shellfish waters becomes markedly murky when extended to recreation and swimming waters. In the first place, the increased illnesses suffered by swimmers as compared to nonswimmers is largely nasopharyngeal in character. The coliform can hardly be expected to provide an estimate of this type of risk. And secondly, the overwhelming epidemiologic basis for setting coliform criteria for drinking and shellfishing waters is almost completely missing from the area of recreation and swimming waters.

Garber (11), in reviewing various state and municipal criteria, observed that coliform values ranged from 50 to 3000/100 ml. Further, the standards were expressed in terms of arithmetic means, geometric means, or monthly medians. A maximum is occasionally specified, and frequently the percentage of samples that may exceed a stated concentration is indicated. The most widely used criterion is that the arithmetic mean coliform density not exceed 1000/100 ml and that this concentration not be exceeded in more than 20% of the samples in any one month. Streeter (12), assumed a 10-ml accidental intake, and concluded that, based upon various coliform-pathogen ratios, the 1000/100 ml value is a compromise between that which would be desirable and that which is practical.

Epidemiological studies conducted in various bathing waters by Stevenson of the United States, Moore of England, and Buczowska of Poland (5) all resulted in important contributions. Stevenson actually found a correlation between illness incidence and coliform counts in fresh waters when the counts were in the range of 2300–2700/100 ml. His salt water study data, however, failed to show any such tendency, an observation that was verified by Buczowska. The Moore studies—the most extensive of the three—concluded that bathing in sewage-polluted seawater carries only a negligible risk to health and where the risk is present it is probably associated with chance contact with aggregates of infected fecal material.

The next major input to the swimming water quality criteria scene was a recommendation of the Committee on Water Quality Criteria for the U.S. Secretary of the Interior (13). This committee recommended the application of a single set of criteria for fresh, estuarine, and marine waters. For all waters, it recommended an average not to exceed 2000 fecal coliforms per 100 ml and a maximum of 4000/100 ml. For waters classified for recreational purposes other than primary contact (swimming, etc.), it recommended a log mean of 1000/100 ml with no more than 10% exceeding 2000/100 ml. For primary contact recreation, it recommended a log mean of 200/100 ml with not more than 10% of the samples exceeding 400/100 ml.

Henderson (14), recognizing the lack of epidemiological support for these criteria, attacked the recommendations. Geldreich (15), however, in applying modern fecal coliform concepts to the Stephenson findings in a retroactive manner, supports the committee's recommendations.

The success of the coliform criteria applied to drinking waters in controlling waterborne illness is without question due in large part to the close relationships between coliforms and the types of organisms (intestinal pathogens) which the coliform criteria are designed to protect against. Among swimmers, eye, ear, nose, and throat ailments represent more than half of all illnesses. In the case of *pool* users, these illnesses represent as high as 68% of illnesses. In applying classic sanitary engineering doctrine to such waters, namely, that water quality should be judged by the combination of a sanitary survey and bacteriological criteria, it would appear that something is needed in addition to an intestinal indicator. It appears obvious that accidental ingestion of intestinal pathogens represents only one route by which a swimmer can become ill. Morbidity studies made thus far indicate that an upper respiratory pathogen indicator would be much more important. But which? Of approximately equal importance to the gastrointestinal group of illnesses are organisms that cause dermatitis. Here, again, an indicator might serve a useful role.

D. Food Production and Processing. The previously mentioned National Technical Advisory Committee on Water Quality Criteria (13) reviewed the research and experience that had been accumulated on irrigation water quality. They suggested that the most likely situation to cause trouble is where the contaminated water is applied to crops by overhead sprinkling. They also condoned the use of accepted indicator organisms to judge the bacteriological quality of irrigation waters. In their review, they recommended "guidelines" which would be subject to research confirmation. The guidelines are as follows:

The monthly arithmetic average density of the coliform group of bacteria shall not exceed 5000 per 100 ml and the monthly average density of fecal coliforms shall not exceed 1000 per 100 ml. Both of these limits shall be an average of at least two consecutive samples examined per month during the irrigation season and any one sample examined in any one month shall not exceed a coliform group density of more than 20,000 per 100 ml or a fecal coliform density of more than 4000 per 100 ml.

Geldreich and Bordner (16) examined the relationship of the occurrence of fecal coliforms and *Salmonella*. They found that for values under 1000 fecal coliforms per 100 ml, *Salmonella* occurrence was 62.5%; above this fecal coliform value, the occurrence was 93.8%. They concluded that this provided good evidence of support for the National Technical Advisory Committee recommendations. More importantly, perhaps, they do not overlook the "physical" aspects and remind us that "Practical reduction of the public health hazard can be accomplished only through a concept of multiple barriers to pathogen exposure and retention," and that, as part of this concept, "from harvest to consumption of raw produce, all applications of water to freshen these foods must be of drinking water quality."

It is definitely possible, when water is utilized in the preparation or processing of foods, that the severest demands are being made upon its bacteriologic quality. If *Salmonella* or enteropathogenic *E. coli,* for example, are present in a domestic supply and some housewives utilize the water to wash vegetables for salads, add some "good" (bacteriological media) dressings, and then hold the salads at room temperatures for several hours prior to serving, the relatively few organisms initially present could conceivably develop into the type of inoculum needed to initiate an illness. For straight drinking purposes, however, the same quality supply might not provide a "pathogenic dose" to more than an occasional person and then only very infrequently. Also, the small dose of pathogens may stimulate rather than overwhelm the body's natural protective mechanisms.

Many a veterinarian has lamented the lack of concern that some stockmen express for the bacteriological quality of the water served their animals. The National Technical Advisory Committee on Water Quality Criteria (13) reviewed the information available on pathogens in livestock water supplies but make no specific recommendations. In the previously mentioned section on irrigation waters, however, the committee notes that the coliform criteria ". . . are especially applicable where the tops or roots of the irrigated crop are to be consumed directly by man or livestock."

REFERENCES

1. E. E. Geldreich, *Sanitary Significance of Fecal Coliforms in the Environment,* Water Pollution Control Research Series, Publ. WP-20-3, FWPCA, USDI, Cincinnati, Ohio (1966).
2. *Standard Methods for the Examination of Water and Wastewater,* 12th ed., American Public Health Association, New York, 1965.
3. D. C. Jackson, *J. Infect. Diseases* **8,** 241 (1911).
4. J. J. Hinman, Jr., *J. Amer. Water Works Assoc.* **7,** 821 (1920).

5. J. E. McKee and H. W. Wolf, *Water Quality Criteria,* Publ. No. 3A, State Water Quality Control Board, Sacramento, Calif., 1963.
6. *Public Health Service Drinking Water Standards,* PHS Publ. No. 956, USDHEW, Washington, D.C. (1962).
7. G. M. Fair and J. C. Geyer, *Water Supply and Wastewater Disposal,* Wiley, New York, 1954.
8. B. B. Berger et al., *J. Sanit. Eng. Div. Amer. Soc. Civil Eng.* **SA1**(7112), 111 (1970).
9. C. B. Kelly, *Technical Conference on Fish Inspection and Quality Control,* Topic VI, Technical Note: Food and Agriculture Organization of the United Nations (1969).
10. L. S. Houser, Ed., *National Shellfish Sanitation Program Manual of Operations, Part I—Sanitation of Shellfish Growing Areas,* PHS Publ. No. 33, USDHEW, Washington, D.C. (1965).
11. W. F. Garber, *Sewage Ind. Wastes* **28**, 795 (1956).
12. *Bacterial Quality Objectives for the Ohio River,* Ohio River Valley Water Sanitation Commission (1965).
13. *Report of the Committee on Water Quality Criteria,* USDI, Washington, D.C. (1968).
14. J. M. Henderson, *J. Sanit. Eng. Div. Amer. Soc. Civil Eng.* **SA6**, 1253 (1968).
15. E. E. Geldreich, 89th Annual Conference, American Water Works Association, San Diego, Calif. (1969).
16. E. E. Geldreich and R. H. Bordner, *Bacteriol. Proc.* (1970).

15 The Detection of Enteric Viruses in the Water Environment

H. I. Shuval and E. Katzenelson,
Department of Medical Ecology,
Environmental Health Labora-
tory, Hebrew University-Hadas-
sah Medical School, Jerusalem,
Israel

15-1. Introduction

The possibility that water might serve as the vehicle for the transmission of certain virus diseases, particularly those whose infectious agent is excreted through the enteric tract, has been considered feasible for some time. Mosley (1) has pointed out that over 50 documented water-borne epidemics of infectious hepatitis have been recorded over the years. Apart from infectious hepatitis, poliomyelitis and viral gastroenteritis were the only other viral infections that caused epidemics suspected of being transmitted by water. In most cases, the epidemiological evidence was inconclusive. An exception is possibly a small polio epidemic in Nebraska in 1952 (2) where strong evidence suggested that it was caused at least partially by a water-borne virus. The possibility that viral gastroenteritis may be water-borne on occasion cannot be ruled out, however.

The massive waterborne epidemic of infectious hepatitis in Delhi, India, in 1955 (3) in which some 30,000 persons became infected by the contaminated municipal drinking water which had undergone what is generally considered complete and adequate treatment, including chlorination, emphasized the need to develop new methods of monitoring water supplies for viruses. Over the years, evidence has pointed to the fact that the usual bacterial parameters of water purity, particularly the coliform group, may not provide an adequate index as to the safety of water from a virological point of view. The need to monitor water specifically for viruses of enteric origin presents many problems, and the developments in this field will be reported here.

15-2.   The Presence of Enteric Viruses in Water

Over 100 virus types are known to be excreted from humans through the enteric tract and may find their way together with sewage into sources of drinking water. Many of these viruses are known to cause disease in man. However, the critical question is whether these viruses can survive long enough and in high enough concentration to cause disease in people consuming such contaminated water. The concentration of enteric viruses in sewage and polluted water is an important factor to consider. Clarke and Kabler (4) calculated a theoretical average number of enteric virus in infectious units in sewage and found it to be about 500/100 ml. In our own studies we have found the enteric virus concentration in the sewage of five communities in Israel to average 100/100 ml (5). Based on these figures it can be assumed that the virus concentration in polluted river water would range from 0.1–1 viral infectious units/100 ml as a result of physical dilution only. The number is lower during the cold months and somewhat higher in the late summer and early fall due to seasonal variation of enteric virus diseases. It can be assumed that this concentration will be further reduced both by processes of natural die-away and by water treatment, imperfect as they may be in the removal and inactivation of viruses. Under normal circumstances only a relatively small number of infective units will, at worst, penetrate a water supply system which derives its raw water from a heavily contaminated river.

Simultaneous infection of a large number of people is therefore rather improbable under normal conditions with modern water treatment methods. Sporadic infections, however, are possible, at least theoretically. The latter becomes true particularly in the light of Plotkin and Katz's claim (6) that "one infective dose of tissue culture is sufficient to infect men if it is placed in contact with susceptible cells." They were able to reach this conclusion as a result of studies on attenuated polio viruses, respiratory viruses, agents of ocular diseases, viruses of animals, and other agents. This might mean that even when virus concentrations as low as one virus infectious unit per 1000 ml are present in water, a certain number of individuals might well become infected by consuming the normal daily intake of 1–2 liters per capita.

With this in mind, one may form a picture of water as playing a small but important role in the spread of viral diseases in man in areas provided with modern treatment facilities. The effect of such slightly contaminated water may lead to sporadic cases of disease dispersed over a large area. However, these occasional cases may, in turn, act as foci and through food or personal contact cause epidemics which may involve much larger numbers of people. In other cases when heavily contaminated water reaches large population groups without adequate treatment, explosive mass epidemics have occurred and may well occur in the future. It is therefore obvious that the development of methods for the detection of viruses in water are required to allow for an adequate evaluation of the virological safety of water supplies and treatment processes. Bacterial evaluation of water as an indicator of contamination cannot replace such methods since it became apparent that viruses are not as sensitive as bacteria to hostile environmental factors or to standard purification procedures, and they may be present in water even when bacterial counts are at acceptable standards (7, 8).

15-3. Types of Viruses in Water

Water being used for drinking and bathing can act as a vehicle for the transmission of most viruses. The Picorna group of viruses is the one most commonly found in sewage; it includes the polio, coxsackie, and echo viruses. Adeno viruses which cause respiratory and eye infections, and sometimes diarrhea, are commonly found in feces.

Infectious hepatitis is actually the only disease for which a water-borne infection has been proven beyond any doubt. However, its viral characteristics are not yet clear. There are some claims that the responsible virus has been isolated from suspected cases of hepatitis, but most virologists feel that these claims are as yet insufficiently established. It must be remembered, however, that in spite of the latter there is strong evidence supporting the view that this disease is actually caused by a virus (9).

15-4.   Isolation and Identification of Enteric Viruses

Viruses can only multiply inside living cells and therefore live organims, such as animals, chick embryos, or tissue culture must be used for their isolation in the laboratory. For the enteric viruses, tissue cultures which may be of two types are generally used: primary tissue cultures and continuous cell cultures. Primary tissue cultures are usually prepared according to the method of Enders et al. (10). This method is based on the fact that 0.25% trypsin acts on small cuts prepared from a tissue (usually a kidney) by separating the cells from each other. When put inside a suitable glass or plastic flask, tube, or plate together with a tissue culture nutrient medium, these cells attach to the wall of the vessel and multiply. As a result, a monolayer of cells is formed on the wall. Continuous cultures are very similar, but instead of a tissue from an organ, a tissue culture is used as a source for the cells. Initial isolation of enteric viruses is usually done on primary tissue cultures prepared from monkey kidneys which typically have a higher sensitivity than most cell lines. However, any isolated virus can be adapted to cultures of the continuous type.

After inoculation of a virus into a tissue culture some of the cells become infected. The virus multiplies within these cells and spreads to the neighboring cells. At the same time, the infected cell usually undergoes morphological and biochemical changes and dies. The result is a slow process of destruction of cells in the culture, a phenomenon known as the cytopathic effect (CPE). The process of viral spread from cell to cell can be slowed down by adding a layer of agar together with tissue culture medium over the cells. As a result, instead of being rapid and confluent, the CPE will be limited to a smaller area which looks macroscopically like a hole in the monolayer of cells. These holes are also known as "plaques." A single plaque usually originates from a single infected cell which may be caused by a single virus infectious unit. This method is used for quantitation of enteric viruses in tissue cultures in the same manner as agar plates are used in bacteriology for the determination of bacterial counts. The term "plaque forming unit" or PFU was given to the lowest concentration of viruses that form one plaque on a monolayer of cells.

Different viruses cause cytopathic effects which differ morphologically.

Also, plaques may be of different sizes and shapes. However, this phenomenon cannot be used for the final identification of the isolates since some viruses, belonging to different groups, cause identical CPE.

Final identification can only be achieved with specific antisera. Here the identification is based on the fact that specific antiserum will neutralize the effect of the virus against which it was prepared.

15-5. Quantitation of Enteric Viruses

Two methods are available for the quantitative determination of enteric viruses in a given sample of material being assayed, both of which give accurate results. Selection of the method to be used is usually based on the experience and resources of the laboratory. In the first one, the tube assay method, serial dilutions of the virus suspension to be tested are prepared. Groups of tissue culture tubes are inoculated with each dilution. After proper incubation at 37°C, the inoculated tubes are examined for CPE. Quantitation is obtained by finding the lowest dilution of the virus suspension that caused CPE in 50% of the tubes. The figure obtained is known as the $TCID_{50}$ (tissue culture infectious dose—50%) value of the virus suspension. Using this same method, it also is possible to calculate the virus concentration as a most probable number (MPN).

In the second, the plaque assay method, quantities of 0.3–1.0 ml of virus dilutions are inoculated into plates or bottles, the cells of which are later covered with an agar overlay. After proper incubation, usually at 37°C in a humid atmosphere containing 5% CO_2, the inoculated tissue cultures are examined for the presence of plaques. When plaques are present, they are counted and their number for each of the dilutions is determined. The number of PFU in the original virus suspension is then calculated and the virus concentration is reported as PFU/ml or other unit of volume.

15-6. Detection of Viruses in Water

The isolation of viruses from water is principally the same as from any other source. However, the main difficulty is the usually low concentration of viruses in water. We have already seen that this number may be 1 PFU/1000 ml of water or even less. Considering the amount of 0.1 ml of inoculated material for a tube, 0.3 ml for a plate, and about 1 ml for a bottle, it is obvious that for the detection of a single virus unit in a 1000 ml sample at least 1000 bottles or 10,000 tubes should be inoculated. It is not surprising that in the past it has been difficult to detect the presence of viruses in water, even during epidemics.

The story of virus detection in water is therefore a story of the development of methods for virus concentration. The actual inoculation of specimen and the quantitation procedures are essentially the same as in clinical virology.

15-7. Methods for the Concentration of Viruses from Water

A. Gauze Pad Method. One of the first methods used to detect viruses in water was the gauze pad method. Here a modification of the "Moore" swab technique (11) for recovery of bacteria from water is exploited. Gauze pads or pads filled with cotton or plastic foam sponges are placed in the water and left for varying periods, usually from one to several days. During that period viruses are adsorbed or entrapped in the pad from the flowing stream of water leading to a concentration of viruses. The pads are often treated with dilute sodium hydroxide to increase the pH of the soaked up water to 8.0. This facilitates the elution of viruses from the pads (12). Liquid expressed from the pads is tested for viruses. Further concentration of the virus from the expressed liquid increases the sensitivity of the method.

This method has been used by many workers for the detection of viruses in water and sewage. Melnick *et al.* (12) were able to demonstrate its superior sensitivity compared to a grab sample. In their work, when equal volumes of expressed fluid and of grab sample were compared, more isolates were obtained from the pads. In our own studies of this phenomenon it appears that the liquid eluted from pads suspended for 1 or 2 days in sewage or contaminated water contained from 10–50 times as many viruses as detected in parallel grab samples of the same water. However, the mode of action and the degree of concentration have not as yet been determined. Although the pad method is not quantitative, it has proved itself to be a useful and effective procedure for detecting viruses in water.

B. Sample Incorporation Method. In this method the conventional culture medium is so concentrated as to allow for the incorporation in it of 10 to 60 ml of the sample to be assayed (13). According to another approach, large volumes of a maintenance medium, prepared from the water that is being tested, are inoculated onto cell cultures (14). No concentration of virus from the water prior to inoculation is attempted in these cases. Nevertheless, by this method virus detection is enhanced since with the same tissue culture tube or bottle a volume of sample 10–20 times larger than that normally inoculated can be assayed.

C. Ultracentrifugation. The centrifuge is mostly used for the concentration of small suspended particles from fluids. Being extremely small particles in the size range of 20–200 mμ, viruses are no exception, but because of their small size, relatively high forces of the order of 60,000 × g for 1 hour are required. These are obtained using an ultracentrifuge. In the usual procedure, the water sample is first centrifuged at a relatively low speed to reduce the number of larger particles including bacteria. The supernatant is then centrifuged at high speed. The sediment obtained contains the virus, and is resuspended in a small volume of tissue culture medium. A high concentration factor may thus be reached. When accurate results are required, restirring of the sediment has to be avoided. A "trap" of 2% gelatin, layered on the bottom of the centrifuge tube, has been used. In this way, the detection of a few virus units in 1 liter of sample water is possible (15).

Anderson *et al.* (16) described the development of a complex centrifugation system for the isolation and separation of small numbers of virus particles from large volumes of fluid. A high performance continuous-flow centrifuge was constructed which removed over 95% of suspended virus at a flow rate of 2–3 liters/hr. In their rotor the flowing stream moved over a stationary density gradient. Trapped virus particles were banded isopycnically in the gradient and were never pelleted. The recovered material could be further fractionated on the basis of sedimentation rate by zonal centrifugation.

The main disadvantage of ultracentrifugation is that it requires very expensive apparatus, especially when Anderson's technique is used. This may explain why the method has not been used extensively for the detection of viruses in water under field conditions.

D. Membrane Filter Adsorption. This technique is based on the fact that viruses are adsorbed onto the matrix of membrane filters of the "Millipore" or "Gelman" type, even when the pore diameter of the filter is 10–20 times larger than the virus. Addition of $MgCl_2$ to the virus suspension usually increases the adsorption. A suspension containing virus is passed through a membrane filter and the viruses are eluted from the membrane with either serum or gelatin (17, 18), or with sodium lauryl sulfate, a surface active agent (19). One of the problems involved with testing large volumes of water with this method is clogging of the filter system. In practice, using coxsackie virus A9 in tap water and membranes of 0.45 μ porosity, it appeared that there was at least a 50% probability of detecting virus at a level of two plaque-forming units per liter (20). Berg (21) reports complete recovery of enteroviruses and 50–80% of reovirus 1 from large volumes of distilled water with a 3% solution of beef extract after intensive sonication of the membrane filter. This technique was not effective with tap water, however.

E.   Soluble Ultrafilters.   The water to be tested is passed through a soluble filter which concentrates the viruses. The filter is then dissolved in a solvent and the suspension containing the viruses is inoculated into tissue culture. Filters of aluminum alginate gel and a 3.8% solution of sodium citrate as a solvent have been used (22). With this method it was possible to recover 10 $TCID_{50}$ virus from 10 liters of water.

F.   Hydroextraction.   A sample of water is placed in a cellulose dialyzing bag which is then immersed in a hydrophylic agent, polyethylene glycol (23–25). Water is then adsorbed from the bag while viruses, like other molecules of high molecular weight, remain inside. A concentration of 100 or more may thus be obtained, with an efficiency of recovery somewhat under 50%.

G.   The Phase Separation Method.   This method is based on the discovery of Albertsson (26) that the result of certain mixtures made from two polymers such as dextran sulfate and polyethylene glycol in an aqueous solution leads to the formation of a two-phase system. Introduction of particles and macromolecules into this system will result in the partition of the particles in the two phases, depending on their size and surface properties (27). Viruses show a nearly one-sided distribution and the method may therefore be used for their concentration (28, 29). The concentration is accomplished by adding polymer solutions to a virus suspension in such proportions that almost all the virus particles are collected in a small volume bottom phase and may be drained off separately. In this way, the virus may be concentrated 100–200 times, depending upon the kind of phase system used. Purification of viruses is also obtained since other substances, such as protein and cell fragments, distribute in a different way than viruses in the phase system. Dextran sulfate may be easily removed from the virus suspension with barium or potassium ions. This allows for repetition of the concentration procedures (30, 31). Using this relatively simple method Shuval *et al.* have shown that a concentration factor of 500 can be achieved using the two-step phase-separation (PS) procedure and as few as 1–2 virus infection units per liter of sample could be efficiently detected (24, 32). The usefulness of this method in detecting enteric viruses in sewage, water, and seawater has been demonstrated. On one occasion during the investigation of a waterborne epidemic it was possible to detect enteroviruses in contaminated well water with the two-step PS method (5).

H.   Electrophoresis.   Viruses are usually negatively charged at neutral pH values (33) and will therefore move toward the cathode when a virus suspension is placed in an electric field. This principle was used by Bier for concentrating bacteriophages in water (34). A simple procedure was developed by which electrophoretic transport was used to bring about adsorption of bacteriophages on dialyzing membranes. With this method a sample of water could be processed in a relatively short time, and a concentration factor of 100 achieved. It seems likely that with some adaption this method could be used for the concentration of enteroviruses, too.

I. Adsorption to Particulate Material.   Viruses may be adsorbed from water onto a variety of particulate material and then be eluted with a much smaller volume of liquid. Methods that involve the use of precipitates of inorganic salts such as calcium phosphate (35–37), cobalt chloride (38), aluminum hydroxide (36, 39), and ammonium sulfate (39) or the use of insoluble polyelectrolytes (41–43), ion exchange resins (44, 45), and passive hemagglutination (23, 46) were described. The adsorbent is first added to the tested water and later separated from it by centrifugation or filtration. Resuspension is carried out in a small volume. Concentration factors of 20–100,000 times have thus been obtained, depending on the method used. Because of its low cost, simplicity, and relative high efficiency in recovering viruses, this method holds much promise.

15-8.   Comparative Studies of Virus Concentration Methods

The great variety of methods which have been described for concentrating viruses in water indicate how different approaches can lead toward the same goal, the detection of a small number of viruses in a large volume of water. However, this variety emphasizes the lack of one general method of proved efficiency, low enough in cost to be acceptable by all workers in the field. It is as yet impossible to say which method is preferable although considerable progress has been made in recent years in improving techniques for the detection of viruses in water. A systematic comparative study of all the available methods is still to be made. However, a limited number of comparative studies have been published which may throw some light on the problem.

Gibbs and Cliver (23) compared three different methods: passive hemagglutination, hydroextraction, and ultracentrifugation. The virus they used was reovirus. They found passive hemagglutination specific for this virus and ineffective for others. The hydroextraction method had a disadvantage as the concentrated fluid resulting from this procedure was somewhat toxic to cell cultures. Ultracentrifugation was superior to the other methods for the demonstration of viruses in food extracts.

Shuval and coworkers compared the PS method which they had adapted and developed with hydroextraction ˙(24) and found the former to be about 10 times more sensitive in detecting enteroviruses in clarified sewage, normal saline solution, distilled water, and phosphate buffer solution. The efficiency of recovery of the PS method approaches 100%, while that of the hydroextraction method is between 40 and 50%. In later studies they demonstrated the effectiveness of the PS method in detecting viruses in seawater as well.

Lund and Hedström (40) also compared the PS method with a combined ammonium sulfate–ultracentrifugation method. They found both methods to be equally efficient for the detection of enteroviruses in sewage. However, the PS method was by far the simpler.

Moore et al. (47) compared membrane filtration with adsorption on aluminum hydroxide and found the latter to be better for the concentration of poliovirus from waste water. They found very high losses of virus with the membrane method. Aluminum hydroxide was also compared with insoluble polyelectrolytes by Wallis et al. (41). In their study polyelectrolytes were more efficient in detecting enteroviruses in sewage.

Graveille and Chin (45) compared three different methods of sewage sample preparation. Their samples were collected by gauze pads which were then inoculated into tissue cultures unconcentrated; concentrated by ultracentrifugation; and concentrated with the resin precipitation method. Evaluation of the methods was based on the number of successful isolates. Their results showed very clearly the superiority of ultracentrifugation over the other two methods. A valuable finding of their study was that a combination of two methods of concentration can increase the number of virus isolates from sewage.

15-9.   Evaluation of Virus Concentration Methods

Evaluation of virus concentration methods may be obtained when results from different studies are compared. Selection of criteria for the comparison is based on the assumption that the best method should enable the quantitative detection of very small numbers of virus infectious units in water. This will become possible only if the method allows for utilization of large quantities of water, has a very high concentration factor, and maximum recovery efficiency for as many virus types as possible. Assuming a concentration of viruses in water of 1 virus unit/10 liters, then the method should enable the processing of up to 100 liters of water, and have a concentration factor of about 10,000 with recovery efficiency approaching 100% for as broad as possible a spectrum of viruses.

Table 15.1   A Comparison of Various Methods for Detecting Viruses in Water[a]

	Sample Volume, liters	Concentration Factor	Percent Recovery Efficiency
Gauze pad method	?	10–50	0.5–5
Sample incorporation method	0.06–0.15	10–20	95
Ultracentrifugation	1–10	30–?	20–100
Membrane filter adsorption	10–100	1,000	5–38
Soluble ultrafilters	10	1,000	35–100
Hydro extraction	1	100	24–64
Phase separation (two steps)	5	1,000	35–100
Electrophoresis	0.3	100	100
Adsorption to particulated materials:			
Calcium phosphate	3.0	1,000	50–100
Cobalt chloride	1	1,000	100
Aluminum hydroxide	3.8–19	1,000	50–100
Ion exchange resins	0.1	50–100	20–50
Passive hemagglutination	?	?	80–100
Insoluble polyelectrolytes	1,000	1,000–100,000	35–80

[a] Based on published data.

Comparative results from various studies on virus concentration methods are given in Table 15.1. A rough comparison of some of the methods is thus possible. The figures given were either taken directly from the reports or were calculated on the basis of published data referred to in this review.

Information on the specificity of the methods for detecting various types of viruses was not available since different viruses were used by different workers. Such information, however, is very important for an evaluation of the method. An example of the nature of this problem has been reported in connection with the PS method.

The PS method was evaluated for its ability to recover seven types of enteroviruses. It was found to be highly effective in the recovery of the 3 types of poliovirus and coxsackie virus types B-3 and A-9. It was, however, found to be inhibitory for coxsackie virus B-2, ECHO virus type 6, and influenza A virus (48).

Another example is the passive hemagglutination technique, which is highly specific for a single type of virus (23, 26). In some of the methods such as electrophoresis, only a single type of virus was tested so that no information is available as yet as to its ability to concentrate a broad range of viruses, although it may well be capable of doing so.

The work on the concentration of viruses with aluminum and calcium salts included an evaluation of its efficiency with many types of viruses.

Recovery efficiencies of Herpes, Pox, Adeno, Papova, Myxo, Paramyxo, Reo, Entero, Arbo, and Rhino viruses were compared. Apart from Reo virus, which showed a very poor recovery, the others showed recovery efficiencies in the range of 50–100% (47).

Wallis *et al.* (43) showed that they were able to recover about 40% of the poliovirus added to natural water by passing 300 gallons through a filter composed of a thin layer of insoluble polyelectrolyte. The virus concentration in the water was 2–12 PFU/gallon. Even though the efficiency of recovery is somewhat low, this method may prove to be particularly useful due to the large volumes of water that can be tested.

A very important factor in selecting a method for the routine detection of viruses in water is the cost involved. Some of the methods such as Anderson's continuous flow ultracentrifuge technique require the use of very expensive apparatus and may be feasible for research purposes only. Others such as the PS method and concentration on aluminum hydroxide and insoluble polyelectrolytes are much cheaper, and can be used for the routine monitoring of water without special equipment or expense.

15-10. Summary and Conclusions

The possibility that enteric virus diseases, particularly infectious hepatitis, may be water-borne both in epidemic form and at times as sporadic cases, calls for the development of methods for evaluating the safety of water supplies from a virological point of view. The need is especially great in light of ever-increasing evidence that coliform bacteria, the classical water pollution indicator organisms, are less resistant under certain conditions to both natural hostile environmental factors as well as man-made water and waste water treatment processes. Many of the enteroviruses persist longer in water courses and the ocean than do coliforms and are less easily removed or inactivated even by the most rigorous of treatment methods. This may mean that situations can arise where coliform counts are at acceptable low levels either because of natural die-away or because of active intervention by man in the form of treatment while pathogenic enteroviruses may still be present in water in concentrations high enough to cause disease in man. This is apparently what happened in Delhi in 1952, and there is presumptive evidence at least suggesting that it can happen more frequently than we might assume.

A specific method to detect viruses in water is required both for the routine monitoring of water supplies and to enable the evaluation of water and waste water treatment processes under actual field conditions. Such a method should be capable of detecting a few virus infectious units in a relatively large volume of water. The method should ideally be capable of testing at least a 100-liter water sample with the ability to determine with a high degree of reliability whether or not any viruses are present and if so how many. The method should have a virus concentration factor of 1,000–100,000 with a high efficiency of recovery for very low concentrations of viruses. It also must have the ability to concentrate and detect all enteric virus types with equally high efficiency. Last but not least, the method should be relatively inexpensive so that it can be introduced into most routine water monitoring laboratories. Over a dozen different methods have been developed and studied to date, with none of them completely meeting all of the ideal requirements outlined above. Some appear quite promising and approach the ideal specifications listed.

Adsorption to particulate matter, membrane filters, and PS are some of the most promising methods yet reported upon although others may well prove to be of equal or better performance.

Possible combinations of methods may hold promise as well. For example, it may be attractive to test a sample of 10 liters by the PS method using the two-step procedure which provides a concentration factor of about 500–1000. This would result in a final volume of some 10–20 ml to be assayed for viruses. This volume, containing essentially all of the viruses present in the sample, could be assayed in one tissue culture bottle using the sample incorporation method.

Routine monitoring of water supplies for viruses will most likely not be within the reach of most local water supply quality control laboratories in the foreseeable future. However, the possibility does exist that the routine laboratory can concentrate the sample by one of the methods reviewed here which is deemed to be most suitable, freeze the concentrate for later shipment, and then have the assay done at a regional water quality or public health laboratory having complete facilities and staff for virus assay and identification work. Frozen water samples containing concentrated virus can be held at $-20°C$ for weeks before testing without appreciable reduction in virus count. This provides a decided logistic advantage over bacteriological examination of water which usually must be carried out immediately at a local laboratory.

In conclusion, despite the need for considerable further development of existing virus detection methods and the need to run carefully controlled comparative studies on the most promising ones, it can be said that a number of effective methods for detecting low concentrations of viruses in large volumes of water, which are relatively inexpensive and easy to operate, are currently available. Some are being used routinely with good effect for that very purpose. Virus assay methods currently available could play an important role in protecting the public health in many communities if they were utilized more widely. This could become feasible, particularly in programs where the actual virus assay work is done in one regional laboratory servicing a number of local water quality control laboratories.

REFERENCES

1. J. W. Mosley, in *Transmission of Viruses by the Water Route* (G. Berg, Ed.), Wiley-Interscience, New York, 1967.
2. P. M. Bancroft, W. E. Engelhard, and C. A. Evans, *J. Amer. Pub. Health Assoc.* **164**, 836 (1957).
3. J. M. Dennis, J. Amer. Water Works Assoc. **51**, 1288 (1959).
4. N. A. Clarke and P. W. Kalber, *Health Lab. Sci.* **1**, 44 (1964).
5. H. Shuval, in *Developments in Water Quality Research,* Ann Arbor-Humphrey Science Publishers, Ann Arbor, 1970.
6. S. A. Plotkin and M. Katz, in *Transmission of Viruses by the Water Route* (G. Berg, Ed.), Wiley-Interscience, New York, 1967.
7. J. Kjellander and E. Lund, *J. Amer. Water Works Assoc.* **57**, 893 (1965).
8. H. Shuval, S. Cymbalista, A. Wachs, Y. Zohar, and N. Goldblum, *Proceedings of the 3rd International Conference on Water Pollution Research,* W.P.C.F. Wash., D.C., 1967.
9. R. E. Kissling, in *Transmission of Viruses by the Water Route* (G. Berg, Ed.), Wiley-Interscience, New York, 1967.
10. J. F. Enders, T. H. Weller, and F. C. Robbin, *Science* **109**, 85 (1949).
11. B. Moore, *Monthly Bull. Ministry Health and Public Health Lab. Service* **7**, 241 (1948).
12. J. L. Melnick, J. Emmons, E. M. Opton, and J. H. Coffey, *Amer. J. Hyg.* **59**, 185 (1954).
13. B. D. Rawall and S. H. Godbole, *Environ. Health* **6**, 234 (1964).
14. G. Berg, D. Berman, S. L. Chang, and N. A. Clarke, *Amer. J. Epidemiol.* **83**, 196 (1966).
15. D. O. Cliver and J. Yeatman, *Appl. Microbiol.* **13**, 387 (1965).
16. N. G. Anderson, G. B. Cliver, W. W. Harris, and J. G. Green, in *Transmission of Viruses by the Water Route* (G. Berg, Ed.), Wiley-Interscience, New York, 1967.
17. D. O. Cliver, *Appl. Microbiol.* **13**, 417 (1965).
18. C. Wallis and J. L. Melnick, *Bull. World Health Organ.* **36**, 219 (1967).
19. C. Wallis and J. L. Melnick, *J. Virol.* **1**, 472 (1967).
20. D. O. Cliver, in *Transmission of Viruses by the Water Route* (G. Berg, Ed.), Wiley-Interscience, New York, 1967.
21. G. Berg, Proceedings of the National Specialty Conference on Disinfection, American Society of Civil Engineers, New York, 1970.
22. H. Gastner, in *Transmission of Viruses by the Water Route* (G. Berg, Ed.), Wiley-Interscience, New York, 1967.
23. T. Gibbs and D. O. Cliver, *Health Lab. Sci.* **2**, 81 (1965).
24. H. I. Shuval, S. Cymbalista, B. Fattal, and N. Goldblum, in *Transmission of Viruses by the Water Route* (G. Berg, Ed.), Wiley-Interscience, New York, 1967.

25. D. O. Cliver, in *Transmission of Viruses by the Water Route* (G. Berg, Ed.), Wiley-Interscience, New York, 1967.
26. P. A. Albertsson, *Biochim. Biophys. Acta* **27**, 373 (1958).
27. P. A. Albertsson, *Fractionation of Cell Particles and Macromolecules in Aqueous Two-Phase Systems,* Almquist and Wilksells, Uppsala, 1960.
28. P. A. Albertsson, *Nature* **182**, 709 (1958).
29. G. Frick and P. A. Albertsson, *Nature* **183**, 1070 (1959).
30. T. Wesslin, P. A. Albertsson, and L. Philipson, *Arch. Ges. Virusforsch.* **9**, 510 (1959).
31. L. Philipson, P. A. Albertsson, and G. Frick, *Virology* **11**, 553 (1960).
32. H. I. Shuval, B. Fattal, S. Cymbalista, and N. Goldblum, *Water Res.* **3**, 225 (1969).
33. C. C. Brinton and M. A. Lauffer, in *Electrophoresis* (M. Bier, Ed.), Academic Press, New York, 1959.
34. M. Bier, G. C. Bruckner, F. C. Casper, and H. B. Roy, in *Transmission of Viruses by the Water Route* (G. Berg, Ed.), Wiley-Interscience, New York, 1967.
35. J. B. Salk, *Proc. Soc. Exp. Biol. Med.* **46**, 709 (1941).
36. J. Taverne, I. H. Marshall, and R. Fulton, *J. Gen. Microbiol.* **19**, 451 (1957).
37. C. Wallis and J. L. Melnick, *Am. J. Epidemiol.* **85**, 459 (1966).
38. N. Grossowitz, A. Mercado, and N. Goldblum, *Proc. Soc. Exp. Biol. Med.* **103**, 872 (1960).
39. C. Wallis and J. L. Melnick, in *Transmission of Viruses by the Water Route* (G. Berg, Ed.), Wiley-Interscience, New York, 1967.
40. E. Lund and C. E. Hedström, in *Transmission of Viruses by the Water Route* (G. Berg, Ed.), Wiley-Interscience, New York, 1967.
41. C. Wallis, S. Grinstein, J. L. Melnick, and J. E. Field, *Appl. Microbiol.* **18**, 1007 (1969).
42. S. Grinstein, J. L. Melnick, and C. Wallis, *Bull. World Health. Organ.* **42**, 291 (1970).
43. C. Wallis, J. L. Melnick, and J. E. Field, *Water Res.* (In Press).
44. S. M. Kelly, *Amer. J. Public Health* **43**, 1532 (1953).
45. C. R. Gravelle and T. D. Y. Chin, *J. Infect. Diseases* **109**, 205 (1967).
46. J. E. Smith and R. J. Courtney, in *Transmission of Viruses by the Water Route* (G. Berg, Ed.), Wiley-Interscience, New York, 1967.
47. M. Moore, P. P. Ludovici, and W. S. Jetes, *J. Water Pollution Control Fed.* **42**, R21 (1970).
48. J. Grindsod and D. W. Oliver, *Arch. Ges. Virusforsch.* **28**, 337 (1969).

Part VI
Microbiological Approaches to Pollution Control

16 A Critical View of Waste Treatment

J. W. M. la Rivière, International Courses in Hydraulic and Sanitary Engineering, Laboratory of Microbiology, Delft University of Technology, Delft, Netherlands

16-1. Introduction

In this chapter an attempt is made to discuss waste treatment from the microbiological point of view. As the subject is too vast for exhaustive treatment within the scope of this book, this chapter focuses on principles and concepts mainly illustrated by domestic waste treatment processes.

Treatment of water-borne wastes serves the purpose of controlling water quality of surface and ground waters at the level that is desired for the purposes these waters have to serve, either directly, for example, for recreation, fishing, or wild life preservation, or indirectly as raw water sources for industry or community water supply. Treatment may occur (*a*) immediately after the waste is collected and before discharge occurs, (*b*) in the surface or ground water itself through natural and assisted natural purification, and (*c*) in the production of drinking or industrial water. Since in all these processes parts of the waste are removed or transformed, they should be viewed together as one integrated process.

Since more and more often the raw water source for the water supply of one city is the diluted waste of another upstream city the integrated treatment procedure must be placed on a regional basis with a river basin, a delta, or other practical entity as one unit. For some waste categories, for example, pesticides and heavy metals, it even becomes necessary to consider the entire biosphere.

As water and waste treatment procedures have much in common, both are dealt with together in one section of this chapter, to be followed by a brief discussion of waste management in surface waters. After some considerations on the impact of waste treatment on the biosphere and the possibilities for waste utilization, the chapter is concluded by an examination of the question of how microbiologists can rationally approach waste treatment problems.

16-2.   Objectives

The objectives of treatment, that is, removal of any kind of pollutant down to an acceptable level, are dictated by the uses to be made of the treated water and can be fairly rigorously laid down in standards in cases like drinking and industrial waters. But even here standards have to be considered ephemeral as they are liable to alteration and amplification by (a) more knowledge on toxicology of known pollutants, for example, as to long-term effects or synergism; (b) the appearance of new industrial chemicals; and (c) the locally prevailing economic situation as well as the cultural history. With tight budgets it is clearly preferable to serve many with safe water of lower quality than only a few with high grade water. In parts of continental Europe a much higher hardness is tolerated by the public than in the United Kingdom or some areas of the United States.

For the quality of waste treatment plant effluents it is impracticable to prescribe categorical standards other than as a first guideline. A much more refined approach adjusted to the case in question is required. Nature and quantity of the individual pollutants rather than just total amount of BOD or COD should be taken into account. Additive effects of various discharges as well as dilution factors and self-purifying powers of the receiving waters should be considered in conjunction with the use the community wants to make of the receiving water.

It should be realized that the setting of a standard is an important economic act. After all, water can be made as pure as we want it but, particularly at low concentrations, removal of an extra 0.1 mg/liter of a pollutant can be quite costly. We literally pay for our ignorance when, for lack of data, a standard is set too high. In cases where the technology involved permits this at little extra cost, an additional safety barrier should, of course, always be included in the standard. This situation is illustrated by the bacteriological standards for drinking water which are probably stricter than necessary, as they are determined by what is cheaply technically attainable rather than by rigorous medical rationale. On the other hand, standards for bathing water on beaches, where remedial technology is almost completely lacking, must depend more heavily on the medical evidence. The statistics of disease incidence as related to beach-water quality and the question of how many cases of disease can be accepted per 10,000 bather hours make this a difficult and as yet unsolved problem. As indicated elsewhere in this book, the problem is particularly difficult for water-borne viral diseases. The situation becomes even more complicated when the economic advantage of polluting, for example, a sea arm down to a minimum of 5 instead of 7 mg O_2/liter, has to be balanced against the loss of categories of marine life. While the costs of the extra purification can be easily computed, the value of preserved genes is hard to estimate, especially when the region in question is not biologically unique.

After this relativism it will be clear that the data on standards for BOD and coliform organisms presented in Table 16.1 merely serve to provide

Table 16.1   BOD$_5$^{20} and Coliform Content of Various Categories of Water and Waste Water (1, 2)

	BOD$_5$^{20}, mg/liter	Coliform Bacteria/ml
Domestic waste water	>400	>80,000
Sewage treatment effluent	20	8,000
Raw water suitable for water supply	4	5,000
Drinking water	3	0.01

the order of magnitude of the changes in quality that various forms of treatment must effect, and thus help delineate the framework within which waste treatment takes place.

The examples given in Table 16.1 deal with the most common types of pollution, those with organic matter and with pathogens. As discussed elsewhere in this book, the former, even when harmless by itself, is objectionable because of the consequences it has for the O_2 balance and for undesirable growth of microorganisms, for example, in pipelines.

16-3.   Treatment of Organic Wastes

A.   General.   At the three points where treatment occurs, that is, in waste treatment proper, natural purification, and drinking and industrial water production, microbiological processes play important if not dominant roles. Almost without exception, these processes are based on (*a*) intensified and controlled mineralization(*e.g.,* activated sludge treatment, trickling filters, slow sand filters) or (*b*) methane fermentation (anaerobic digestion), the two processes having in common that their main end products are either harmless (CO_2, H_2O) or easily removable (cells, CH_4). Since cells of pathogens in an environment foreign to them also constitute organic matter subject to mineralization, biological treatment leads to a reduction of numbers of pathogens as well as of organic matter.

Table 16.2 gives the composition of settled domestic sewage in Steven-

Table 16.2   Composition of Settled Domestic Sewage[a]

Settled Sewage		Comparable Medium	
$BOD_5{}^{20}$	370 mg/liter	0.05% glucose	$BOD_5{}^{20}$:370 mg/liter
		$(28 \times 10^{-4}\ M)$	
Organic C	219 mg/liter		Organic C:207 mg/liter
NH_3-N	46 mg/liter	0.025% NH_4Cl	Total N: 68 mg/liter
Organic N	22 mg/liter		
Organic P	7 mg/liter	0.004% K_2HPO_4	P: 7 mg/liter

[a] Other minerals and growth factors present in abundance.

age (3, 4) and a crude "translation" of this into microbiological language, showing that this settled sewage conforms to an approximate $10\times$ diluted laboratory medium. From this composition several general conclusions can be drawn:

1.   STOICHIOMETRY OF MINERALIZATION REACTION; KEY ROLE OF O_2. The O_2 requirements per liter of sewage obviously exceed by a factor of 20 to 40 the maximum amount of O_2 that could be dissolved in it. Unstirred sewage, therefore, very quickly becomes anaerobic; its mineralization requires aeration. Thus the low solubility of O_2 makes it the key substance in present day biological waste treatment. Unless the waste is diluted 80 to 100 times with air-saturated water the resulting mixture, if not aerated, will soon contain less than 5 mg O_2/liter and as a result the fish will die.

It is therefore not surprising that organic matter is often quantitatively measured and expressed by the quantity of oxygen required for its microbial oxidation under certain conditions.

Since BOD is discussed elsewhere in this book, it suffices here to state that the oxygen demand of any organic compound can be easily computed from the reaction equation, a procedure always to be preferred to any other method in cases where the composition of the waste is known. Table 16.3 simplifies these computations and provides an easy opportunity

Table 16.3 Aliphatic Organic Compounds Ranked According to Redox Status and Number of C Atoms per Molecule

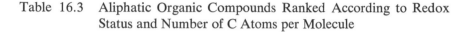

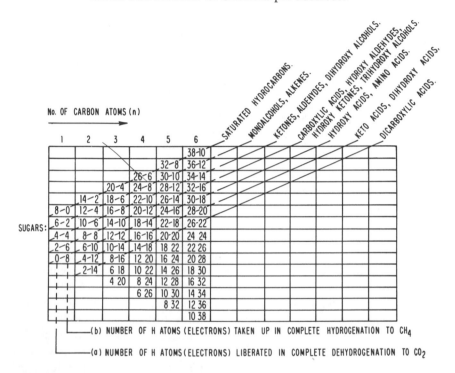

(b) NUMBER OF H ATOMS (ELECTRONS) TAKEN UP IN COMPLETE HYDROGENATION TO CH_4

(a) NUMBER OF H ATOMS (ELECTRONS) LIBERATED IN COMPLETE DEHYDROGENATION TO CO_2

Reactions of inorganic H acceptors:

$$O_2 + 4H \rightarrow 2H_2O$$
$$2HNO_3 + 10H \rightarrow N_2 + 6H_2O$$
$$HNO_3 + 8H \rightarrow NH_3 + 3H_2O$$
$$H_2SO_4 + 8H \rightarrow H_2S + 4H_2O$$
$$H_2CO_3 + 8H \rightarrow CH_4 + 3H_2O$$

APPLICATIONS: (a) Computation of theoretical demand for H acceptors, including BOD, with corresponding production of reduced H acceptor. (b) Computation of maximum CH_4 yields in digestion and CH_4/CO_2 ratios. (c) Computation of number of H atoms involved in the transformation of one organic compound into another.

PROCEDURE: Select square representing compound in question. $a =$ demand for H acceptor in number of H atoms/molecule of compound $=$ moles of H/mole of compound. Total theoretical BOD per mole is $\dfrac{a}{4} \times 32$ g O_2.

$$BOD_5{}^{20}: \frac{68}{100} \times \frac{a}{4} \times 32 \text{ g } O_2$$

$$\frac{na}{a+b} = \text{maximum } CH_4 \text{ yield in moles } CH_4/\text{mole of compound}$$

$$\frac{a}{b} = \frac{CH_4}{CO_2}, \text{ ratio of } CH_4 \text{ and } CO_2 \text{ produced in mole/mole}$$

EXAMPLE: Butyric acid (fourth row, square 20 12) produces 20 H atoms per molecule when completely transformed to CO_2. This corresponds with 5 moles of O_2, 4 moles of HNO_3 (when reduced to N_2), or 2.5 moles of H_2SO_4 per mole of butyric acid. In the latter cases 2 moles of N_2 and 2.5 moles of H_2S are formed, respectively. One mole of butyric acid yields $(4 \times 20)/32 = 2.5$ moles of CH_4; the CH_4/CO_2 ratio is 5/3. Transformation of butyric acid to a tetrose (16 16) would require removal of $20 - 16 = 4$ moles of H atoms/mole butyric acid.

for comparing different compounds as to overall redox-status. It also shows the yields of complete methane fermentation.

Obviously the O_2 demand values derived from these calculations are those for complete oxidation. In practice considerable transformation into cell material, up to 60% of the substrate, will take place, which then by itself may become a substrate yielding CO_2, H_2O, and new cells, after which the process repeats itself again and again. This intervention of assimilation does not, of course, change the final O_2 demand but it does have an effect on the rate of O_2 consumption, which for practical purposes in some instances may be equivalent to removal of organic matter.

In view of the average composition of microbial cells, it is clear that neither N nor P will become limiting factors. On the contrary, we may expect a fully mineralized effluent to be an excellent medium for algal growth because of the N and P that remain in it.

Only those anaerobic processes can be efficient for BOD reduction where complete dehydrogenation of the organic matter is coupled with release of reduced products like H_2 and CH_4. The only fermentation useful for the purpose is the methane fermentation, as the only other possible alternative is excluded by the composition of the medium: there is no stoichiometric amount of nitrate available.

The C–N–P ratio as found in the medium renders it impossible to use resting cell methods in oxidative mineralization. The only two factors that could be used to arrest growth are the supply of O_2 and organic matter, which are the reactants in the very reaction we want to perform. Hence substrate removal and growth are irrevocably coupled with one another. It should be borne in mind that this need not be the case in industrial wastes with deviating C–N–P ratios.

2. ECOLOGICAL SUCCESSION OF MICROBIAL POPULATIONS. ROLE OF ODOR OF DECOMPOSITION PRODUCTS. When a medium is left to anaerobic degradation, volatile reduction products like H_2S and NH_3, supplemented by fatty acids and protein degradation products, will emerge. In this way waste products render themselves identifiable by smell.

The earliest motivations for sanitary behavior must have been purely physiological and instinctive; because of the fact that the odor of waste products registers in our brain as repulsive, man stayed away from decaying organic matter and thus was saved from massive infections by the powerful defense mechanism of his olfactory genes.

The empirical discovery that the odor of oxidized decay products is absent or not obnoxious opened the way for eliminating bad odors by maintaining a surplus of O_2 (air) in the stored waste, which then also led, of course, to its mineralization. Even today this principle is almost the only alternative to storing the waste and letting it decay anaerobically in such a way that the odors are contained. The first principle has led to the activated sludge and trickling filter methods; the latter to anaerobic digestion. It is clear that an inexpensive method for the destruction or masking of odors would profoundly affect the motivations and techniques in sanitary engineering.

The fact that oxidative degradation supersedes anaerobic decomposition and its odor production is often taken for granted. As pointed out by Stumm (5), the sequence of these processes is readily explained by the equilibrium constants (or redox potentials) of the reactions involved which dictate the sequence of reactions as given in Table 16.4. Neat evolutionary dovetailing has led to a distribution of properties among the

Table 16.4 Equilibrium Constants for the Principal Microbial Dissimilation Reactions at 25°C and pH 7(5)

	H Donor	H Acceptor	Log K
Aerobic respiration	CH_2O	O_2	22.0
Denitrification[a]	CH_2O	$NO_3{}^-$	21.0
Fermentation[b]	CH_2O	CH_2O	14.4
Sulfate reduction	CH_2O	$SO_4{}^{2-}$	4.5
Methane fermentation	CH_2O	$CO_3{}^{2-}$	4.0

[a] Reduction to N_2.
[b] Dismutation to CH_3OH and $CHOOH$.

groups of microorganisms involved that corresponds to a large extent to the thermodynamic scheme. Examples are: (*a*) sulfate reducers and methane bacteria are obligate anaerobes; and (*b*) the organic substrates to which methane bacteria almost exclusively restrict themselves are the final degradation products of the fermentation and sulfate reduction processes that have to precede methane fermentation.

By manipulation of O_2 concentration the system can be subjected to aerobic or to anaerobic degradation. With the latter, nitrate will disappear before sulfate, and sulfate will disappear before methane fermentation develops. This state of affairs is exploited in the process of microbiological N removal. After complete nitrification a period of anaerobiosis is allowed for, and this immediately leads to nitrate reduction provided enough organic substrate is available. This method would not exist if nitrate reduction were last in the sequence and followed methane fermentation.

Similarly, production of sulfur via H_2S from sulfate reduction in sewage enriched with sulfate, as described by Butlin, Selwyn, and Wakerley (6), is only possible if methane fermentation is last in the sequence. It should also be noted that it is possible to operate an activated sludge plant with or without nitrification. Also here, O_2 concentration is one of the decisive factors (7).

Thus the nature of the predominant process during treatment or natural purification can be manipulated by controlling the concentration of H acceptor, that is, O_2, nitrate, or sulfate, but always within the inexorable thermodynamic framework of the sequence discussed above. In no way do these methods differ from the time-honored Beijerinckian methods applied to crude elective cultures in the laboratory.

In contrast to beliefs occasionally expressed among sanitary engineers, inoculation is not a matter of genuine concern in microbiological treatment since the lack of asepsis guarantees continuous inoculation with practically any conceivable microbe. In due course a diverse microbial population fitting the created niche will establish itself.

It is, of course, effective to use inocula from similar, actively operating installations if one wants to start up a freshly built plant quickly. This is particularly true in the case of anaerobic digestion. Slow sand filters and trickling filters need a working-in period to allow for "ripening," that is, development of the microbial film on the solid substratum.

3.   KINETIC CONSIDERATIONS. THE KEY ROLE OF ORGANIC MATTER CONCENTRATION AS A LIMITING FACTOR.   Obviously, for economic reasons, we want the mineralization reactions to proceed in continuous culture rather than in batch culture. Flow rates should be as high as possible in order to reduce detention time, which is equivalent to reactor volume and

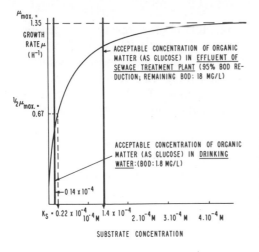

ACCEPTABLE CONCENTRATION OF ORGANIC
MATTER (AS GLUCOSE) IN <u>EFFLUENT OF
SEWAGE TREATMENT PLANT</u> (95% BOD RE-
DUCTION; REMAINING BOD: 18 MG/L)

ACCEPTABLE CONCENTRATION OF ORGANIC
MATTER (AS GLUCOSE) IN <u>DRINKING
WATER</u>: (BOD: 1.8 MG/L)

Figure 16.1 Standards for organic matter content of waste treatment plant effluents and of drinking water compared with growth limiting concentrations of glucose for *Escherichia coli* under aerobic conditions at 37°C (4).

hence equivalent to construction costs/ton of water/hour treated. Although entirely comparable to processes in the microbiological industry, the present process differs considerably in its constraints: the medium composition cannot be freely chosen; lack of too costly asepsis precludes free choice of organisms; and the price of the product, treated water, is governed by complex sociopolitical mechanisms that are hard to control, particularly in the case of waste treatment.

Since a high reaction rate is essential, the only factor that can be manipulated, the O_2 concentration, must be given a value such that it does not become rate limiting and remains above 2–3 mg O_2/liter. Thus the design of aerobic treatment plants is based on (*a*) adequate O_2 input and (*b*) mineralization rates as determined by the organic matter concentration prevailing in the operating reactor, P and N being present in excess. In continuous culture this concentration is *not* dictated by the concentration in the incoming waste (S_R) but by the value of the organic matter concentration (S) chosen as acceptable for the effluent. This value is a key factor in determining the size of the aeration tank. The value of S_R is, of course, the main determinant for the capacity of the aeration equipment, which should provide O_2 equivalent to $S_R - S$.

Figure 16.1 offers a basis for roughly assessing the framework within which such continuous cultivation takes place. It presents the maximum allowable concentrations of organic matter in drinking water and in waste treatment effluents, expressed as glucose, compared with growth-limiting concentrations of glucose for *Escherichia coli* plotted against specific growth rate μ. Obviously, our mineralizing microbial populations are not pure cultures of *E. coli,* nor is domestic waste a single organic substrate like glucose. Hence the following conclusions must be relative and qualitative rather than absolute and quantitative:

1. If we consider the simplest reactor of the open, homogeneous type—the single stage stirred reactor in Herbert's (8) classification— satisfactory removal of the limiting organic substrate appears possible at a reasonable rate with a corresponding detention time somewhere between 0.5 and 3 hours without practicing recirculation of cells. The latter is an additional factor to be used for manipulation but entails the technical problem of separating the cells (see below).

2. In view of the range of standard values for the BOD of effluents on the Monod curve, it is evident that the choice of the actual standard value greatly influences the maximum feasible flow rate, that is, the economy of the process: a small decrease of the standard value requires a drastic decrease in flow rate.

3. For the preparation of drinking water, only very low flow rates would be applicable in the reactor type under consideration.

4. Although a direct consequence of continuous culture theory, it seems appropriate to emphasize that in the case of a toxic degradable organic substrate like phenol the concentration of this compound in the waste is of no concern; even if it is above the toxic level, elimination continues as long as the concentration in the reactor is below the toxic level. There are evident difficulties in starting up such a plant and risks in operating it under such conditions.

5. Because of dilution in the receiving water and the possibility of O_2 becoming the limiting factor, natural purification growth rates will generally be lower than those in treatment plants.

Against the background of the general microbiological framework presented above, some of the most important technical methods will be discussed in somewhat more detail, particularly with respect to their microbiological mechanisms.

Figure 16.2 presents the major types of microbiological treatment pro-

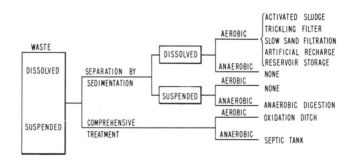

Figure 16.2 Main categories of microbiological treatment processes for organic wastes.

cesses [for a detailed description see Fair *et al.* (1)]. Treatment can be aerobic or anaerobic. The waste to be treated may be mixed, that is, dissolved and suspended, or either suspended or dissolved after separation by sedimentation. For domestic waste the suspended matter may amount to approximately 50% of the total BOD.

B. Aerobic Processes. The *activated sludge* process is carried out in reactors, classified as single stage, stirred reactors with feedback of cells (8). Rigorous mathematical treatment would be possible provided one pure culture and one single organic substrate were involved. Although this is not the case, sanitary engineers tacitly assume that the same methodology also holds for mixed cultures on multiple substrates and in their calculations empirical constants are used in "pure culture formulae." Notwithstanding the scientific illegitimacy of this, extensive practice has provided useful "constants" for design purposes for some standard types of organic waste under given climatic conditions.

The microbiologist, at first sight, might question the necessity of feedback of cells in this system. Without it one would expect sufficiently low BOD values with reasonable detention times for the aqueous phase of the effluent. Since, however, part of the dissolved organic waste is transformed into cells, more than half of the waste might be discharged into the receiving water as freely suspended microbial cell material if no separation were practiced.

Cell removal necessitates operation at low growth rates so as to obtain settleable flocs, and feedback of flocs has to be applied as compensation for the low growth rate. Thus the design suffers from an additional constraint: it has to reconcile rapid mineralization (high growth rate) with cell removal by sedimentation (floc formation requires low growth rates). The seriousness of this constraint is demonstrated by the work of Tenney and Stumm (9) who studied the possibility of removing suspended, unflocculated cells by chemical coagulation so as to create freedom for optimal design of the mineralization step.

An alternative to the Tenney and Stumm proposal would be discharge of cells with the effluent. From the medium composition given, one would expect the cells to possess little if any endogenous carbonaceous reserves that would cause considerable and immediate O_2 consumption in the receiving water. The O_2 demand exerted when the cells themselves die and are decomposed would, of course, remain, but there are reasons to expect that this might well turn out to be a slow process because (*a*) oxidative mineralization of activated sludge requires excessively long detention times, and (*b*) algal cells produced in oxidation ponds are usually released in surface waters although they theoretically constitute organic matter equivalent to almost the entire BOD value of the waste entering the pond.

It therefore appears that the rate of oxygen uptake of resting or dying microbial cells, whether bacterial or algal, might be so small as to become acceptable for some receiving waters. An additional benefit to treatment plant simplification would be an increased self-purification in the receiving water: the dissolved organic matter that has remained in the effluent probably consists of various compounds, each in low concentration, that are very resistant to degradation under the conditions of the aeration basin. The "tubular piston-flow reactor with initial inoculation," that is, the cell material discharged into it, is ideally equipped for removing such compounds. For many waters, however, especially lakes and long rivers, there are great dangers in the release of bacterial and algal cells since they must eventually exert their full oxygen demand as well as release the inorganic nutrients that are contained in them. In such cases the idea of separate cell removal by other means than sedimentation might well be considered.

Notwithstanding many studies during the past decades, the microbiological nature of activated sludge is not well known. Understandably, the composition of the mixed population varies with the substrate, flow rate, pH, temperature, etc. A great number of different bacterial species have been isolated and counted, but this is not enough to answer the important practical question of how one can manipulate the process so as to obtain heavy (well settleable) flocs of maximum oxidative activity. Under certain conditions the sludge becomes too light for proper settling within the designed detention time ("bulking sludge"), and this phenomenon can often be associated with the appearance of filamentous organisms like *Sphaerotilus natans* or *Geotrichum candidum*. By what mechanisms these organisms come to the fore is still unclear; since the oxidative capacity of such sludge is not impaired, one wonders if another, possibly cheaper, way of cell removal could not be used if the process was made to produce such light flocs continuously.

The wish to control floc properties has led to studies on the mechanism of floc formation. In many of these, attempts were made to make bacterial model suspensions flocculate by means of pH control and/or addition of various flocculants, including bacterial polysaccharides. It is not yet clear to what extent the results will lead to improved floc control. Meanwhile, it might pay to realize that in the actual treatment process no two-stage process, that is, growth of freely suspended bacteria followed by their flocculation, occurs but only a proliferation of flocs. In addition, it is evident that the system selects flocs rather than free bacteria since the former are recirculated. Furthermore, it would be of interest to know what percentage of the floc consists of inert material. From average data (10, 11) on oxygen uptake rates for bacterial suspensions, one would conclude that activated sludge is less than 10% as active per dry weight unit when compared to a *Pseudomonas* suspension, for example. One might, therefore, well suspect that a considerable part of the floc consists of insoluble organic matter that is hard to degrade under treatment conditions; microbial "bones or ashes," that is, undigested microbial cell components like cell walls might well qualify for this matrix material, since such substances would automatically be selected for by the recirculation process of the settled material.

As pointed out previously (4), the system must be expected to select for multivorous organisms, that is, for organisms that can simultaneously oxidize a great number of substrates. In the continuous culture system they have a selective advantage both over those that can adapt to only a few substrates and those that can attack many substrates but only sequentially. The picture is further complicated by the numerous interactions between different microbial species that must be expected to establish themselves on ecological grounds.

As long as a constant flow rate is maintained, all different organisms present have their own steady-state density but all have the same growth rate to which the concentrations of their respective limiting substrates are related. Since growth-factor requiring organisms have a selective advantage over their less exacting counterparts, as—finally—so elegantly shown by the Zamenhoffs (12), it is clear that the stage is set for the development of dual, triangular, and more complicated interdependecies at many different assimilatory and dissimilatory levels. These, then, at the same time constitute beautiful regulatory mechanisms that keep the proportions of the different parts of the microbial population more or less constant. Besides nutritional mechanisms involving waste substrates, factors like slime forming capacity have an important selective role. Finally, the bacterial cells are themselves substrates for each other, bacteriophages, *Bdellovibrio* species, protozoa, etc.; the density of the members of this secondary group is clearly regulated by direct interdependency with those of the primary group.

This state of affairs in activated sludge should tantalize any ecologist as an ideal model for quantitative study, and further knowledge should lead to improved control of the practical process. Research at Stevenage has already led to the possibility of manipulating nitrification (7) and the activity of protozoa (13).

In judging the performance of treatment plants, capacity for BOD removal and effluent BOD should not be the sole points of concern. It becomes more and more necessary that attention be paid to the nature of the substances that slip through the treatment plant and thus enter the sources of water supply. Viruses, phosphates, heavy metals, and "recalcitrant" micropollutants are examples.

The failure of treatment plants to eliminate these compounds has contributed to serious environmental disorders which in turn have triggered preventive and remedial measures. Among the latter, so-called tertiary treatment methods play an increasingly important role. Chlorination of effluents, polishing of activated sludge effluents by trickling filters, chemical phosphate removal and microbial nitrogen elimination are examples. This development underlines the natural limitations of microbiological treatment methods in an industrialized society; these methods selectively eliminate in an artificially intensified manner the same compounds that are also readily degraded in the environment itself. Detailed analysis of effluents should help to further define these limitations and also throw an interesting light on the microbiological processes going on in the aeration tank.

Comprehensive chemical waste treatment processes, like chemical coagulation combined with carbon adsorption, have the advantage that compounds not retained by such processes are not necessarily also refractory to microbial degradation in the receiving water. Against this advantage the problem of safe sludge disposal and any environmental hazards involved in the production of the chemicals have to be balanced.

Of the many variations that exist of the activated sludge process, the so-called *oxidation ditch* (14) merits special attention. Here dissolved and suspended organic matter are treated simultaneously in an ellipsoidal ditch provided with brush aeration. In its simplest form the ditch is at the same time aeration tank (brushes on) and sedimentation tank (brushes off), and operates as a stirred reactor with feedback (partial retention) of cells. Instead of a few hours, the detention time is between 1 and 3 days, depending on temperature; this leads to formation of a small amount of sludge, which, in contrast to ordinary activated sludge, can be easily dewatered on drying beds without risk of bad odors. Full nitrification which permits complete N removal by intermittent aeration can be easily obtained. If the rotors are stopped, anaerobiosis and nitrate reduction quickly commence.

Another comprehensive way of treating whole waste is by means of *oxidation ponds*. These are gaining widespread importance as simple and cheap devices in areas where land is inexpensive and much light is available. As their theoretical basis is still badly formulated (15) and many variations of the system occur, it is difficult to recognize even the basic principles. One design objective could be predominantly anaerobic treatment by methane fermentation in which a top layer of algal growth replaces costly covers, otherwise necessary to contain offensive odors. Another design objective, often found in the literature, is the transformation of the organic waste via oxidative bacterial mineralization into algal cell material, the algae providing the oxygen required for the mineralization step, which in turns provides CO_2 for algal growth. Since some anaerobic digestion will always take place in practice, it is likely that most systems will operate somewhere between the two extreme conditions given. In this case the net change in BOD may be small when most of the waste is transformed into organic matter contained in algal and bacterial cells. It must be admitted that transformation of organic matter into a less rapidly degradable form may constitute a genuine solution in instances where postponement or more even distribution in space and time of O_2 consumption can meet the local needs, as is conceivable for short rivers. But the method would be greatly improved, and include P removal, if the algal cells were harvested; this would become feasible if the algal cell mass could be economically used. This would depend on the extent of possible ecological manipulation of the predominant algal species (cf. 16).

Besides the activated sludge systems which use homogeneous, stirred reactors, heterogeneous two-phase reactors of the *packed-tower type* can be applied to aerobic degradation of organic wastes, but this is only possible for dissolved material.

In these reactors removal is secured by extensive contact between a moving water film and a layer of sessile organisms. In the so-called slow sand filter (SSF) used for drinking water production, no provision for aeration is made because the O_2 concentration in the raw water is high enough (or if necessary, previously adjusted) for mineralizing the organic matter present. As this is impossible with waste, the high concentration counterpart of SSF, the so-called trickling filter (TF), has three phases instead of two, air and water passing simultaneously through the reactor. Accordingly, the grain size of the material that supports the film is much smaller in SSF (0.15–0.35 mm) than in TF (6–10 cm).

In these reactors successive elimination of solutes occurs, the sequence being determined by the growth rates of the microbes involved and the concentrations of the substrates. A substance S, present in high concentration and oxidized by a bacterium with low K_s and high μ_{max}, will lead to formation of a film of its specific population closer to the point of entrance in the reactor than a substance with the opposite characteristics, for which the specific microbe can only competitively succeed in gaining a hold after the concentration of substance S has been reduced, that is, at a point further removed from the entrance. In view of the low μ_{max} of nitrifiers, one would expect nitrification to occur in one of the last stages.

Growth necessarily continues in both reactors to the point where the film comes off; hence a true steady state is not attainable. In practice, almost permanent steady performance is obtained as cell removal is balanced by growth. SSF differs from TF in that it is deep enough to permit mineralization of cells that are produced in it, cells released in the upper layers being sequentially mineralized in the lower ones. As a result, the total bacterial count in the effluent is much lower than that in the raw water.

As shown by Figure 16.1, activated sludge treatment of water up to the standards of drinking water would require very long detention times, and indeed, one might well consider the quality improvement on storage in a reservoir as the logical counterpart of SSF treatment just as activated sludge treatment is the counterpart of TF treatment.

Slow sand filters require more space than installations for chemical treatment with coagulation followed by rapid (true) sand filtration, and during the last few decades SSF has carried the undeserved label of being old-fashioned. In most countries SSF has been superseded by chemical treatment methods, but recently the SSF method has begun to gain renewed popularity. In developing countries it has many advantages when cheap space is available (17), and in industrialized countries where ground water supplies tend to become exhausted, artificial recharge with surface water is more and more practiced. The processes in ground water recharge resemble those in SSF to a large extent.

The microbiology of the packed-tower type reactors has not yet been exhaustively studied, which is undoubtedly caused in part by the name "filter," as mentioned above. It appears that there are still possibilities for increased efficiency and wider applications. Examples illustrating this development are the experimental use of plastic materials to support the microbial film and the use of TF in combination with activated sludge treatment for final polishing of the effluent.

C. Anaerobic Processes.

Anaerobic treatment by methane fermentation (digestion) is practiced on a large scale. Comprehensive treatment of dissolved and suspended organic matter takes place in the well-known septic tanks while treatment of suspended matter by this method is routinely carried out in conjunction with activated sludge or trickling filter mineralization. There is as yet no method in operation designed specifically for anaerobic treatment of dissolved organic wastes, nor has the comprehensive treatment been developed to a scale larger than that of the septic tank.

Historically, the treatment of suspended organic solids consisted of utilization as fertilizer or land filler. This is only possible where cheap transport and potential users are available. Ordinarily transport of untreated sludges does not occur because of their very high water content and tendency to produce repulsive smells.

Anaerobic microbial treatment is most effective in reducing the volume of the sludge and renders it easily susceptible to dewatering without risk of odors. However, drying followed by burning is a serious competitor in some socioeconomic frameworks.

Since sludges do not lend themselves well to stirring and aeration, it is readily understandable that aerobic treatment of sludges alone is not methodically being attempted and that the classical treatment developed from mere storage in an enclosed space so as to avoid the nuisance of smells is used. If the methane produced by the resulting fermentation is collected, it may well be sufficient to supply 70–90% of the energy requirements of the entire treatment plant.

Because of the intractability of methane bacteria and the heterogeneous nature of the medium, rational design of these digesters has been strongly retarded; detention times between 10 and 20 days are considered quite acceptable.

Present studies aim to improve the process by working in the thermophilic rather than the mesophilic range, by better mixing, and by separating the process into the two steps of introductory fermentation and methane fermentation proper. If both possess different parameters for optimum performance, combining them in one reactor, as is done now, cannot be expected to produce the best results. In addition, more fundamental studies by Lawrence and McCarty (18) have been carried out on the methane fermentation itself, which appears to be the rate-limiting step. These hold promise for more rational design and also perhaps for anaerobic treatment of dissolved organic wastes, which would have the great advantage that cell densities could be increased without a corresponding requirement for stepping up the O_2 input. The authors have shown that microbial populations capable of both required fermentations can be retained within packed anaerobic reactors while the water phase flows through.

An excellent review of all aspects of anaerobic digestion has been prepared at the National Institute for Water Research in Pretoria (19–22).

16-4. Waste Management in Surface Waters

It was felt until recently that natural purification was a process of nature that could not be tampered with by man. However, led by the unpleasant consequences of overtaxing this essential microbiological process, that is, phenomena like anaerobiosis in rivers and hypolimnia of lakes, engineers have made attempts to assess quantitatively the self-purifying powers of surface waters by predictive equations. These eventually led to remedial measures to increase the self-purifying capacity which has now been recognized as an economic commodity of high value.

Since failure of natural purification first manifests itself in the oxygen balance, most efforts are directed towards aeration of streams at strategic points with air or oxygen, by the building of weirs, or even by the addition of nitrate. As shown elsewhere in this book, the art of reservoir operation, for example, by means of destratification devices, algicidal treatment, and application of thin films to reduce evaporation losses, is rapidly developing. In rivers as well as lakes, thermal pollution is adding its own demands for assessment of heat balances and the ecological impact of temperature modification. In coastal areas special and costly measures for rapid mixing are being carried out to prevent local anaerobiosis caused by waste discharge. Outfalls several kilometers long and equipped with diffusers have been constructed in many coastal cities.

So far O_2 concentration and the concentration of organic matter seem to be the main parameters chosen for manipulation. Since substrate removal rate is proportional to the product of growth rate, fixed by substrate concentration, and microbial mass, the latter might also qualify as a parameter. In the first stretch of the oxygen sag curve, for instance, substrate removal is slow because cell density is low, and therefore very little oxygen is taken up from the air. If cell density was already high at the point of discharge, the first stretch of the river downstream of the outfall could also be used intensively for substrate removal. In this way the shape of the sag curve could be influenced by one more means. Addition of cells at the outfall would change a river from a plain tubular reactor to a "tubular reactor with feedback of cells" which could easily be realized by means of a small aerated basin switched parallel to the sewer just before the outfall. Similar utilization of the many miles of sewers for pretreatment of the waste could be considered; the anaerobic conditions prevailing in the sewage would certainly guarantee high O_2 transfer rates if adequate facilities were provided.

The large-scale measures taken to combat incidental oil pollution (see Chapter 7) in surface waters are mostly designed to emulsify or to sink the oil rather then to promote its microbial mineralization. The obvious bottleneck for this process in this case is not lack of O_2 but a severe imbalance in the C–N–P ratio. Nontoxic emulsification combined with dosage of suitable P and N sources constitutes a sound microbiological approach that until now has not received the attention it deserves.

The examples given above illustrate the fact that management of surface water quality is progressing beyond mere manipulation of waste inputs. This is a natural outgrowth of the much older technology of quantitative water management as practiced by hydraulic engineers in the construction of canals, irrigation systems, artificial lakes, etc. Quantity and quality cannot be regulated separately, and in current large hydraulic projects an integrated approach is gradually emerging.

Understandably, the scientific basis for the new technology of surface water quality management is still small, but the economic pressures have become so strong that sophisticated predictive calculations are being made by computer with constants no biologist would underwrite.

The unavailability of biological data rather than mathematical methodology appears to be the limiting factor in the development of predictive models for water quality management. Since the time and space scale of the phenomena renders quantitative studies on large water-bodies very costly and time consuming, laboratory models are becoming increasingly important. Examples are those of Wuhrmann (see Chapter 6) for shallow rivers and of Gates, Marlar, and Westfield (23) for deep rivers.

16-5.   Impact of Waste Treatment on the Biosphere; Waste Utilization

Waste treatment and particularly its microbiological aspects must not only be considered from the technological point of view, but also against a general evolutionary background.

Waste products are important factors in evolution. The waste product oxygen, for instance, had a profound effect on primordial anaerobes when it first arose. It has given rise to predators of the very organisms that produce it, and by its incorporation in the cycles of the biosphere has become the most important gaseous food for man.

Until recently man has lived by gradually intruding upon the biosphere as if its resources and capacity as a sink for wastes were unlimited. Now we discover by the feedback effect of our own waste production that the earth is finite. This forces us to recognize that Nature and the niche of man are no longer separate entities but that all that is left of Nature is man and his household. Our house is nothing less than the entire biosphere, and the need to put it in order is creating a widespread response.

The very fact that environmental control has become possible on a micro- as well as a macroscale imposes upon us the awesome responsibility to control the human niche in a way that is optimal for human evolution as judged by ourselves. This leads directly to the conclusion that we must wean ourselves from the notion that environmental control is a defensive activity aimed at restoring ecosystems to their natural state. For even if such a state could unequivocally be defined, we must realize that any ecosystem man does elect not to destroy or modify has become a synthetic entity produced by man's conscious manipulation. Now that we can control concentrations of oxygen, phosphorus, and other nutrients, as well as the temperature of large bodies of water, we possess a new liberty to choose from many other options in creating new environments which offer a wide scope to fantasy and experiment.

In this trend of thought wildlife preservation is no longer based on cosmic emotion but on drive for self-preservation and, accordingly, wildlife reservations acquire the status of gene-banks and laboratories for evolutionary experimentation. It is no longer a matter of whether or not to preserve a rare bird but of how many and where.

The same trend of thought leads to the duty of integrating waste production in the physiology of the biosphere by finding a constructive use for our waste materials. An additional impetus to waste utilization is given by the realization that the biosphere is finite, not only in its capacity for waste absorption but also in its capacity to produce food.

Waste utilization is, taken by itself, nothing new, for the best forms of waste treatment are the processes that do not carry this label and produce valuable commodities from raw materials that would be waste products if unused. The status of a substance within the spectrum ranging from obnoxious waste to costly raw material depends upon socioeconomic standards; these are now rapidly adapting themselves by introducing recycling of many materials still considered wastes.

Thus the microbiologist also has the task of finding new ways or improving old methods for utilizing domestic and industrial wastes. The possibilities are numerous and range from methane production, microbial S production and composting, to fish pond farming, algal farming, and sea fertilization. As shown by the inspiring review by Allen (24), several of these methods are being used more extensively in developing nations than in the industrialized world. Instead of being condemned as old fashioned, they might well be considered prototypes for modern processes. Our eutrophication and thermal pollution problems, for instance, could be substantially altered if predominance of algae suitable as a food for a desirable fish could be assured at more elevated temperatures. And if no such satisfactory alga–fish system exists today, there is nothing to prevent the biologist from attempting to create it.

In cases where we fail to mobilize microbial systems into destruction or utilization of a given waste product, we will have to adapt the waste to actual microbial potentials. Cases in point are plastics and nonbiodegradable pesticides which cannot be used indefinitely because of the serious impact they exert on the entire biosphere.

It is an important preventive task for the microbiologist to help design molecular structures for such substances that will conform to the additional constraint of acceptable biodegradability in the biosphere and to develop biological pest control methods based on the use of insect pathogens.

16-6. Conclusions

In the previous discussion a sanitary engineer might have some difficulty in recognizing the description of equipment he continually designs and operates. This is caused by the difference in outlook, semantics, motivation, and background that exists between engineers and microbiologists. This diagnosis does not constitute a complaint from the side of the microbiologists; if anyone, the sanitary engineer would have legitimate grounds for reproach for having been left to fend for himself as a practicing microbial ecologist by running the largest microbiological industry in the world without the benefit of assistance from the microbiological profession.

In the outcry raised by biologists against environmental pollution, microbiologists have been remarkably silent. It is true that microbes are ubiquitous and versatile and that the threat of extinction of rare microbial species is at first sight less serious than that for animals and plants. Also it is understandable that microbiologists are attracted by many other areas of rewarding activity like molecular biology, genetics, and industrial microbiology.

Still, they should realize that they carry a very special responsibility in environmental affairs because of their preoccupation with the kingdom of the Protista, which in mass surpasses that of the animals and plants. Where zoologists and botanists have the duty to sound the alarm, the microbiologist is in a position to assist in offering remedial and preventive measures since, after all, waste treatment until very recently was exclusively taken care of by microbial mineralization throughout the entire biosphere. In view of its involvement in industry, microbiology is easily amenable to engineering applications in applied ecology. And, finally, microorganisms promise to be excellent tools for model laboratory studies in ecology, just as they are already in biochemistry and genetics.

But in order to fully exploit the environmental benefits held in store by the microbial kingdom, the microbiologist has to overcome his aversion to working with mixed cultures and to reading the sanitary engineering literature. The existing experience with natural communities and semi-natural populations such as those of activated sludge has made it abundantly clear that reproducible behavior is not the exclusive prerogative of pure cultures. One has to realize, however, that the parameters valid for description of pure culture systems cannot be applied to mixed culture systems without modification and amplification.

In pure culture experiments, at least in those of short duration, the genetic material is limited and constant, and only its phenotypical expression is liable to variation by environmental conditions. In natural communities, subject to constant inoculation with all kinds of microorganisms, there is not only phenotypic variation by the environment but also selection of the relevant genetic material itself from an almost complete pool of microbial genes. This makes the system much more complex but not so capricious as to be completely refractory to study and interpretation. The substrates in the medium give rise to sequences of enzymes which can be distributed in different combinations over a multitude of different microbes which through manifold interrelationships come to behave like one organism, within which feedback and other regulatory mechanisms operate in much the same way as they do in one single cell of a pure culture. Thus we have lost, in comparison with the pure culture, one of the equations that describe the system, that is, *genetic material = constant*, but we have gained another condition that was not valid for the pure culture but which must be fulfilled for the natural community: *no ecological niche can remain unexploited*.

While in pure culture work our interest is organism-oriented and the changes in the medium are only considered relevant in their relation with the organism under study, work with natural communities should be substrate oriented since the substrates are the determinants of changes in the population.

These considerations tend to show that mixed cultures under constant inoculation will be more reproducible with respect to steady-state substrate concentrations than with respect to microbial constitution because one and the same total potential of enzymes can be distributed over different microbial species in different ways. Uncontrolled fluctuation in species distribution would render waste utilization, for example, by cultivating a specific desirable alga, quite difficult, but that this is not impossible is suggested by the perennial persistence of *Spirulina* in Lake Tchad and of *Oscillatoria rubescens* in the Vierwaldstättersee, for example. It is probable that besides nutrient manipulation, massive inoculation can be employed to control the composition of the population, provided the desired organism is a potential majority organism for the system in question. Thus we could exploit the effect of sequence of introduction of genes into the ecosystem just as it is done so successfully in agriculture.

In cases where our main environmental interests are concerned with the disappearance of a given waste, the exact composition of the mineralizing population is far less important than knowledge of the parameters determining maximum degradation rates.

Thus the reproducibility tacitly assumed, for example, for activated sludge processes, by the sanitary engineer may well have a sound scientific basis after all when we choose descriptive parameters as disappearance rates of substrates, steady-state concentrations of substrates and of categories of enzymes, etc. The work by the National Institute for Water Research in Pretoria on the determination of DNA, dehydrogenase, protease, amylase, cellobiase, and phosphatase in various sludges appears to be a promising approach in this direction (25–28). It is likely that systems under constant inoculation will yield more information than aseptic model systems set up with a small number of pure cultures. Although the latter permit more rigorous study, they do not differ from pure cultures in that for them the condition that every niche should be filled is also invalid.

The above digression merely intends to point out that there is a conceptual as well as technological challenge in applied microbial ecology. The fact that so little has been done is not the only basis for some promise of success. While we have succeeded in the large-scale cultivation of desirable plants and animals on vast areas of land, it is extremely unlikely that every large-scale attempt to manipulate the vast resources offered by the protistal kingdom would be doomed to failure.

Still, it would be unwise to expect too much too soon. Sanitary engineers must not be brought to believe that employing microbiologists will automatically solve their problems. Environmental microbiology still has a long way to go as it moves along the path from diagnosis towards curing, preventing, and exploiting pollution. Nor must we forget that other technologies such as chemical engineering have important roles to play in waste treatment. But this must not hold back the microbiologist from making up for lost time.

REFERENCES

1. G. M. Fair, J. C. Geyer, and D. A. Okun, *Water and Wastewater Engineering: Vol. 2, Water Purification and Wastewater Treatment and Disposal,* Wiley, New York, 1968.
2. K. E. McKee and H. W. Wolf, *Water Quality Criteria,* State Water Quality Control Board, Resources Agency of California, Sacramento, Calif., 1963.
3. H. A. Painter, M. Viney, and A. Bywaters, *J. Proc. Inst. Sewage Purif.* 302 (1961).
4. J. W. M. la Rivière, *Proceedings of the 3rd International Conference on Global Impacts of Applied Microbiology,* Bombay, 1971 (In Press).
5. W. Stumm and J. J. Morgan, *Aquatic Chemistry, Wiley-Interscience,* New York, 1970.
6. K. R. Butlin, S. C. Selwyn, and D. S. Wakerley, *J. Appl. Bacteriol.* **19**, 3 (1956).
7. A. L. Downing, H. A. Painter, and G. Knowles, *J. Proc. Inst. Sewage Purif.* 130 (1964).
8. D. Herbert, in *Continuous Culture of Micro-organisms,* S.C.I. Monograph No. 12, London, 1961, p. 187.
9. M. W. Tenney and W. Stumm, *Proc. Purdue Univ. Indust. Waste Conf.* **19**, 518 (1964).

10. H. W. van Gils, Report No. 32. Research Institute of Public Health Engineering, T.N.O., Delft, 1964.
11. H. G. Schlegel, *Allgemeine Mikrobiologie,* Thieme, Stuttgart, 1969.
12. S. Zamenhoff and P. J. Zamenhoff, Paper Presented at the 10th International Congress for Microbiology, Mexico City, 1970.
13. C. R. Curds and J. M. Vandyke, *J. Appl. Ecol.* **3**, 127 (1966).
14. J. K. Baars, *Bull.World Health Org.* **26**, 465 (1962).
15. E. F. Gloyna, *Waste Stabilization Pond Concepts and Experiences,* Preliminary report for W.H.O., Geneva, 1966.
16. W. J. Oswald and C. G. Golueke, in *Single-Cell Protein* (R. I. Mateles and S. R. Tannenbaum, Eds.), M.I.T. Press, Cambridge, Mass., 1968.
17. L. Huisman, *Preliminary Report on Slow Sand Filtration,* Technological University, Delft, 1969.
18. A. W. Lawrence and P. L. McCarthy, *J. Water Pollution Control Fed.* **41**, R1 (1969).
19. D. F. Toerien and W. H. J. Hattingh, *Water Res.* **3**, 385 (1969).
20. J. P. Kotzé, P. G. Thiel, and W. H. J. Hattingh, *Water Res.* **3**, 459 (1969).
21. W. A. Pretorius, *Water Res.* **3**, 545 (1969).
22. G. G. Cillie, M. R. Henzen, G. J. Stander, and R. D. Baillie, *Water Res.* **3**, 623 (1969).
23. W. E. Gates, J. T. Marlar, and J. D. Westfield, *Water Res.* **3**, 663 (1969).
24. G. H. Allen, *Proceedings of the FAO Technical Conference on Marine Pollution and Its Effects on Living Resources and Fishing,* Rome 1971 Doc. No. R-13.
25. W. H. J. Hattingh and M. L. Siebert, *Water Res.* **1**, 197 (1967).
26. G. Lenhard and L. D. Nourse, J. K. Baars, ed. *Proceedings of the Second International Water Pollution Research Conference,* Pergamon, New York, 1965.
27. G. Lenhard, *Hydrobiologia* **19**, 67 (1967).
28. P. G. Thiel and W. H. J. Hattingh, *Water Res.* **1**, 191 (1967).

17 New Approaches to Water Quality Control in Impoundments

J. E. Ridley, Metropolitan Water Board, Limnological Laboratories West Molesey, Surrey, U.K. and J. M. Symons, Bureau of Water Hygiene, U.S. Department of Health, Education and Welfare, Cincinnati, Ohio

This chapter deals with some of the water quality problems experienced during operation of eutrophic impoundments (see also Chapters 1, 2, and 3). The problems are accentuated if the water mass stratifies thermally, and in recent years attention has been focused on low-cost engineering systems for partial, or complete, destratification of impoundments with holding capacities of up to 50,000 mg.

Some of the methods in current use are described, with particular reference to improvements in biochemical and biological quality within the impoundment that are advantageous in the management of hydroelectric schemes, fisheries, river regulation, and waterwork supplies.

389

17-1. The Eutrophic Impoundment

A.   Function and Size.   The construction of large impoundments is an inevitable consequence of urban and industrial expansion. In many areas the water supply requirements for large cities can no longer be satisfied by underground resources, and an increasing need exists for exploitation of surface waters that may be polluted by domestic and industrial effluents referred to in earlier chapters.

The quantity requirements may be solved by construction of dams across valleys as a means of retaining winter floodwaters. An alternative solution is to provide a series of pumped-storage reservoirs along the length of a river as a means of regulating river flows or for direct abstraction by waterworks.

The valley impoundment is frequently designed for multipurpose usage such as river flow regulation, waterworks abstraction, hydroelectric power supplies, fisheries, and possibly recreational amenities. More recently, an interest in estuarine barrages as a long-term solution of the quantity problems has developed, although these schemes require many miles of aqueducts or overground mains for distribution of the impounded water to areas of shortage.

The following examples illustrate the range of areas and volumes involved:

Valley Impoundments	Area, acres	Volume, mg (Imperial)
Grafham Water (U.K.)	1,570	13,000
Chew Valley Lake (U.K.)	1,210	4,500
Derwent (U.K.)	1,000	11,000
Inniscarra (U.K.)	2,000	12,500

The United States has 87 reservoirs with volumes of 300,000 mg (U.S.) or greater.

Pumped Storage Reservoirs	Area, acres	Volume, mg (Imperial)
Queen Mary Reservoir (U.K.)	707	6,700
Wraysbury Reservoir (U.K.)	500	8,000

The United States has 22 major pumped storage projects with ultimate capacities of 122 to 1,403 MW of electric power.

Estuarine Barrages	Area, acres	Volume, mg (Imperial)
Plover Cove (Hong Kong)	3,063	41,000
Morecambe Bay (U.K.)	22,000	65,000

This type of impoundment is not yet popular in the United States.

The profound effect on the biotic components of the eutrophic waters caused by impoundments has been known for almost a century. For example, Frankland (1) demonstrated that storage of sewage-polluted river water led to a marked decrease in its content of pathogenic bacteria. If *Escherichia coli* is used as an index of fecal pollution, at least 90% reduction in the numbers of this organism will occur if polluted waters are stored for periods of about 10 days (see also Chapter 14). This improvement in bacteriological quality is frequently used to advantage in water supply schemes.

As the quantity design of any impoundment scheme is often based on reliable daily yields of about 1 to 2% of the volume held in storage, the theoretical retention period of about 50 to 100 days will usually be in excess of the minimum period required for reduction in the numbers of fecal organisms. Thus, an impoundment may fulfill the dual functions of waterworks quantity reserve and as a first barrier in the destruction or removal of undesirable bacteria or viruses (see also Chapter 16).

Other biotic associates of the feeder streams will also be modified during passage through an impoundment because the organisms have been transferred from a fluviatile to a lacustrine environment. The types and quantities of autotrophic and heterotrophic organisms that flourish in the impoundment will be determined by numerous factors, including the concentrations of abiotic substances passing in from the feeder streams. Geographical location, local topography, and the rate at which water passes through the impoundment are also significant. Physical factors such as solar energy and the effects of wind action on the water surfaces are important regulators of primary production, by algae and by macrophytes, within the impoundment. In turn, prolific zooplankton populations will flourish in the presence of suitable algae, bacteria, and organic detritus. These planktonic animals, together with the invertebrate fauna of the littoral and benthic zones, are a major source of food for young fish.

Thus the eutrophic impoundment can be described as an enriched ecosystem where communities of plants and animals produce significant changes in very short periods and the rates of change will be greatly influenced by physical phenomena such as thermal-density layering of the water column.

B. Physical, Chemical, and Biological Conditions. The physicochemical and biological conditions in natural lakes have been studied by limnologists for more than a century. The reader is referred to Hutchinson (2, 3) for detailed information. The thermal characteristics of lakes were studied by Whipple as long ago as 1895, using an electrical thermometer, while Drown discovered summer stratification of oxygen in Massachusetts lakes in 1891. In the United Kingdom, thermal stratification and depth distribution of dissolved oxygen in natural lakes and waterworks reservoirs were studied in detail from about 1930, although the first records of thermal gradients in Scottish lochs were obtained by Jardine as long ago as 1812.

Despite the availability of data from studies of numerous natural lakes, the designers of man-made impoundments in the early part of this century made little or no allowance for controlling thermal stratification, other than attempting to insure a good circulation (4). As a consequence, a maximum water depth of 40 feet for reservoirs in temperate climates became accepted as the means of limiting the extent of thermal stratification and deoxygenation of the lowest layers, on the assumption that midsummer wind forces would still be adequate for maintaining near-isothermal conditions.

During cyclonic summers this proved to be correct, but during drought summers the combination of excessive solar heating and the reduction of flow from feeder streams resulted in thermal gradients that persisted throughout summer until fall overturn. In recent years the need to conserve land resources has led to construction of impoundments where the water depth is considerably more than 40 feet, so that thermal stagnation may persist until early winter.

If a eutrophic impoundment is assumed to contain nitrate-N concentrations of about 2 mg/liter, and phosphorus-P at 0.2 mg/liter, as well as adequate quantities of other nutrients such as SiO_2, Mg, Ca, Na, K, and HCO_3, the potential crops of algae are extremely high. In terms of numbers of algal cells, blooms of the order of 5000 to 50,000 cells/ml are to be expected at intervals between spring and fall if the physical conditions of the environment are favorable to growth and reproduction. The number of cells may be less important than the type of algae involved, however, and the blue-green algae are often regarded as the ultimate stage of eutrophication (see also Chapter 10).

Death, settlement, and decomposition of successive algal blooms may then deoxygenate the lowest layers of water in the impoundment, and unless the anaerobic layers are rapidly dispersed into the upper and well-oxygenated layers, a progressive deterioration will occur that will depend upon the intensity of thermal-density layering of the water column.

During midsummer stratification in temperate climates, the physical, chemical, and biological profiles in a eutrophic reservoir may be as shown in Table 17.1

Although the density differences between the upper layers (epilimnion) and lowest layers (hypolimnion) of a thermally stratified impoundment

TABLE 17.1   Midsummer Stratification in Temperate Climates

	Depth below Water Surface, ft			
	3	25	40	60 (Bottom)
Temperature, °C	25+	20	10	10
Dissolved oxygen (percentage saturation)	150+	100	25	0[a]
pH Value	8.5	8.2	8.0	7.5
SiO_2, mg/liter	1	1	6	8
Ammonia-N, mg/liter	0.05	0.10	0.80	3.00
Phosphorus-P, mg/liter	0.02	0.05	0.20	2.00
Algae, cells/ml	20,000+	5,000	1,000	20,000[a] (decomposing)

[a] Possibly hydrogen sulfide present, 10 mg/liter.

may be small, they effectively reduce the capability of the light midsummer winds to promote vertical circulation of the water column. As a consequence, algae in the topmost layers will flourish until specific nutrients are depleted, and at the same time a release of nutrients in the lowest layers, where sedimented organisms are decomposing, will occur.

As the two zones are separated by the thermocline, that is, the plane of maximum rate of decrease in temperature with depth, mixing will not occur under natural conditions until heat is lost from the upper layers in late summer and the thermocline is depressed by increased wind forces of the fall.

Thus, in midsummer the total quantity of well-oxygenated water available is limited if excessive quantities of algae are to be avoided. This is of some embarrassment to waterworks because, at the same time, the bottom water will be unusable unless the treatment facilities include means of dealing with unacceptable tastes and odors and the various reducing substances that interfere with conventional chlorination procedures.

These facts are well documented in standard textbooks and in numerous scientific publications (2, 3, 5, 6) together with various techniques for chemical control of algal blooms in impoundments, and also the wide range of treatment facilities available to the water works industry for dealing with low quality waters.

While the accentuation of quality problems in deep impoundments caused by thermal stratification has been clear for many years, design engineers tended to concentrate on treatment facilities for correcting any quality deficiency after the water had left the impoundment. Only within the past 20 years has any serious attempt been made to consider modification of the thermal conditions in large impoundments. The following sections of this chapter deal with some of the engineering methods that have proved effective in many areas.

17-2.   Impoundment Destratification

The design of systems for specifically preventing or destroying thermal stratification resulted from a variety of objectives. For example, Hooper, Ball, and Tanner (7) pumped hypolimnion water to the surface of West Lost Lake, Michigan (area 3.6 acres, depth 20 to 42 ft) as a means of cooling, deepening, and fertilizing the epilimnion—primarily in connection with fisheries management.

Riddick (8) used a floating aerator, in effect, a modified air-lift pump, to insure even distribution of milk of lime in a water supply reservoir (area about 20 acres) and in so doing he eliminated a thermal difference of 11°F between the epilimnion and hypolimnion within 7 days. Heath (9) discharged compressed air at the bottom of a eutrophic lake in Sweden that was devoid of oxygen and incapable of supporting plant or fish life.

Laurie (10) reported the use of air-lift pumps ("Bubble-guns") at the Inniscarra reservoir (area 2000 acres; depth at dam, 103 ft) in Ireland for transferring hypolimnion water to the epilimnion at the rate of 20 million gallons per day (mgd). The pumping resulted in marked increases in dissolved oxygen content at turbine discharge depths about 65 to 75 feet below the surface, thereby reducing the risk of fish mortality downstream of the reservoir. Ford (11) described the effects of compressed air injection at Lake Wohlford, California (area 222 acres, maximum depth 80 ft) where a temperature difference of 8°F in the water column was virtually eliminated in 80 hours. In his experiment, the compressor output was 210 cu ft/min, and the lake was subsequently kept isothermal by intermittent use of the injector. Many other examples are reported in the literature, but the main effect of every system has been destruction of density layering of the water column or the reduction of the density gradient so that relatively light forces would be capable of maintaining isothermal conditions.

A.   Mechanical Pumping.   Water supply systems in lowland areas invariably include pumping stations for lifting water from rivers and discharging it into storage reservoirs. This facility insures some degree of turbulent disturbance of thermal stratification although designers tend to introduce the water at a low velocity in order to insure deposition of suspended solids.

For controlling thermal stagnation in waterworks reservoirs in the United Kingdom, Cooley and Harris (12) designed inlet systems based on the jet discharge principle for reservoirs of 300 to 500 acres in area and with depths of 50 to 70 ft. Their experiments showed that thermal-density layering in midsummer resulted almost entirely from solar heating, but that a reservoir with a capacity of 4000 mg and a depth of 60 ft could be prevented from stratifying if about 100 mgd of incoming water from the river could be discharged through a series of inclined jet inlets, at a velocity of about 8 ft/sec. The jet inlet system was designed to entrain about 10 times the volume of water being pumped into the reservoir and at the same time to promote two-directional rotation of the water mass. This new approach to reservoir engineering design proved effective in maintaining near-isothermal water columns throughout summer, and another system of pumping was designed for use in reservoirs where it would not be possible to inject river water. This second system was intended for water masses used mainly as drought reserves, but which would stratify thermally every summer (13).

The advantages of jet entrainment systems for destratifying waterworks reservoirs and reducing the problems at treatment works have been recorded in Reports of the Director of Water Examination, Metropolitan Water Board, London, since 1963.

At a reservoir now under construction for the Metropolitan Water Board, water area 500 acres and depth 70 ft, the engineering design includes a triple system of stratification control (14). The incoming water can be discharged through a series of six jets placed a few feet above the floor of the reservoir. The jet orifices are 36 inches in diameter and the total inflow volume will range from 100 to 200 mgd at a discharge velocity of about 10 ft/sec. Some of the jet inlets are inclined horizontally, while others are inclined at angles of 22½ and 45° to the horizontal. The direction of the jets, relative to the peripheral embankment of the reservoir, was determined by extensive trials with scale models.

At times when the inflowing water is excessively turbid, the high-velocity jet inlets described above can be by-passed and the water introduced with diminished momentum through inlets of large cross-sectional area (180 ft^2). Both the above systems rely upon the availability of river water for injection, but when river abstraction is restricted by dry weather flows, control of stratification can be continued by a system of shore-mounted recirculating pumps. Three submersible pumps will be used for transferring water down from strata 30 to 50 ft below the water surface and discharging it between 3 and 20 ft above the floor of the reservoir. The rate of subsurface transfer will be 50 mgd, and the jet discharge velocity will be 10 ft/sec to insure entrainment of at least 10 times the pumped volume.

In the United States comprehensive studies of natural and artificial destratification of impoundments have been in progress for many years. For examples, see Symons (15) and Symons, Carswell, and Robeck (16). In 1964, studies were begun by the U.S. Public Health Service to determine if artificial destratification of lakes and reservoirs could prevent adverse reactions and thereby maintain good quality water vertically throughout the impoundment during a summer season. Theoretical analysis showed that little work was required to lift a unit volume of cold water to the surface of a stratified lake, that favorable density currents would develop during such an operation, and all of the cold water initially present would be drained to the deepest portion of the reservoir. Based on these favorable concepts, artificial destratification was attempted, and the results are summarized below.

The first destratification method investigation involved mechanical pumping of cold water from the bottom of a stratified water body and discharging it at the surface. A 12-in. mixed-flow pump driven by a gasoline engine was floated over the deepest part of the study lake from an anchored raft. The suction line extended to within 1 or 2 ft of the bottom and drew water from this depth for discharge at the surface.

1. 1964 EXPERIENCE.   In September 1964 this apparatus was placed in Vesuvius Lake (100 acres, 30 ft deep at the dam) in south-central Ohio. Temperature and dissolved oxygen (DO) data show that between September 4 and 17 the lake was made nearly isothermal and some DO appeared at all depths. Temperature profile data taken after 2.5 days of pump-operating time and at the end of the test (8.5 days of pump-operating time) at four locations in the lake show that, except for a pocket of deep, cold water trapped near the dam by a hump in the lake bottom, the entire lake could be influenced from one pumping location.

2. 1965 EXPERIENCE.   In August 1965 the apparatus was placed in Boltz Lake, in northern Kentucky. The lake has a surface area of 96 acres and a volume of about 2,900 acre-ft. Bullock Pen Lake, a lake with relatively similar size and shape (142 surface acres and about 3,200 acre-ft volume), a few miles from Boltz Lake, was chosen as a "control" on the experiment. Important water quality parameters measured weekly at various depths at the pump site in the test or mixed lake and near the dam in the control lake before the mixing of the test lake began in August 1965 showed that water quality of the two lakes was behaving in similar fashion. The test lake was artificially destratified by pumping bottom water to the surface for 5 weeks during August and September 1965. This artificial destratification improved the water quality in the test lake.

B. Diffused-Air Pumping.   Although the pumping operation described above was effective in destratifying the test lake and improving its water quality, the energy input per unit volume was somewhat higher than originally expected. In an effort to provide the same improvement in water quality with reduced energy input, the test was repeated with a diffused air-pump during the spring and summer of 1966. In addition to a change in mixing equipment, the 1966 test investigated the technique of water quality maintenance by early and periodic mixing instead of water quality improvement by one late mixing.

1. 1966 EXPERIENCE. The equipment used for this operation was a 22.4-kW-hour portable air compressor that delivered about 115 cfm air at about 30 to 40 psi. The air was released through 16 porous ceramic diffusers spaced at 3-ft intervals in a cross pattern. This gave a cylinder of rising air, approximately 40 ft in diameter, that increased to about 100 ft in diameter at the surface. The 1966 destratification was started in the spring rather than in midsummer, and mixing was performed periodically throughout the summer (called multiple mixing). This technique maintained good quality in the mixed lake throughout the summer.

At least 28 other reservoirs in the United States have been mixed by diffused-air pumping for the purpose of water quality control, but two are particularly worthy of note because of their size. One is Allatoona Reservoir in Georgia, mixed by the U.S. Army Corps of Engineers (17), and the other is Eufaula Reservoir, mixed by the Federal Water Quality Administration (18). The former has a capacity of 367,000 acre-ft while the latter has a volume of 570,000 acre-ft in the central pool of the large, multiarmed reservoir that has a total capacity of 2,800,000 acre-ft.

The Allatoona project was started in the spring of 1968 and the five 60-hp compressors operated continuously until fall. This operating scheme was repeated in 1969, but the compressors were started in March before stratification had begun. The results of this project are currently being evaluated (1970), but a definite improvement in DO concentration occurred in the lower waters when 1968 and 1969 data were compared to data taken prior to mixing (1966) or after mixing (1970).

The Eufaula project was started in late July 1967 and operated for 25 days. While the size of the unit and the time of operation were not adequate for complete circulation of the central pool, 65,000 acre-ft of water below a depth of 22 ft were aerated by the twenty-fifth day of operation. In 1968 the aeration device was moved to within 750 ft of the dam. Under these circumstances mixing the water had very little influence up-reservoir from the dam, but the DO content of the low level power discharges was increased.

C. Evaluation of Equipment and Design Performance. Because mechanical equipment is used for artificial destratification, some method is needed for designing the most effective apparatus. Calculation of the "oxygenation capacity" and "destratification efficiency" as defined by Equations 1 and 2 is recommended.

Oxygenation capacity (OC) =

$$\frac{\text{Net change in Oxygen balance from } t_1 \text{ to } t_2 \times \text{hypolimnion volume in billion gallons}}{\text{Total energy input from } t_1 \text{ to } t_2} \qquad (1)$$

where t_1 = time of start of mixing, t_2 = time of end of mixing, and oxygen balance = $[(DO + (-S_{0_2}{}^-) + (-Fe_{0_2}{}^{2+}) + (-Mn_{0_2}{}^{2+})]$ where DO = pounds of dissolved oxygen per million gallons below thermocline, $S_{0_2}{}^-$ = pounds of oxygen equivalent of sulfide (concn. $\times$ 2) per million gallons below thermocline, $Fe_{0_2}{}^{2+}$ = pounds of the oxygen equivalent of reduced iron (concn. $\times$ 0.29) per million gallons below thermocline, and Mn_{0_2} = pounds of the oxygen equivalent of reduced manganese (concn. $\times$ 0.29) per million gallons below thermocline.

Dividing the hypolimnion into 5- or 10-foot vertical sections and using the volume and chemical analysis for each layer is a convenient method of calculating the oxygen balance. The units of OC are pounds of DO transferred per kilowatt-hour.

Destratification efficiency (DE)

$$= \left[\frac{\text{Net change of stability from } t_1 \text{ to } t_2}{\text{Total energy input from } t_1 \text{ to } t_2} \times 100\right] \qquad (2)$$

where t_1 = time of start of mixing, t_2 = time of end of mixing, and, stability = minimum energy needed to mix the lake.

Stability can be calculated by dividing the lake or reservoir into 5- or 10-foot layers and calculating the weight of each, using the existing temperature profile. The gain or loss of energy is determined as each layer is arithmetically "moved" to the depth of the isothermal center of gravity. These gains and losses in energy are algebraically summed to obtain a net energy in the stratified condition. The calculation is then repeated, taking the body of water as isothermal. The isothermal net energy, which should be smaller than that above, is subtracted from the stratified net energy and the resulting difference is the stability, the energy needed to mix the lake and make it isothermal.

17-3. Effects of Impoundment Destratification on Water Quality

A. Biodegradation of Synthetic Organics. Large-scale laboratory experiments in which river water treated with synthetic organic compounds was placed in 500-gal, 3-ft diameter, 10-ft deep aluminum tanks were performed to determine the influence of stratification on the biodegradability of these materials (19). The water in the experimental apparatus was thermally stratified through the use of immersion heaters to maintain the upper water at a temperature about 23°C, while cooling coils maintained the lower ⅓ of the volume at approximately 15°C. The lower water was maintained in an anaerobic condition through the periodic addition of glucose, while the upper portion was maintained aerobic through contact with the air at the surface of the water column and by gentle mixing.

Three synthetic organic materials were tested; two synthetic detergents and a herbicide. These materials, sodium lauryl sulfate, linear alkylate sulfonate, and 2,4-dichlorophenoxyacetic acid, were added to the river water placed in the apparatus described above. One series of experiments was run with sterilized water as a control, while the other tank contained the natural organisms in the river water. Periodic sampling for these three compounds showed that in all cases that the rate of degradation was slower under the cool deoxygenated conditions of the simulated hypolimnion when compared to the rate in the warm aerobic simulated epilimnion. This would indicate that artificial destratification of a natural reservoir, so as to maintain warmer aerobic conditions throughout the water depths, would enhance the biodegradation of organic materials.

Unfortunately, only the relatively insensitive COD test was performed during the field experimentation with artificial destratification so that confirmation of the laboratory data cited above could not be made. Future work should include measurements of the detergent and pesticide levels in artificially destratified reservoirs to show whether or not the laboratory data are confirmed in a natural situation.

B.   Inorganic Substances

1. DISSOLVED OXYGEN.   Figures 17.1 and 17.2 show that artificial destratification both by the mechanical pump and the diffused-air pump method raised the concentration of DO at the lower depths of the 62-ft deep test or mixed lake. These data also show that the mechanical pump was not too effective in providing DO at the 45-ft depth during 1965, and that during the month of August 1966 the diffused-air pump could not cope with the oxygen demand at the 0.7 depth (45 feet). The decline in DO concentration at the 5-ft depth during the fourth mixing in 1966, as shown in Figure 17.2, shows that when large quantities of oxygen-demanding materials are suddenly raised into the epilimnion, a rather severe drop in DO concentration can occur. A slow, continuous mixing might avoid this undesirable decline in DO concentration.

In the United Kingdom, destratification of waterworks reservoirs has

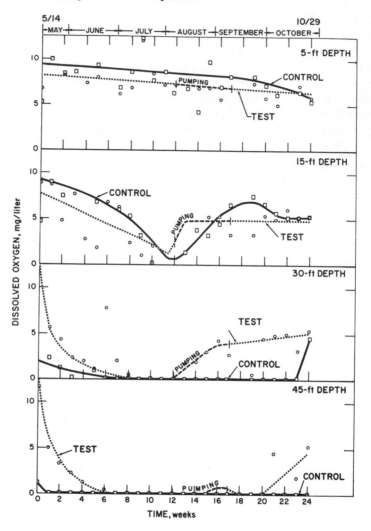

Figure 17.1 Comparison of changes in dissolved oxygen concentration with time at four depths in test and control lakes.

also resulted in marked improvements in dissolved oxygen concentrations in the depths of water columns. At one site the continuous injection of river water through jet inlets ensures oxygen levels ranging from 80 to 100% saturation through a 58-ft water column from spring until late fall. If the destratification system is not used, anaerobic conditions prevail in the bottom layers throughout summer.

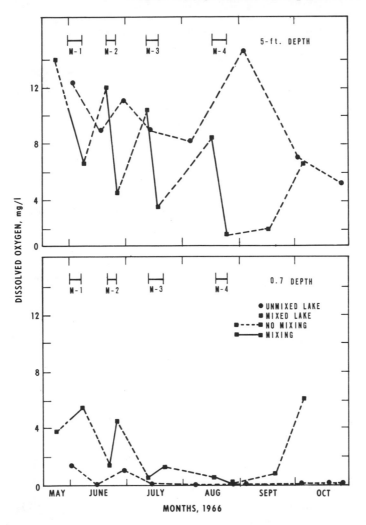

Figure 17.2 Influence of diffused-air pumping on dissolved oxygen.

At another site, static recirculation pumps are used to break down the thermal gradients that are well established by midsummer, and improved oxygen conditions in the depths occur within 2 weeks of commencing the destratification operation (13).

2. SULFIDE. Figures 17.3 and 17.4 show the influence of oxygenation of the lower waters during the 1965 and 1966 experiments. In this case, sulfide was nearly completely eliminated by artificial destratification. This

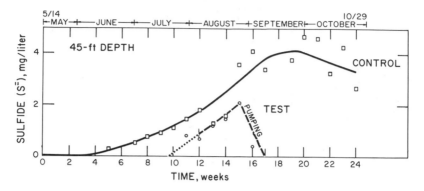

Figure 17.3 Comparison of changes in sulfide concentration with time at 45-foot depth in test and control lakes.

occurred during both years, but because of the periodic mixing that occurred during 1966, relatively high concentrations of sulfide did not develop in the test or mixed lake.

Sulfide concentrations of 10 mg/liter will occur in the bottom layers of a stratified eutrophic reservoir for several weeks during midsummer. Continuous destratification now inhibits sulfide formation, while intermittent

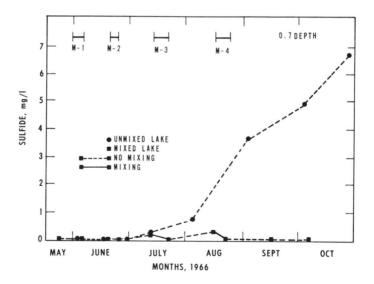

Figure 17.4 Influence of diffused-air pumping on sulfide.

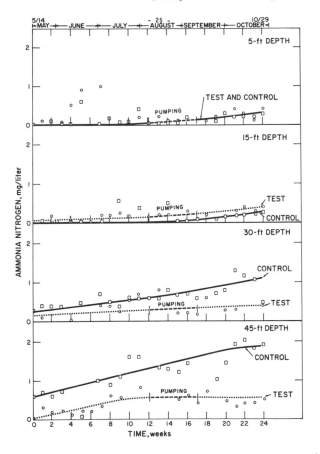

Figure 17.5 Comparison of changes in ammonia nitrogen concentration with time at four depths in test and control lakes.

use of the pumping equipment reduces the sulfide concentrations to tolerable levels.

3. AMMONIA NITROGEN. Figures 17.5 and 17.6 show the influence of mixing on the ammonia nitrogen concentration during the 2 years of experimentation. In both cases the concentration of ammonia nitrogen in the lower levels of the test or mixed lake was reduced by the mixing processes. This may have been caused by the provision of dissolved oxygen at these depths, which might have prevented the anaerobic decomposition of bottom muds with the subsequent release of ammonia nitrogen from the protein material. The rise in ammonia nitrogen concentration

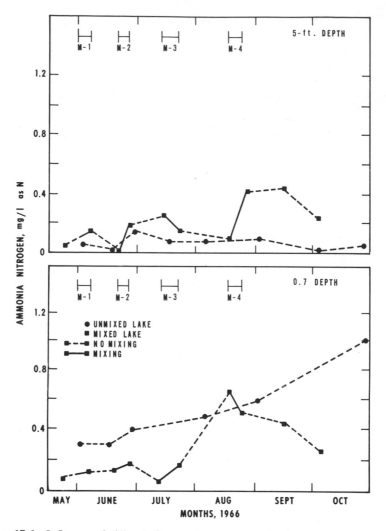

Figure 17.6 Influence of diffused-air pumping on ammonia nitrogen.

that occurred at the 5-ft depth during the fourth mixing of the 1966 experiment is unexplained at this time (1970).

An inverse relationship seems to exist between the concentrations of dissolved oxygen and of ammonia nitrogen in the lower layers of a stratified impoundment. As modern waterworks treatment relies upon minimal amounts of ammonia nitrogen for effective chlorination, any reduction within the impoundment is of hygienic significance. Destratification techniques effectively increase the depth-volume of water containing acceptable concentrations of ammonia nitrogen.

4. MANGANESE. Figures 17.7 and 17.8 show that the manganese concentration in the lower portions of the test or mixed lake were reduced both by the mechanical mixing operation performed during 1965 and the periodic diffused-air mixing carried out during 1966. The data taken at the 0.7 depth during the fourth mixing in 1966 are particularly significant. Temperature was used as the criterion for cessation of mixing during this experiment, and the data on manganese taken during the fourth mixing show that significant quantities of manganese remained at the 0.7 depth when mixing stopped. If the manganese concentration had been used as the control parameter, this mixing would have progressed much longer and the quality of the lake water would have been improved much more than it actually was. This shows the importance of choosing the proper parameter for controlling the beginning and end of any artificial destratification operation.

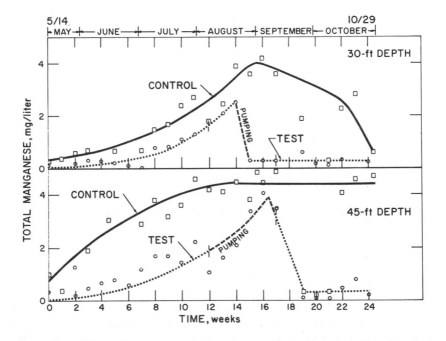

Figure 17.7 Comparison of changes in total manganese concentration with time at two depths in test and control lakes.

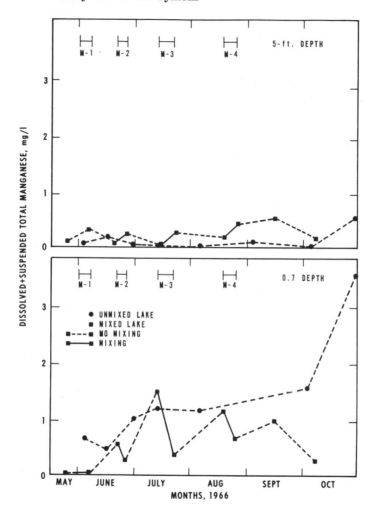

Figure 17.8   Influence of diffused-air pumping on total manganese.

C.   Plankton Populations.   Figure 17.9 compares the plankton population in the control lake and the test lake during the 2 years of study.
These data indicate that in both 1965 and 1966, a decline in plankton
organisms occurred at the 5-ft depth in the study lake during the artificial
destratification operation. Analysis of plankton populations at the deeper
depths of the test lake showed that these declines were not merely the
result of dilution, because similar population declines occurred at all depths.
These declines were not, however, permanent. As shown by the data, regrowth occurred after the cessation of mixing.

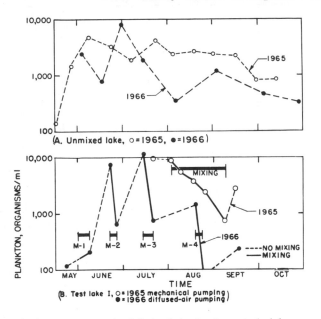

Figure 17.9　Plankton counts at the 5-ft depth in the three study lakes.

Figures 17.10 through 17.13 show the change in composition of the standing crop that occurred in the artificially destratified lakes when compared to similar data in the control lake, in which blue-green algae predominated in August and September. A change in this predomination pattern occurred both during 1965 and 1966 in the lake that was artificially destratified.

Although the precise reasons for these changes in algal predominance patterns, or in actual quantities of algal particles, are at present unknown, similar effects have been reported from destratified reservoirs (4).

Reduced to its simplest terms, the outflow from a eutrophic impoundment can be regarded as a suspension of particles, predominantly algal in origin, which have to be removed by a waterworks. If a destratification technique in any way reduces the total numbers of algae in a reservoir, or merely changes the dominant type of algal particle to one that is more easily filtered, some economy must result.

Stratified eutrophic reservoirs produce successive crops of algae from early spring until fall, although short periods always exist when the water is of high clarity, usually when zooplankton are actively grazing. A filtration works receiving water from such a source may, therefore, be periodically subjected to gross overloading with algae for several months, interspersed with short periods of underloading.

Reservoirs that are continuously destratified tend to produce smaller

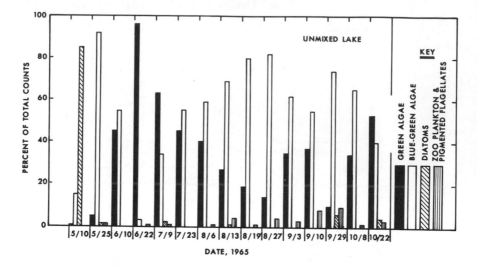

Figure 17.10 Composition of plankton standing crop in unmixed lake in 1965.

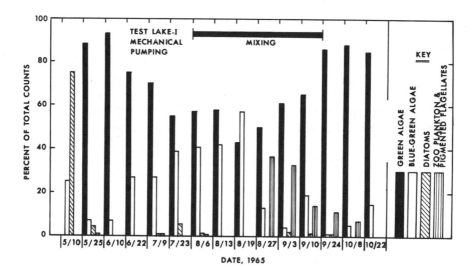

Figure 17.11 Composition of plankton standing crop in test lake 1 during 1965.

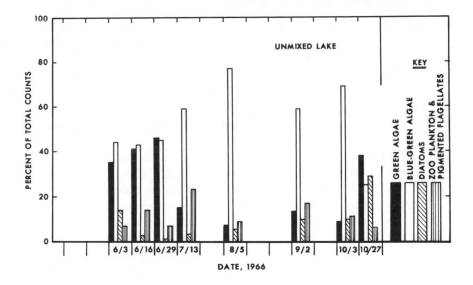

Figure 17.12 Composition of plankton standing crop in unmixed lake in 1966.

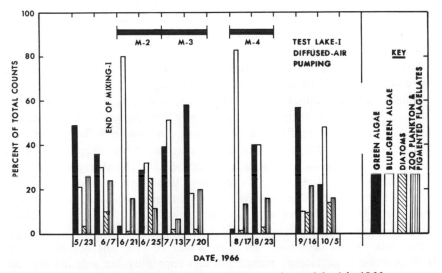

Figure 17.13 Composition of plankton standing crop in test lake 1 in 1966.

409

standing crops of algae, provided that the depth of water is about 40 ft, and the possible mechanisms are discussed later in this chapter. With this type of source, waterworks filters are subjected to tolerable loadings for much longer periods and are thus able to operate at optimal efficiency.

At the present state of knowledge more than speculation on the mechanisms responsible for any advantageous changes in the biotic patterns of destratified impoundments would be unwise, although some factors are more obvious than others.

For example, if destratification prevents the establishment of excessively warm layers near the surface, this may adversely affect proliferation of certain blue-green algae that prefer temperatures in excess of 20°C.

Also the increased rate of movement of water particles within the impoundment may be a factor. As simple planktonic particles, some algae will be unable to resist the induced water currents that will periodically remove them from the photic zone. This intermittent transfer from areas of light saturation to the darkness of the depths may affect rates of carbon assimilation by the algae, and also expose them to grazing invertebrates —as well as bacteria and fungi—which normally colonize the lower layers of a reservoir. In this instance, the depth of the water column through which the algal cell must travel will be critical because the time interval between removal and return to the photic zone will determine the amount of carbon gained or lost by the cell.

Additionally, possibly destratification transfers some unknown substances from the bottom layers of a reservoir that inhibit algal growth or reproduction. In a stratified water column, these substances—if they exist —would remain isolated from the algal zone until the fall overturn.

Furthermore, in the quiescent conditions of a stratified reservoir, particles of silt and organic detritus tend to settle fairly rapidly. If destratification currents retain these particles in suspension, they will then absorb some proportion of the light available to the algae.

Finally, if a destratification operation is regarded as a means of superimposing an artificial turbulence, that is, over and above the normal movements of water that result from convection currents and the interchange of energy between the wind and a water surface, the effects will be to produce changes in the environmental microclimate by disruption of density layering. Therefore, simple organisms that have a short life-history would be unable to adapt readily to the changed physical and chemical conditions. If a particular alga was unable to adapt to light limitation, its rate of carbon assimilation would fall, and if this balanced the respiration rate of the cell, then the alga would survive but not reproduce. If respiratory requirements exceed the cell capability for assimilation, the cell must die unless it has previously stored reserve products.

This is an oversimplification of an extremely complex environmental situation, but it could account for observed sequences during destratification operations when the standing crop of algae first ceases to increase and then rapidly declines. If the sequence is interrupted by stopping destratification when the algae are at the carbon assimilation-respiration balance point, then some proportion of the surviving algae will be capable of renewed growth and reproduction. This might explain some of the algal quantity patterns which result from intermittent use of destratification equipment, although it must be stressed that a considerable amount of fundamental research will be needed before any of the comments in the preceding paragraphs can be justified. However, the factors mentioned above are useful starting points for future research concerning biotic changes.

In conclusion, artificial destratification cannot as yet be regarded as the panacea for every water quality problem in an impoundment, although it is probably a major breakthrough in the management of eutrophic impoundments.

The authors hope that designers of new impoundments will consider the merits, and deficiencies, of the various destratification techniques, and that operators of existing installations will continue to collect fundamental data and relate the information to their field experiences.

REFERENCES

1. P. Frankland, *Proc. Inst. Civ. Eng.* **127**, 83 (1896).
2. G. E. Hutchinson, *A Treatise on Limnology,* Vol. 1, Wiley, New York, 1957.
3. G. E. Hutchinson, *A Treatise on Limnology,* Vol. 2, Wiley, New York, 1967.
4. J. E. Ridley, "Symposium on Eutrophication," *J. Soc. Water Treat. Exam.* **19**, 374 (1970).
5. W. O. Skeat, Ed., *Manual of British Engineering Practice,* 4th ed., Volumes 1–3, Heffer, Cambridge, England, 1969.
6. W. S. Holden, Ed., *Water Treatment and Examination,* 8th ed., Churchill, London, 1970.
7. F. F. Hooper, R. C. Ball, and H. A. Tanner, *Trans. Amer. Fish. Soc.* **82**, 222 (1952).
8. T. M. Riddick, *Water and Sewage Works* **104**, 231 (1957).
9. W. A. Heath. *Water and Sewage Works* **108**, 200 (1961).
10. A. H. Laurie, *Water and Waste Treat. J.* **8**, 363 (1961).
11. M. E. Ford, *J. Amer. Water Works Assoc.* **55**, 267 (1963).
12. P. Cooley and S. L. Harris, *J. Inst. Water Eng.* **8**, 517 (1954).
13. J. E. Ridley, P. Cooley, and J. A. P. Steel, *Proc. Soc. Water Treat. Exam.* **15**, 225 (1966).
14. P. Cooley, *Proc. 2nd Conf. Wtr. Qual. Technol.,* Hungarian Hydrolog. Soc., Budapest, 1970.

15. J. M. Symons, *Water Quality Behavior in Reservoirs—A Compilation of Published Research Papers,* PHS Publ. 1930, Cincinnati, Ohio, 1969, 616 pp.
16. J. M. Symons, J. K. Carswell, and G. G. Robeck, *J. Amer. Water Works Assoc.* **62**, 322 (1970).
17. F. H. Posey, Jr., and J. W. DeWitt, *J. Power Div., Proc. ASCE* **96**, 173 (1970).
18. L. E. Leach, W. R. Duffer, and C. C. Harlin, Jr., *Pilot Study of Dynamics of Reservoir Destratification,* FWPCA, Ada, Okla., 1968, 24 pp.
19. J. DeMarco, J. M. Symons, and G. G. Robeck, *J. Amer. Water Works Assoc.* **59**, 965 (1967).

Index